Hearing

ITS PHYSIOLOGY
AND PATHOPHYSIOLOGY

Hearing

ITS PHYSIOLOGY
AND PATHOPHYSIOLOGY

Aage R. Møller, Ph.D.

Program in Communication Disorders
School of Human Development
The University of Texas at Dallas
Dallas, Texas

ACADEMIC PRESS

A Harcourt Science and Technology Company

San Diego   San Francisco   New York   Boston   London   Sydney   Tokyo

Academic Press
A Harcourt Science and Technology Company
525 B Street, Suite 1900, San Diego, California 92101-4495, USA
http://www.academicpress.com

Academic Press
Harcourt Place, 32 Jamestown Road, London NW1 7BY, UK
http://www.hbuk.co.uk/ap/

Library of Congress Catalog Card Number: 00-101601

International Standard Book Number: 0-12-504255-8

PRINTED IN THE UNITED STATES OF AMERICA
00 01 02 03 04 05 SB 9 8 7 6 5 4 3 2 1

CONTENTS

SECTION **I**

The Ear

3 Physiology of the Cochlea

4 Electrical Potentials in the Cochlea

SECTION II
Auditory Nervous System

5 Anatomy of the Auditory Nervous System

6 Representation of Frequency in the Auditory System

7 Is Temporal Code or Place Code the Basis for Frequency Discrimination?

8 Coding of Complex Sounds

9 Hearing with Two Ears

10 Electrical Potentials in the Auditory Nervous System

11 Far-Field Auditory Evoked Potentials

SECTION **III**

Acoustic Reflexes

12 Acoustic Middle Ear Reflex

SECTION **IV**

Disorders of the Auditory System and
Their Pathophysiology

15 Auditory Nerve and Central Auditory Nervous System

16 Tinnitus, Hyperacusis, and Phonophobia

INTRODUCTION

Progress in treatment of conductive hearing loss has benefited from understanding of the function of the normal middle ear and the diseased middle ear. Understanding the physiology and pathophysiology of the middle ear is important for the surgeon who operates on the middle ear and the diagnostician who does the work-up of the patients before the operation and follow-up after treatment. The function of the normal and the pathologic middle ear, including the basis for recording of the ear's acoustic impedance, is described in this book in a manner that assumes minimal knowledge of physics. The function of the acoustic middle ear reflex and its use in the diagnosis of disorders of the auditory system are described in detail. Pathologies that affect the cochlea, such as noise induced hearing loss and hereditary degeneration of hair cells, have no present cure but a better understanding of the pathophysiology is important for diagnosis and management of deafness and severe hearing loss. Cochlear implants have changed the treatment of severe hearing loss but optimal use of such hearing instruments requires understanding of basic auditory physiology regarding coding and processing of complex sounds such as speech sounds. We are now in an era where disorders with symptoms such as tinnitus and hyperacusis are becoming amendable to treatment. While most disorders of the auditory system have detectable morphologic abnormalities, these disorders often lack detectable morphologic changes and other objective signs are often absent. It has become evident that symptoms such as tinnitus, hyperacusis, and phonophobia are often caused by physiological abnormalities that are located in the central nervous system and are a result of neural plasticity that result in changes in synaptic efficacy. A complex series of events seems to be necessary so that such pathologies become manifest. The importance of the nonclassical auditory nervous system has become evident through studies of disorders such as tinnitus. This little known and rarely mentioned part of the brain may have much wider importance than previously known. Progress in our understanding and management of such disorders requires a thorough understanding of the function of the normal ear and the auditory nervous system as well as that of the pathologic auditory system. I discuss such information and also cover less common disorders such as bilirubinemia and cortical lesions and acoustic tumors and their diagnosis. The recent developments in understanding pathologies of the auditory system have made it imperative that the clinician has a thorough understanding of the basic function of not only the ear but also the auditory nervous system. *Hearing: Its Physiology and Pathophysiology* provides a thorough description of coding of complex sounds in the auditory system and discussion of the analysis that occurs in the ear and the auditory nervous system. The effect of pathologies of the cochlea and the auditory nervous system on processing of complex sounds is discussed in a way that is applicable to providing up-to-date health care. Relevant information about the anatomy and the physiology of the auditory nervous system serves to build the basis for understanding the function the auditory nervous system. I also describe the various electrical potentials that are generated in the auditory nervous system, what anatomical structures generate the different components of such far field potentials as the BAEP and the MLR, and how they change as a result of different types of pathologies.

Hearing: Its Physiology and Pathophysiology not only prepares the clinician and the clinical researcher for the challenges of the modern clinical auditory discipline but the knowledge it provides about the pathophysiology of the auditory system is also essential to individuals engaged in basic research in the auditory field. It is my hope that such knowledge can guide research efforts in basic auditory research into clinically relevant questions. The text thus aims at cross-fertilization between clinicians, clinical researchers, and basic scientists.

I am indebted to many people for their assistance and support while writing this book. I received valuable help from many individuals: Karen Pawlowski, who edited the manuscript, greatly improved the book and Pritesh Pandya helped with the illustrations. I am grateful to my students at the Callier Center for Communication Disorders. They provided important feedback on earlier versions of the text used as teaching material. I also extend thanks to Craig Panner and the editorial staff of Academic Press in San Diego. Special thanks go to the copyeditor, directed by Kelly Ricci, who made my manuscript into the finished product. The University of Texas at Dallas has given practical support during the writing phase.

I thank my wife, Margareta B. Møller, MD, DMSc, for unfailing support and constant encouragement during the manuscript preparation. I am also grateful for her helpful comments on earlier versions of several chapters of the book.

Aage Møller

The Ear

The ear as a sensory organ is far more complex than other sensory organs. The sensory cells are located in the cochlea, but the cochlea not only serves to convert sound into a code of neural impulses in the auditory nerve, it also performs the first analysis of sounds that prepares sounds for further analysis in the auditory nervous system. This analysis consists primarily of separating sounds into bands of frequencies before coding them into a discharge pattern in the individual auditory nerve fibers. The separation of sounds is accomplished by the properties of the basilar membrane and the sensory cells that are located along its length. The cochlea is more frequency selective for weak sounds than louder sounds, which facilitates detection of weak sounds. The cochlea also compresses the amplitudes of sounds, which makes it possible to code sounds within the very large range of sound intensities that is covered by normal hearing. Without such amplitude compression the ear could not detect and analyze sounds in the intensity range of normal hearing.

While the gross anatomy of the cochlea has been known for many years, recent studies of its morphology and function have produced what seems to an endless series of surprising and intriguing results. In fact, the sensory transduction in the cochlea has attracted more research effort than any other part of the auditory system and the function of the auditory receptor organ is better known than that of any other sensory system.

The cochlea is fluid filled, which means that sounds must be converted into vibrations of fluid in order to activate the sensory cells. Direct transfer of sound to a fluid is very ineffective. The middle ear facilitates such transfer by acting as a transformer.

ANATOMY

When describing the anatomy of body parts such as the ear, it is important to have unambiguous and clear definitions of directions and planes of the body. Several different methods are commonly used. In this book we use the ones illustrated in Fig. I.1. The direction from head to tail is caudal (which means "pertaining to the tail") and the opposite is rostral ("relating to the beak"). Ventral

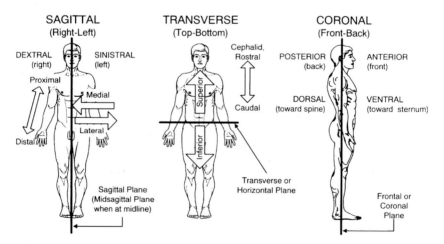

FIGURE I.1   Common anatomical planes with their names, orientations, and directions (from Gelfand, S. A. (1997). "Essentials of Audiology." Thieme, New York).

and dorsal give the directions from the belly to the back. The direction from the midline and out is called lateral, and the opposite direction is medial. A plane extending in the caudal-rostral direction and oriented ventral-dorsally is the saggital plane. The saggital plane that divides the body into two identical halves it is called the mid-saggital plane. A rostral-caudal plane that is perpendicular to the saggital plane is the coronal plane. A plane that is perpendicular to the saggital and coronal planes is the transverse or horizontal plane. It is common in medicine and surgery to use names like posterior and anterior and superior and inferior, which for humans is equivalent to dorsal and ventral, but that terminology becomes ambiguous when used in animals. The description given above can be used for animals as well as for humans.

PHYSIOLOGY

The function of the conductive apparatus of the ear has been extensively studied and is now known in detail. The middle ear is the only part of the entire auditory system where medical or surgical interventions can remedy hearing loss from disease processes or trauma.

During the past decade or so, our understanding of the function of the cochlea has changed in a fundamental way and its function now appears far more complex than perceived earlier. Earlier it was believed that the basilar membrane was a linear system where the properties determined at one sound intensity were directly applicable to all sound intensities. More recently, it has become evident that the frequency selectivity of the basilar membrane depends on the sound intensity. Earlier it was believed that the function of hair cells was limited to transducing the vibration of the basilar membrane into a neural code. The discovery that hair cells also can change their length in response to sound and thus interact with the vibration of the basilar membrane in addition to being transducers radically changed our perception of the function of the cochlea. The best description of the function of outer hair cells is that they act as "motors" that counteract the frictional losses of energy in the cochlea. This function of outer hair cells increases the sensitivity of the ear by about 50 dB and it explains how the loss of outer hair cells causes hearing loss. The interaction between the hair cells and the basilar membrane makes the cochlea more complex than that of other sensory systems, but extensive research during many years has resulted in more knowledge being accumulated about the function of the cochlea than any other sensory organ.

Anatomy of the Ear

ABSTRACT

1. The ear consists of the outer ear, the middle ear, and the inner ear.
2. The outer ear consists of the pinna and the ear canal.
3. The skin of the ear canal is innervated by four cranial nerves: the trigeminal, the facial, the glossopharyngeal, and the vagus nerves.
4. The middle ear consists of the tympanic membrane and three ossicles: the malleus, incus, and stapes.
5. Two muscles are attached to the ossicles, the tensor tympani to the manubrium of malleus and the stapedius to the stapes.
6. The tensor tympani muscle is innervated by the trigeminal nerve and the stapedius muscle is innervated by the facial nerve.
7. The cochlea in humans has a little more than 2-1/2 turns.
8. The cochlea has three fluid-filled compartments, the scala tympani, scala media, and scala vestibuli.
9. The ionic composition of the fluid in the scala tympani and scala vestibuli is similar to that of extracellular fluid (high contents of sodium, low contents of potassium), while the fluid in the scala media is similar to intracellular fluid (high contents of potassium, low contents of sodium).

10. The basilar membrane separates the scala media from the scala tympani and Reissner's membrane separates the scala vestibuli from the scala media.
11. The fluid space of the scala tympani and scala vestibuli communicates with the cerebrospinal fluid space through the cochlear aqueduct.
12. The endolymphatic fluid space communicates with the endolymphatic sac.
13. Hair cells are organized along the basilar membrane in one row of inner hair cells and three to five rows of outer hair cells.
14. The hair cells of the cochlea are transformed epithelium cells that differ from vestibular hair cells in that they lack a kinocilium.
15. Each inner hair cell is innervated by many (Type I) auditory nerve fibers, while each (Type II) nerve fiber innervates many outer hair cells.
16. Efferent nerve fibers terminate directly onto outer hair cells while other efferent fibers terminate on the dendrites of the Type I fibers that innervate the inner hair cells.

INTRODUCTION

The ear consists of three parts, the outer ear, the middle ear, and the inner ear. The inner ear consists of two parts, the vestibular apparatus for balance and the cochlea for hearing (Fig. 1.1). The outer ear and the middle ear conduct sound to the cochlea, which separates sounds with regard to frequency before they are transduced into a neural code in the fibers of the auditory nerve.

OUTER EAR

The different parts of the external ear, "the auricle," have specific names (Fig. 1.2). The groove called the concha is acoustically the most important. The outer ear enlarges in older individuals, especially in men.

EAR CANAL

The ear canal has a length of about 2.5 cm and a diameter of about 0.6 cm. It has the shape of a "lazy S." The most medial part is a nearly circular opening in the skull bone, and the outer part is cartilage. The outer cartilaginous portion of the ear canal is also nearly circular in young individuals but with age the cartilaginous part often changes shape and attains an oval shape. In addition to changing its shape with age, the lumen of the ear canal often becomes smaller with age, and in avid swimmers, it may become very narrow (see Chapter 13).

A

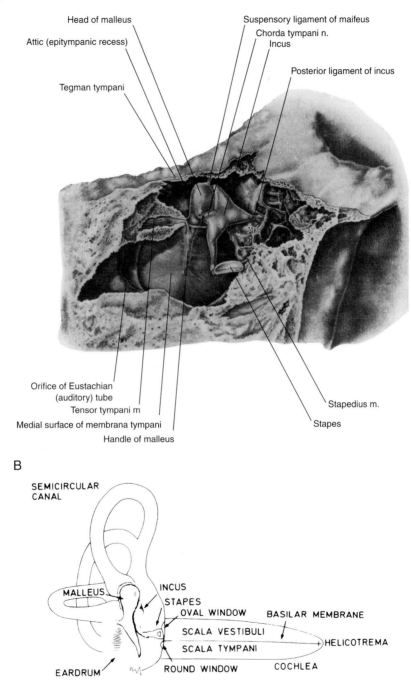

Head of malleus

Suspensory ligament of maifeus

Chorda tympani n.
Incus

Attic (epitympanic recess)

Posterior ligament of incus

Tegman tympani

Orifice of Eustachian
(auditory) tube

Tensor tympani m

Medial surface of membrana tympani

Handle of malleus

Stapedius m.

Stapes

B

SEMICIRCULAR
CANAL

MALLEUS

INCUS

STAPES

OVAL WINDOW

BASILAR MEMBRANE

SCALA VESTIBULI

SCALA TYMPANI

HELICOTREMA

EARDRUM

ROUND WINDOW

COCHLEA

FIGURE 1.1 (A) Cross section of the human ear (from Brodel, M. (1946). "Three Unpublished Drawings of the Anatomy of the Human Ear." W. B. Saunders, Philadelphia); (B) schematic drawing of the ear [96].

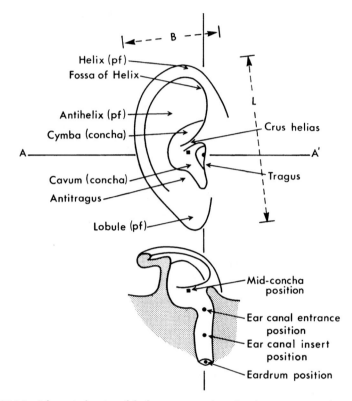

FIGURE 1.2 Schematic drawing of the human external ear showing components that are important for sound conduction: pinna flange, helix, antihelix, lobule, concha (cymba and cavum), and ear canal (from Shaw, E. A. C. (1974). The external ear. *In* W. D. Keidel, and W. D. Neff (Eds.), "Handbook of Sensory Physiology" (Vol. V(1), pp. 450–490). Springer-Verlag, New York).

The ear canal is covered by skin that secretes cerumen (wax) and it has hairs on its surface. There are no sweat glands in the ear canal. Since the skin is not rubbed naturally as skin on other parts of the body it must self-clean dead cells and cerumen. Two types of cells contribute to secretion of cerumen, namely sebaceous cells, located close to the hair follicles, and ceruminous glands. The sebaceous glands cannot secrete actively but form their secretion by passive breakdown of cells. Two kinds of cerumen exist, dry and wet. The dry type is found in people in Asia while the wet cerumen is found in Caucasians, Africans, and Latinos. Accumulation of cerumen in the ear canal to an extent that it becomes occluded is a common cause of hearing loss. Cerumen may also cover the tympanic membrane, which causes hearing loss. Cerumen is often pushed deeper into the ear canals by individuals who attempt to clean their ear canals using cotton swabs. The cerumen is supposed to become dry and leave the ear canal. The secreted cerumen has a slight antibacterial and antifungal property and it may act as an insect repellant.

The outer layer of the skin (epidermis) in the ear canal, together with that of the tympanic membrane, migrates outward. The migration helps heal small injuries and move scars outward as well as transports cerumen out of the ear canal. It has been suggested that failure in this migration of the epidermis may cause several kinds of pathology such as development of cholesteatoma and it may play a role in inflammation of the ear canal.

The skin of the ear canal has an unusual nerve supply. Its sensory receptors (including bare axons) are innervated by four different cranial nerves, namely the sensory portion of the trigeminal nerve cranial nerve (CN V), the facial nerve (CN VII), the glossopharyngeal nerve (CN IX), and the auricular branch of the vagal nerve (CN IX), which supplies the posterior wall of the ear canal. This nerve branch is a part of Arnold's nerve, which also receives contributions from the glossopharyngeal nerve. The innervation of the ear canal by the glossopharyngeal nerve explains why many people cough when the skin of the inner part of the ear canal is touched. The innervation by the glossopharyngeal and the vagal nerve explain why mechanical stimulation of the ear canal can affect the heart and blood circulation and cause sensitive individuals to faint when the ear canal is cleaned for wax.

MIDDLE EAR

The middle ear consists of the tympanic membrane, which terminates the ear canal (Fig. 1.3), and the three small bones (ossicles), the malleus, the incus, and the stapes (Fig. 1.4). Two small muscles, the tensor tympani muscle and the stapedius muscle, are also located in the middle ear. The manubrium of the malleus is imbedded in the tympanic membrane and the head of the malleus is connected to the incus, which in turn connects to the stapes, the footplate of which is located in the oval window of the cochlea. The chorda tympani is a branch of the facial nerve (or nervous intermedius) that travels across the middle ear cavity (Fig. 1.4). It carries taste fibers. The eustachian tube connects the middle ear cavity to the pharynx.

TYMPANIC MEMBRANE

The tympanic membrane (Fig. 1.5) is a slightly oval, thin membrane that terminates the ear canal. It is cone-shaped, with an altitude of 2 mm with the apex pointed inward. Seen from the ear canal, the membrane is slightly concave and is suspended by a bony ring. Normally it is under some degree of tension.

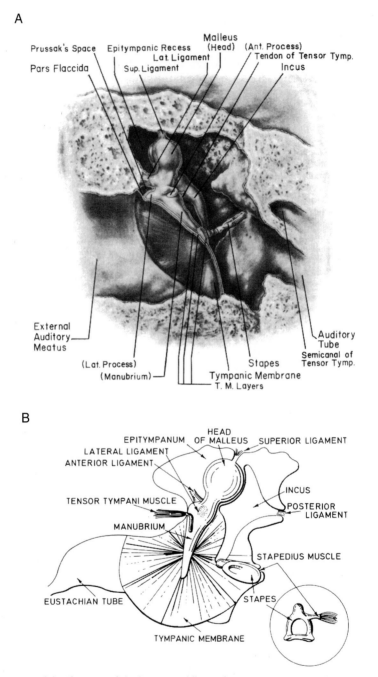

FIGURE 1.3 (A) "Close-up" of the human middle ear from Durrant, J. D., and Lovrinic, J. H. (1984). "Bases of Hearing Science." Williams & Wilkins, Baltimore. (B) schematic drawing of the human middle ear from inside the head (from [92]).

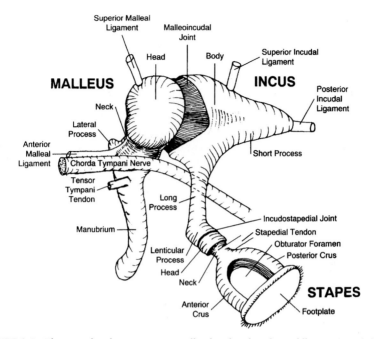

FIGURE 1.4 The ossicular chain as it is normally placed within the middle ear cavity (adapted from Gelfand, S. A. (1997). "Essentials of Audiology." Thieme, New York).

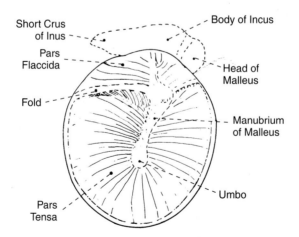

FIGURE 1.5 The tympanic membrane and the position of the malleus and incus (from Durrant, J. D., and Lovrinic, J. H. (1984). "Bases of Hearing Science." Williams & Wilkins, Baltimore).

Its surface area is about 85 mm^2. The main part of the tympanic membrane, the pars tensa, with an area of about 55 mm^2 (Fig. 1.5), is composed of radial and circular fibers overlaying each other. These fibers are comprised of collagen and they provide a lightweight stiff membrane that is ideal for converting sound into vibration of the malleus. A smaller part of the tympanic membrane, the pars flaccida, located above the manubrium of malleus, is thicker than the pars tensa and its fibers are not arranged as orderly as the collagen fibers of the pars tensa. The tympanic membrane is covered by a layer of epidermal cells, continous with the skin in the ear canal. This outer layer of the tympanic membrane migrates from its center outward. That effectively eliminates small injuries and scars and transports small foreign bodies out into the ear canal. The result is that small holes in the tympanic membrane usually heal spontaneously.

OSSICLES

The middle ear bones are suspended by several ligaments (Figs. 1.3 and 1.4). The manubrium of the malleus is embedded in the tympanic membrane, with the tip of the manubrium located at the apex of the tympanic membrane (Fig. 1.5). The head of the malleus is suspended in the epitympanum. The head of the malleus and the incus are fused together in a double saddle joint. The short process of the incus rests in the fossa incudo of the malleus, and it is held in place by the posterior incudal ligament. The long process, also called the lenticular process, of the incus forms one side of the incudo-stapedial joint. The joint between the malleus and the incus is regarded as rigid, while the joint between the incus and the stapes is flexible in the direction that is perpendicular to the pistonlike movement of the stapes, but rigid for pistonlike movements of the stapes. The stapes is suspended in the oval window of the cochlea by two ligaments and one ligament is stiffer than the other.

MIDDLE EAR MUSCLES

Two small muscles are located in the middle ear. One, the tensor tympani muscle, is attached to the manubrium of the malleus and the other, the staped-ius muscle, is attached to the stapes (Figs. 1.3 and 1.4). The tensor tympani muscle extends between the malleus and the wall of the middle ear cavity near the entrance to the eustachian tube. It pulls the manubrium of the malleus inward, displacing the tympanic membrane inward and stretching the mem-brane. The stapedius muscle is the smallest striate muscle of the body. It is attached to the head of the stapes and most of the muscle is located in a bony canal. It pulls the stapes in a direction that is perpendicular to its pistonlike motion, tilting the stapes so that it rotates around its posterior ligament. The tensor tympani muscle is innervated by the trigeminal nerve (the Vth cranial nerve) and the stapedius muscle by the facial nerve (VIIth cranial nerve).

EUSTACHIAN TUBE

The eustachian tube consists of a bony part (the protympanum) close to the middle ear cavity and a cartilaginous part that forms a closed slit where it terminates in the nasopharynx (Fig. 1.6). Optimal functioning of the middle ear depends on keeping the air pressure in the middle ear cavity close to the ambient pressure. That is accomplished by briefly opening the eustachian tube, normally when swallowing. In the adult, the eustachian tube is 3.5 to 3.9 cm long. It follows an inferior (caudal)—medially—anterior (ventral) direction in the head, tilting downward (caudally) by about 45° to the horizontal plane (Fig. 1.6B). The eustachian tube is shorter in young children and it is directed nearly horizontally. The cartilaginous part of the eustachian tube forms a valve that closes the middle ear off from pressure fluctuations during breathing and prevents a person's voice from being transmitted directly to the middle ear cavity. The mucosa inside the eustachian tube (which really is not a tube except for the bony part) is rich in cells that produce mucus and cilia that propel mucus from the middle ear to the nasopharynx. The slit-shaped cartilaginous part of the eustachian tube allows transport of material from the middle ear cavity to the nasopharynx but not the other way. It can be opened by positive air pressure in the middle ear cavity but not by negative pressure, which in fact may close it harder. The most common way the eustachian tube opens is by contraction of a muscle, the tensor veli palatini muscle. The tensor veli palatini muscle is located in the pharynx and innervated by the motor portion of the Vth cranial nerve. This muscle contracts naturally when swallowing and yawning, and some individuals have learned to contract their tensor veli palatine muscle voluntarily.

MIDDLE EAR CAVITIES

The middle ear cavities consist of the tympanum (the main cavity), which lies between the tympanic membrane and the wall of the inner ear (the promontorium), a smaller part (the epitympanum) located above the tympanum, and a system of mastoid air cells. The epitympanum is the location of the head of the malleus (Fig. 1.3). The middle ear cavity and the eustachian tube are covered with mucosa. The total volume of the middle ear cavities is often given to be about 2 cm³, but the size of the middle ear cavities varies considerably from

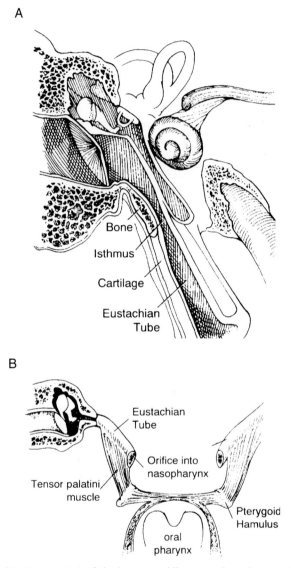

FIGURE 1.6   (A) Cross section of the human middle ear to show the eustachian tube. (B) Orientation of the eustachian tube in the adult. The tensor veli palatini is shown (from Gelfand, S. A. (1997). "Essentials of Audiology." Thieme, New York).

person to person and if the volume of the mastoid air cells is included, the total volume can be as large as 10 cm^3.

INNER EAR

Cochlea

The cochlea is a snail-shaped bony structure that contains the sensory organ of hearing. The cochlea in humans has a little more than 2-1/2 turns (Fig. 1.7). Uncoiled the cochlea has a length of approximately 3.5 cm. The cochlea, together with the vestibular organ, is totally enclosed in the temporal bone, which is one of the hardest bones in the entire body. Together the cochlea and the vestibular organs are often referred to as the labyrinth. The bony structures are known as the bony labyrinth and the content is the membranous labyrinth. The cochlea has three fluid-filled canals, the scala vestibuli, the scala tympani, and the scala media (Fig. 1.8). The scala media, located in the middle of the cochlea, is separated from the scala vestibuli by Reissner's membrane and from the scala tympani by the basilar membrane. The ionic composition of the fluid in the scala media is similar to that of intracellular fluid, thus rich in potassium and low in sodium, while the fluid in the scala vestibuli and scala tympani is similar to that of extracellular fluid (such as the cerebrospinal fluid), thus rich in sodium and poor in potassium.

The scala media narrows toward the apex of the cochlea ending just short of the apical termination of the bony labyrinth. The opening near the apical termination of the bony labyrinth, called the helicotrema, allows communication between the scala vestibuli and scala tympani. In humans, the area of this aperture is about 0.05 mm^2. The basilar membrane (Fig. 1.8) separates sounds according to their frequency (spectrum) and the organ of Corti, located along the basilar membrane, contains the sensory cells that transform the vibrations of the basilar membrane into a neural code.

Organ of Corti

The organ of Corti contains many different cells, of which the hair cells are the most directly involved with the function of the ear. The hair cells, so called because of the hair-like bundles that are located on their top, are arranged in rows along the basilar membrane (Fig. 1.9). The hair cells, transformed cilia cells, have bundles of stereocilia on their top but the hair cells in the mammalian cochlea have no kinocilia. The hair cells are of two main types, outer hair cells and inner hair cells. The human cochlea has about 30,000 outer hair cells arranged in three to five rows along the basilar membrane and about

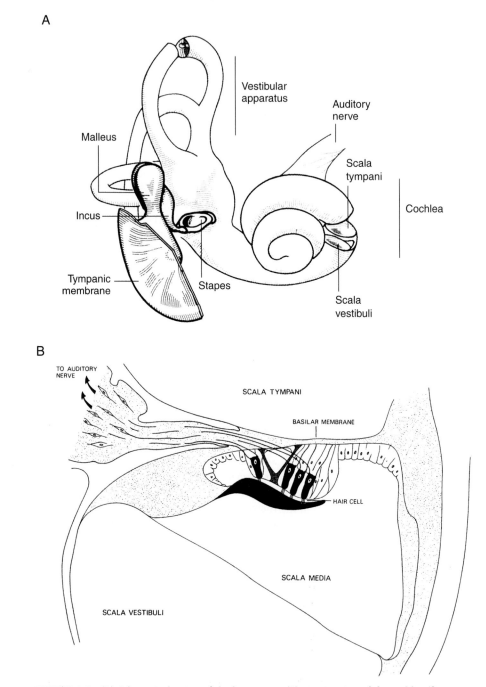

FIGURE 1.7 (A) Schematic drawing of the human ear; (B) cross section of the cochlea (from Møller, A. R. (1975b). Noise as a health hazard. *Ambio* 4:6–13).

A

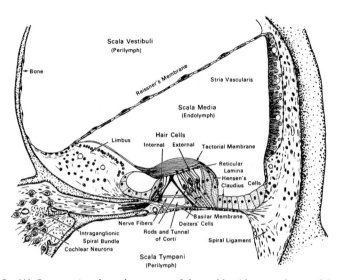

FIGURE 1.8 (A) Cross section through one turn of the cochlea (the second turn of the guinea pig's cochlea) (from Davis, H., Benson, R. W., Covel, W. P., Fernandez, C., Goldstein, R., Katsuki, Y., Legouix, J. P., McAuliffe, D. R., and Tasaki, I. (1953). Acoustic trauma in guinea pig. *J. Acoust. Soc. Am.* **25**:1180–1189). (B) Breschet's drawings of the cochlea, the spiral lamina, and the cochlear nerve. The smaller sketches show differences between the mammalian cochlea and the avian organs of hearing (from Hawkins, J. E. (1988). Auditory physiologic history: A surface view. *In* A. F. Jahn, and J. Santos-Sacchi (Eds.), "Physiology of the Ear" (pp. 1–28). Raven, New York).

10,000 inner hair cells arranged in a single row. The outer hair cells are different from the inner hair cells in several ways. Outer hair cells are cylindrical in shape while the inner hair cells are flask shaped or pear shaped (Fig. 1.10). On each outer hair cell, 50–150 stereocilia are arranged in three to four rows that assume a "W" or "V" shape (Fig. 1.9), whereas the inner hair cells stereocilia are arranged in three to four ascending rows that assume a flattened "U" shape. The stereocilia are linked to each other with specific structures (cross-links) [110]. The tallest tips of the outer hair cell stereocilia are embedded in the overlying membrane, whereas the tips of the inner hair cell stereocilia are not. The outer hair cells in the apical region of the cochlea are longer than in the more basal regions, about 8 μm long in the apical region and less than 2 μm in the base. The diameter of the longest outer hair cell is thus about 1/10th of the diameter of a human hair.[1]

[1] 1 μm = 1 micrometer = 1/1,000,000 of 1 meter, or 1/1,000 of 1 millimeter. A human hair is about 100 μm; a red blood cell is about 7 μm.

B

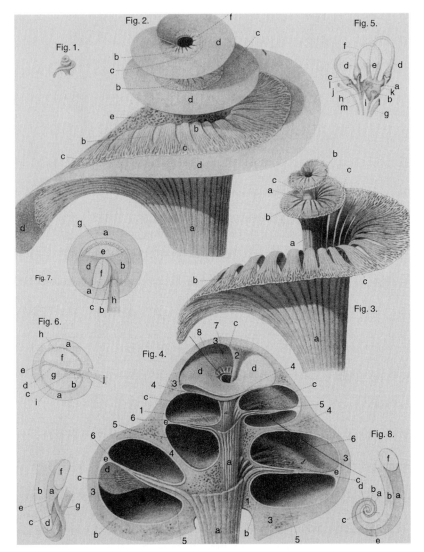

FIGURE 1.8 (*Continued*)

Inner hair cells have about the same dimension in the entire cochlea and all have approximately the same number of stereocilia (about 60). The stereocilia on inner hair cells that are located at the base of the cochlea are shorter

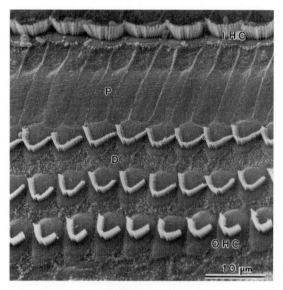

FIGURE 1.9   Scanning electron micrograph of a section of the organ of Corti in a monkey with the tectorial membrane removed to show the organization of the hair cells. One row of inner hair cells (IHC) is visible at the top of the figure and three rows of outer hair cells (OHC) in typical W-shaped formation of stereocilia on the top of the cells are seen (from Harrison, R. V., and Hunter-Duvar, I. M. (1998). An anatomical tour of the cochlea. *In* A. F. Jahn, and J. Santos-Sacchi (Eds.), "Physiology of the Ear" (pp. 159–171). Raven, New York).

than stereocilia of hair cells that are located in the apical region of the cochlea. The stereocilia of inner hair cells are similar to those of outer hair cells, although they are shorter and fatter than the outer hair cell stereocilia. Between the row of inner hair cells and the rows of outer hair cells is the tunnel of Corti, bordered by inner and outer pillar cells (Fig. 1.8).

In addition to hair cells, other types of cells are found in the cochlea. Supporting cells of the organ of Corti are the Deiter's cells and Henson's cells, inner border and inner phalangeal cells. (The rest of the cell types are not mentioned here as they are not thought of as contributing to sound transduction.) The stria vascularis is a structure located between the perilymphatic and the endolymphatic space along the cochlear wall. The stria vascularis has a rich blood supply and its cells are rich in mitochondria, indicating that it is involved in metabolic activity. Many of its intermediate cells have a high content of melanin. The spiral ligament, to which the basilar membrane is attached, supports the stria vascularis.

Basilar Membrane

The basilar membrane consists of connective tissue and it forms the floor of the scala media. It has a width of about 150 μm in the base of the cochlea

A

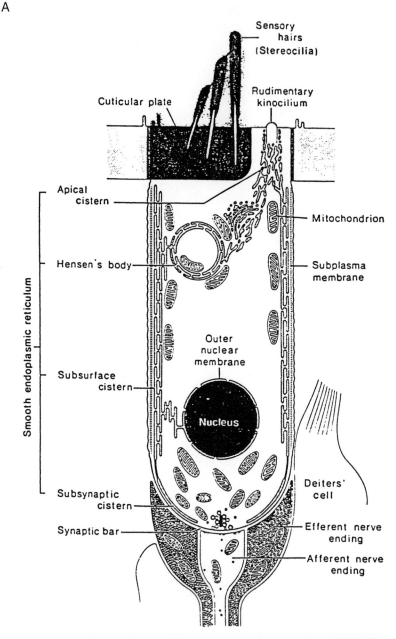

FIGURE 1.10   (A) Schematic drawing of the cross section of an outer hair cell. (B) Schematic drawing of the cross section of an inner hair cell (from Lim, D. J. (1986). Effects of noise and ototoxic drugs at the cellular level in the cochlea: A review. *Am. J. Otolaryngol.* 7:73–99).

B

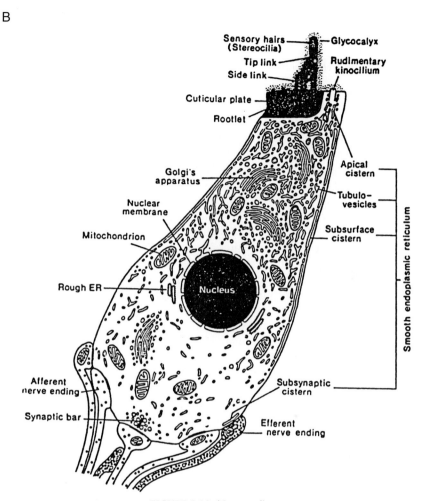

FIGURE 1.10 (*Continued*)

and it is about 450 μm wide at the apex. It is also stiffer in the basal end than at the apex. Due to this gradual change in stiffness, sounds that reach the ear create a wave on the basilar membrane that travels from the base toward the apex of the cochlea. This traveling wave motion is the basis for the frequency separation that the basilar membrane provides before sounds activate the sensory cells that are located along the basilar membrane. As is shown in Chapter 3 the frequency analysis in the cochlea is complex, involving interactions between the basilar membrane, the surrounding fluid, and the sensory cells. The outer hair cells interact actively with the motion of the basilar membrane.

Innervation of Hair Cells

Two types of nerve fibers innervate the haircells, afferent (auditory nerve) fibers and efferent auditory fibers (olivocochlear bundle). The afferent auditory nerve fibers are bipolar cells, the cell bodies of which are located in a spiral, bony canal (Rosenthal's canal) containing the spiral ganglion (Fig. 1.11). The auditory nerve fibers pass through the habenula perforata before they continue as radial fibers to the inner hair cells. In humans, the auditory nerve has about 30,000 afferent nerve fibers. Two types of afferent fibers have been identified. Type I comprises 95% of the auditory nerve fibers and they are myelinated and have large cell bodies. Type II (about 5% of the auditory nerve) are unmyelinated and have small cell bodies. Details about the anatomy of the auditory nerve are given in Chapter 5.

The hair cells connect to the fibers of the auditory nerve via synapses (Fig. 1.12). These connections are different for inner and outer hair cells. Each of the approximately 3,500 inner hair cells in the human cochlea connect to many Type I auditory nerve fibers, while a single Type II auditory nerve fiber connects to many of the approximately 12,000 outer hair cells (Fig. 1.13). It has been estimated that each inner hair cell receives about 20 nerve fibers. The Type II nerve fibers aimed for the outer hair cells cross over the cochlear tunnel to reach the rows of outer hair cells, where each nerve fiber, called an outer spiral fiber, innervates many hair cells and extends apically as much as 0.6 mm (Fig. 1.13) along the outer hair cell region. The inner radial fibers (Type I) are different from the spiral fibers (Type II).

The hair cells also receive efferent (descending) nerve fibers of the olivocochlear bundle (Rasmussen's bundle). Outer hair cells receive the largest number of these connections. The efferent fibers (which number approximately 500–600 in humans [135]) have their cell bodies in the superior olivary nucleus of the brain stem. These fibers are of two kinds. (1) One kind is the medial olivocochlear neurons consisting of large myelinated fibers that originate in the medial superior olivary (MSO) complex, mostly from the opposite side (see Chapter 5). These fibers first travel with the vestibular nerve and then shift to the cochlear nerve. They terminate on outer hair cells. (2) The lateral olivocochlear efferent fibers are small unmyelinated fibers that originate in the lateral part of the superior olivary complex (LSO), mostly on the same side as the ear where they terminate on Type I afferent connections to inner hair cells, near their base. The efferent fibers that reach outer hair cells mainly make presynaptic connections while those reaching the inner hair cells make postsynaptic connections (Fig. 1.12). Many efferent fibers connect with each outer hair cell and each efferent fiber connects to many outer hair cells. Efferent

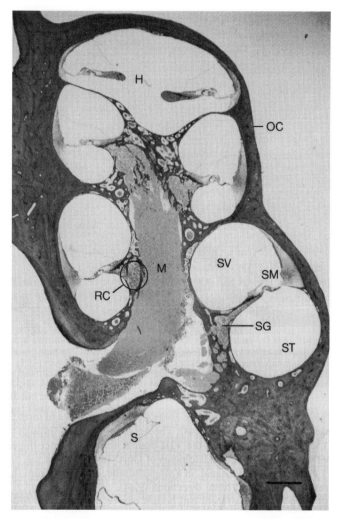

FIGURE 1.11　Spiral ganglion of the auditory nerve shown in a mid-modiolar section of the cochlea showing the otic capsule (OC), modiolus (M), helicotrema (H), the three cochlear scalae [scala vestibulli (SV), scale media (SM), and scala tympani (ST)], and the saccule (S) portion of the vestibular labyrinth. Rosenthal's canal (circled) contains the spiral ganglion (SG). Bar = 500 μm. (from Santi, P. (1988). Cochlear microanatomy and ultrastructure. *In* A. F. Jahn, and J. Santos-Sacchi (Eds.), "Physiology of the Ear" (pp. 173–199). Raven, New York).

fibers connect more sparsely to inner hair cells. It is important to notice that the efferent fibers act directly on outer hair cells while efferent fibers only affect the neural excitation in the nerve fiber that leaves inner hair cells. The

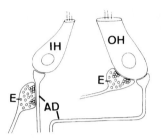

FIGURE 1.12   Schematic drawings of innervation of hair cells by nerve fibers of the auditory nerve. Abbreviations OH, outer hair cells; IH, inner hair cells; AD, afferent dendrite; E, efferent synapse (from Spoendlin, H. (1970). Structural basis of peripheral frequency analysis. *In* R. Plomp, and G. F. Smoorenburg (Eds.) "Frequency Analysis and Periodicity Detection in Hearing" (pp. 2–36). A. W. Sijthoff, Leiden, The Netherlands).

importance of that becomes evident in discussions of the function of outer hair cells compared with that of inner hair cells (Chapter 3).

The cochlea ear also has an autonomic nerve supply. The autonomic fibers, mostly adrenergic sympathetic nerve fibers, mainly innervate blood vessels but they also contact hair cells [28].

FLUID SYSTEMS OF THE COCHLEA

The fluid system of the cochlea is complex. It is shared with the vestibular organ and consists of two distinctly different systems, the perilymphatic system, where the ionic composition of the fluid resembles that of the cerebrospinal fluid, and the endolymphatic system, where the fluid resembles that of intracellular fluid, thus rich in potassium. In the cochlea, the endolymphatic space is separated from the perilymphatic space by Reissner's membrane and the basilar membrane (Fig. 1.8). The ionic composition of the perilymph seems to be important for the function of the hair cells, which are submerged in perilymph.

The fluid space of the perilymphatic system of the inner ear communicates with the cerebrospinal fluid in the skull cavity via the cochlear aqueduct. The endolymphatic system dead-ends in the endolymphatic sac, which is located in the skull space (Fig. 1.15). The pressure in the different compartments is kept in balance by mechanisms that are not entirely known but are thought to involve the function of the endolymphatic sac. An imbalance between the pressure in those two systems can cause hearing loss tinnitus, and vestibular disturbances (see Chapters 14 and 16).

Cochlear Aqueduct

The cochlear aqueduct connects the perilymphatic space with the cranial fluid space (Fig. 1.15). The duct has a very small diameter, 0.05–0.5 mm, and

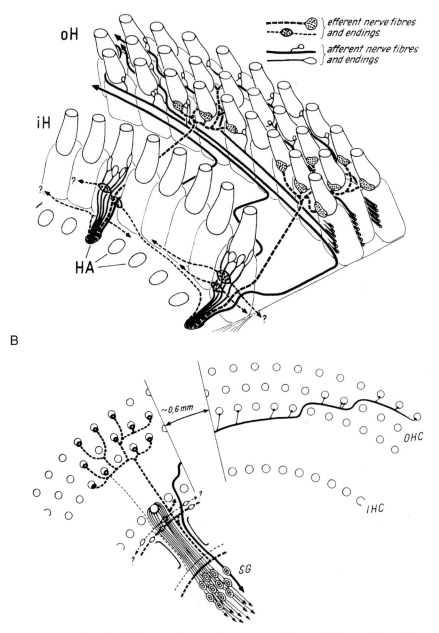

FIGURE 1.13 (A and B) Schematic drawings of innervation of hair cells by nerve fibers of the auditory nerve. Abbreviations: OH, outer hair cells; IH, inner hair cells; SG, spiral ganglion; HA, habenulae openings (from Spoendlin, H. (1970). Structural basis of peripheral frequency analysis. *In* R. Plomp, and G. F. Smoorenburg (Eds.) "Frequency Analysis and Periodicity Detection in Hearing" (pp. 2–36). A. W. Sijthoff, Leiden, The Netherlands).

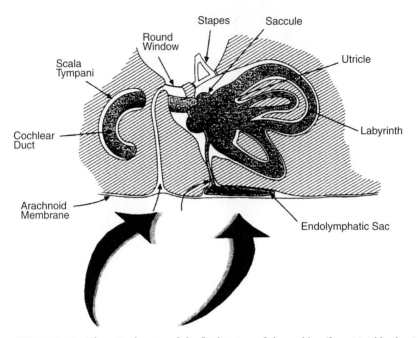

FIGURE 1.14 Schematic drawing of the fluid system of the cochlea (from Marchbanks, R. J. (1996). Hydromechanical interactions of the intracranial and intralabyrinthine fluids. *In* A. Ernst, R. Marchbanks, and M. Samii (Eds.), "Intracranial and Intralabyrinthine Fluids (pp. 51–61). Springer-Verlag, Berlin).

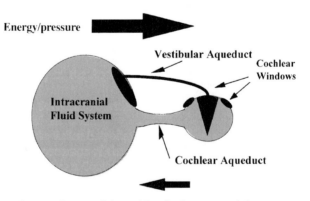

FIGURE 1.15 Schematic drawing of the cochlear fluid systems and their connections with the cerebrospinal fluid in the brain through the cochlear aqueduct (from Marchbanks, R. J. (1996). Hydromechanical interactions of the intracranial and intralabyrinthine fluids. *In* A. Ernst, R. Marchbanks, and M. Samii (Eds.), "Intracranial and Intralabyrinthine Fluids (pp. 51–61). Springer-Verlag, Berlin).

there is evidence that it may not be totally open in many adults. It is known to be open in animals, and in experiments in cats it has been shown that intracranial pressure (ICP) variations are communicated to the perilymphatic space with a short time constant [18]. If the cochlear aqueduct is closed artificially in such animals, changes in the ICP affects the pressure in the perilymphatic space to a much smaller extent and there is a time lag between changes in the ICP and changes in pressure in the perilymphatic space. If the cochlear aqueduct is not open, the perilymphatic pressure will change when the ICP changes because of the endolymphatic sac, but in a more delayed and complex fashion relative to the ICP.

Endolymphatic Duct

The endolymphatic space communicates through the endolymphatic duct with the endolymphatic sac, which lies between two layers of the dura mater. It is located intracranially close to the skull wall near the porus acousticus (the opening of the internal auditory meatus). Reissner's membrane, which separates the endolymphatic space from the perilymphatic space in the cochlea, has a high degree of compliance. Therefore very small changes in pressure can cause large changes in the volume of the endolymphatic space.

Blood Supply to the Cochlea

The arterial supply to the cochlea is the labyrinthine artery. It originates in the anterior inferior cerebellar artery (AICA) and follows the VIIIth cranial nerve in the internal auditory meatus, where it gives off the anterior vestibular artery to the vestibular apparatus (Fig. 1.16). Further into the internal auditory meatus the labyrinthine artery branches to form the vestibula-cochlear artery that supplies parts of the cochlea. The other branch is the spiral modiolar artery that serves as a collateral blood supply to the cochlea [3]. In humans, the labyrinthine artery is much longer than in those animals commonly used in experiments related to hearing. This is because the distance between the brainstem and the cochlea is much longer in humans, partly because the subarachnoidal space is much larger. The auditory nerve, which the artery follows for most of its course, is about 2.5 cm long in humans [71, 72] compared to 0.5–0.8 cm in animals [38].

It is important to note that the labyrinthine artery that runs in the internal auditory meatus is not a single artery but several smaller arteries, almost like an arterial plexus. Such a series of parallel small-caliber arteries attenuate rapid changes in blood flow (pulsation) and thus provide a smooth (constant) blood supply to the cochlea and the vestibular system. The small diameter arteries in connection with a distal reservoir function as a low-pass filter that attenuates

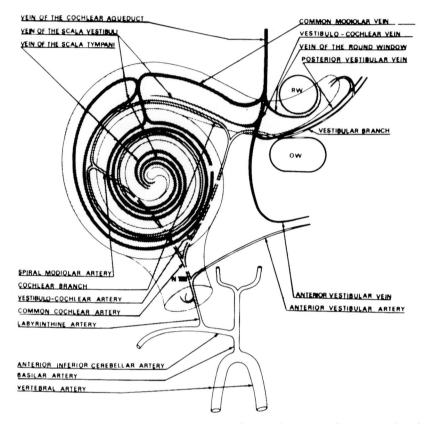

FIGURE 1.16   Arterial blood supply to the cochlea (from Axelsson, A., and Ryan, A. F. (1988). Circulation of the inner ear. I. Comparative study of vascular anatomy in the mammalian cochlea. *In* A. F. Jahn, and J. Santos-Sacchi (Eds.), "Physiology of the Ear" (pp. 295–315). Raven, New York).

fast changes in blood flow. This may be important in avoiding stimulation of the auditory sensory cells from pulsation of the blood supply to the cochlea. This, together with the fact that the labyrinthine artery is an end-artery with little or no collateral blood supply to the cochlea, is important when operating around the VIIIth nerve in the internal auditory meatus, such as in removal of acoustic tumors where the labyrinthine artery may be easily injured causing loss of hearing.

CHAPTER 2

Sound Conduction to the Cochlea

ABSTRACT

1. Sound normally reaches the cochlea via the ear canal and the middle ear, but it may also reaches the cochlea through bone conduction or from sound in the middle ear cavities.
2. The sound pressure at the tympanic membrane depends on the acoustic properties of the pinna, ear canal, and the head.
3. The ear canal acts as a resonator, which causes the sound pressure at the tympanic membrane to be higher than it is at the entrance of the ear canal. The gain is largest near 3 kHz (the resonance frequency) where it is approximately 10 dB.
4. In a free sound field, the head causes the sound pressure at the entrance of the ear canal to be different (mostly higher) than it is when measured at the place of the head without the person being present.
5. The effect of the head on the sound pressure at the entrance of the ear canal depends on the frequency of the sound and on the angle of incidence of the sound (direction to the sound source).

6. The difference in time of arrival of a sound at the two ears is the physical basis for directional hearing in the horizontal plane, together with the difference in intensity of the sound at the two ears.

7. The middle ear acts as an impedance transformer that matches the high impedance of the cochlea to the low impedance of air.

8. The gain of the middle ear is frequency dependent and the increase in sound transmission to the cochlear fluid due to improvement in impedance matching is approximately 30 dB in the mid-frequency range.

9. It is the difference between the force that acts on the two windows of the cochlea that sets the cochlear fluid into motion. Normally the force on the oval window is much larger than that acting on the round window because of the gain of the middle ear.

10. The ear's acoustic impedance is a measure of the tympanic membrane's resistance against being set into motion by a sound.

11. Measurements of the ear's acoustic impedance has been used in studies of the function of the middle ear and for recordings of contraction of the middle ear muscles.

INTRODUCTION

Sound can be conducted to the cochlea mainly through three different routes, i.e., (1) through the tympanic membrane and the ossicular chain, (2) acoustically from sound in the middle ear cavity that reaches the windows of the cochlea directly, and (3) through bone conduction. In the normal ear, the transmission of sound directly to the cochlea is without importance but it is important in some diseases of the middle ear. Bone conduction of airborne sound has little importance for normal hearing but it is important in audiometry where sound applied to one ear by an earphone may reach the other ear by bone conduction (cross transmission).

HEAD, OUTER EAR, AND EAR CANAL

The ear canal, the pinna, and the head influence the sound that reaches the tympanic membrane. The ear canal acts as a resonator and the transfer function from sound pressure at the entrance of the ear canal to sound pressure at the tympanic membrane has a peak at 3 kHz (average 2.8 kHz [111]), where the sound pressure at the tympanic membrane is about 10 dB higher than it is at

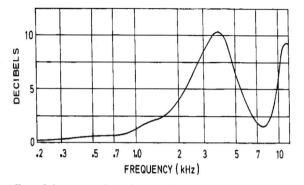

FIGURE 2.1 Effect of the ear canal on the sound pressure at the tympanic membrane. The graph shows the difference between the sound pressure at the entrance of the ear canal and that measured close to the tympanic membrane (from Shaw, E. A. C. (1974). The external ear. *In* W. D. Keidel, and W. D. Neff (Eds.), "Handbook of Sensory Physiology" (Vol. V(1), pp. 450–490). Springer-Verlag, New York).

the entrance of the ear canal (Fig. 2.1).[1] The head acts as an obstacle to the propagation of sound waves. Together the outer ear and the head transform a sound field so that the sound pressure is different (mostly higher) at the entrance of the ear canal compared with the sound pressure that is measured in the place of the head. The effect of the head on the sound at the entrance of the ear canal is related to the size of the head and the wavelength of sound. This means that the "amplification" is frequency (or spectrum) dependent and, therefore, the spectrum of the sound that acts on the tympanic membrane is different from that of the sound field in which the individual is located and it depends on the head's orientation relative to the direction of the sound source.

A solid sphere the size of a head (Fig. 2.2A) has been used as a model of the head in studies of the transformation of sound from a free sound field to that at the tympanic membrane and how that transformation changes when the head is turned at different angles relative to the direction to the sound source. When a sound source is located in front of an observer the sound pressure at the tympanic membrane is approximately 15 dB higher than it is in a free sound field in the frequency range between 2 and 4 kHz (Fig. 2.2B). A dip occurs in the transfer function of sound to the tympanic membrane around 10 kHz. The difference in intensity of the sound that reach the two

[1] Generally, the transfer function of a system is the ratio between its output and its input as a function of the frequency of the input signal. In this specific situation, the transfer function is the ratio between the sound pressure at the tympanic membrane and the sound pressure at the place of the head. When the sound pressure is expressed in decibels, the transfer function is the difference between the (decibel value) of the sound pressure at the tympanic membrane and that at the location of the head.

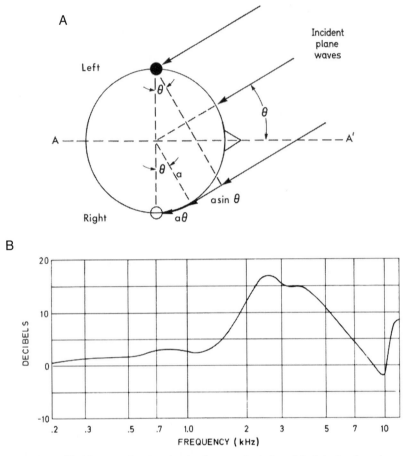

FIGURE 2.2 (A) Schematic drawing showing how a spherical model of the head can be used to study the effect of azimuth on an incident plane sound wave (from Shaw, E. A. C. (1974). Transformation of sound pressure level from the free field to the eardrum in the horizontal plane. *J. Acoust. Soc. Am.* **56**:1848–1861). (B) The combined effect of the head and resonance in the ear canal and the outer ear, obtained in a model of the human head. The difference in sound pressure measured close to the tympanic membrane and a sound pressure in a free sound field with the sound coming from a source located directly in front of the head (from Shaw, E. A. C. (1974). The external ear. *In* W. D. Keidel, and W. D. Neff (Eds.), "Handbook of Sensory Physiology" (Vol. V(1), pp. 450–490). Springer-Verlag, New York).

ears is a result of the head being an obstacle that interferes with the sound field. The head acts as a shield to the ear that is turned away from the sound source, which decreases the sound that reaches that ear. The head acts as a baffle for the ear turned toward the sound source, which increases the sound

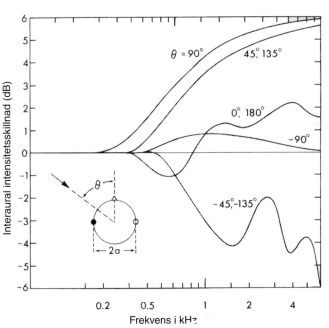

FIGURE 2.3 Calculated differences between the sound pressure (in decibels) at a free field to a model of the head consisting of a hard sphere (Fig. 2.2A). The difference is shown as a function of frequency at different azimuths (from Shaw, E. A. C. (1974). Transformation of sound pressure level from the free field to the eardrum in the horizontal plane. *J. Acoust. Soc. Am.* **56**:1848–1861).

intensity at that ear. Again, these effects of the head on the transfer of sound to the entrance of the ear canal depend on the frequency of the sounds because they are related to the wavelength of sound in relation to the size of the head.[2]

The difference in the sound intensity at the two ears thus depends on the frequency of the sound. It is small at low frequencies because the effect of the head is small for sound of wavelengths that are long compared with the size of the head (Fig. 2.3). In the frequency range between 2.5 and 4 kHz the amplification of sounds by the head and the pinna varies from 21 to 8 dB depending on the angle to the sound source. In a broad frequency range above 1 kHz the intensity of sounds that come from a direction (azimuth) of 45–90°

[2] The wavelength of sound is a function of the propagation velocity of sound and its frequency. It is the propagation velocity divided by the frequency. The propagation velocity of sound is approximately 340 m/s and it depends on the temperature and air pressure. Thus the wavelength of a 1,000-Hz tone is 340/1,000 = 0.34 meter = 34 cm.

relative to straight ahead is approximately 5 dB higher at the entrance of the ear canal than in the area around the listener (Fig. 2.3). The shadow and baffle effects of the head and outer ear contribute to the difference in the sound intensity at the two ears for sounds that do not come from a source located directly in front (0° azimuth) or directly behind (180°) a person.

The results from studies of the effect of the head on the sound pressure at the entrance of the ear canal always refer to the situation where the head is in a free sound field with no obstacles other than the individual on which the measurements are performed. Such a situation occurs in nature with the sound source placed at a long distance and where there is no reflection from obstacles. This is a different situation from an ordinary room where sound reflections from the walls modify the sound field. A free sound field can be artificially created in a room with walls that absorb all sound (or at least most of it) and thus avoid reflection. Such a room is known as an anechoic chamber. Anechoic chambers are used for research such as that of the transformation of sound by the head and the ear canal.

The difference between the sound pressure at the tympanic membranes of the two ears has also been studied using a manniquin equipped with microphones at the place of the tympanic membrane (Fig. 2.4). This model includes the pinna and the results show that the pinna mostly affects transmission of high-frequency sounds. The results of such studies are in good agreement with those using a spherical model of the head. While the studies using a manniquin more accurately mimic the normal situation the results do not include the effect of the absorption of sound on the surface of the normal head. It is also important to point out that the differences in head form and size make results such as those shown in Fig. 2.3 represent the average person only. The transformation of sound from a free sound field to the sound that reaches the tympanic membrane varies between individuals because of differences in the shape of the head.

Because of the ears' position on a human head, sound waves reach each ear at different times (except for sounds from a source located directly in front of the head or directly behind). The difference in arrival time is a nearly linear function of the angle to the sound source (Fig. 2.5). The difference in arrival time is related to the travel time from the sound source and it has a simple linear relation to the azimuth. Sound travels at a speed of about 340 m/s (a little more than 1100 feet per second or about 1.1 feet per millisecond). The maximal difference in arrival time of the two "ears" in the standard model shown in Fig. 2.2A is therefore about 0.6 ms (Fig. 2.5). The values calculated from measurements on a hard spherical model of the head (solid line) agree closely with actual measurements made on a live subject.

Information about the difference in arrival time and the difference in sound pressure at the two ears is used by the central auditory nervous system to decide the direction to a sound source in the horizontal plane (azimuth). It

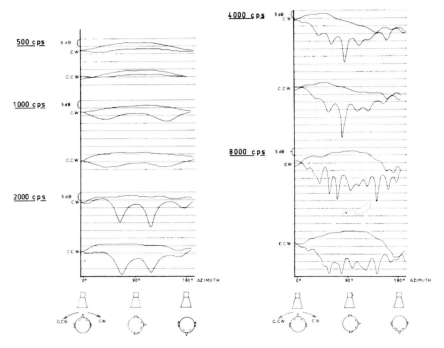

FIGURE 2.4 Sound intensity at the "tympanic membrane" as function of the azimuth measured in a more detailed model of the head (mannequin) than the one shown in Fig. 2.2A. The difference between the sound intensity at the two ears is the area between the two curves (from [105] with permission).

is believed that the intraaural time difference is most important for sounds below 1500 Hz while it is the difference in the intensity that is most important for high-frequency sounds (see Chapter 9).

A change in the direction to a sound source in the vertical plane (elevation) does not cause any change in the intraaural time difference and determination of the elevation must therefore rely on other factors such as changes in the spectrum of broadband sounds which occur as a result of the frequency dependence of the transformation of a sound from the free field to the tympanic membrane. The pinna plays an important role for sound localization in the vertical plane by causing the sound at the tympanic membrane to change as a function of elevation [10]. The effect of the angle to the sound source in the vertical plane on the sound that reaches the two ears is most pronounced above 4 kHz (Fig. 2.6). The sound pressure at the tympanic membrane for 0° azimuth and an elevation of 0° falls off above 4 kHz (solid line in Fig. 2.6). With increasing elevation this upper cutoff frequency shifts toward higher

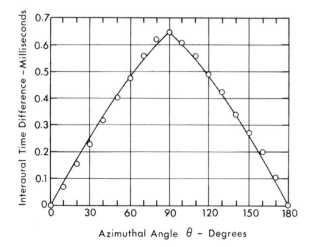

FIGURE 2.5   Calculated intraaural time difference as a function of azimuths for a spherical model of the head (Fig. 2.2A) with a radius of 8.75 cm (solid line) (solid line) and measured values in a human subject (open circles) (from Shaw, E. A. C. (1974). The external ear. *In* W. D. Keidel, and W. D. Neff (Eds.), "Handbook of Sensory Physiology" (Vol. V(1), pp. 450–490). Springer-Verlag, New York).

frequencies (dashed lines). At an elevation of 60° the cutoff is above 7 kHz and at that frequency the sound pressure is more than 10 dB above the value it had at an elevation of 0° [127].

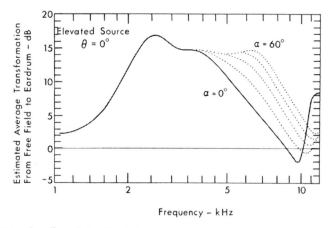

FIGURE 2.6   The effect of elevation (α) on sound pressure at the tympanic membrane (from Shaw, E. A. C. (1974). The external ear. *In* W. D. Keidel, and W. D. Neff (Eds.), "Handbook of Sensory Physiology" (Vol. V(1), pp. 450–490). Springer-Verlag, New York).

Sound Delivered by Earphones

The sound delivered to the ear by earphones is not affected by the acoustic properties of the head. Since the effect of the head, pinna, and ear canal is different for different frequencies, it acts as a spectral filter of sounds. This filter is not effective when earphones are used and that is one reason that music and speech sounds different when listening through ordinary earphones compared to listening in a free sound field. If the sound spectrum that drives earphones is modified (filtered) in a way that imitates the effect of the head, music and speech played through earphones can be made to sound more natural and sound nearly as it does in a (natural) free field [10]. The introduction of such a correction of the spectrum of the input to earphones is the reason sound produced by the common "Walkman" type of tape players sounds natural, giving an impression of "sound space." The effect of turning the head when listening in a free field, however, is absent when listening to earphones.

Supraaural earphones are in general use for testing hearing where one of the requirements is to deliver sound only to one ear at a time. The earphones that are commonly used for audiometric purposes such as the TDH 39 were designed in the 1930s but have remained in use mainly because of the availability of standard calibrations for these earphones. Such supraaural earphones also conduct sound to the other ear, by bone conduction, and the sound delivered to the opposite ear is about 60 dB below that delivered to the ear to which the earphone is applied. That cross-transmission is the reason it is necessary to mask the better hearing ear when testing the hearing in individuals with large differences between hearing thresholds in the two ears. Insert earphones have much less cross-transmission (Fig. 2.7A). For frequencies below 1 kHz the cross-transmission is below 80 dB. Insert earphones have roughly the same frequency characteristics as supraaural earphones but concerns about the accuracy of the calibration remain. Insert earphones also provide much higher attenuation of external noise than supraaural earphones (Fig. 2.7B).

MIDDLE EAR

Two problems are associated with transfer of sound to the cochlear fluid. One is related to the fact that sound is ineffective in setting a fluid into motion because of the large difference in the acoustic properties (impedance) of the two media, air and fluid. Because of the difference in the impedance of the two media, about 99.9% of the sound energy is reflected at the interface between air and fluid and only 0.1% of the energy is converted into vibrations of the fluid. The other problem is related to the fact that it is the difference between the force that acts at the two windows that causes the cochlear fluid to vibrate. Both these problems are elegantly solved by the middle ear. The middle ear acts as an impedance transformer that matches the high impedance of the cochlear fluid to the low impedance of air, thereby improving sound

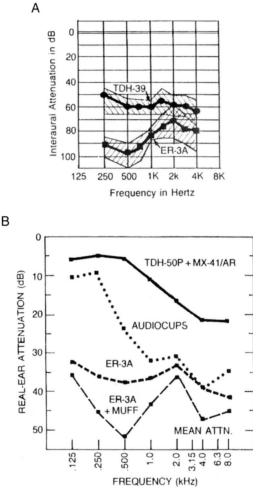

FIGURE 2.7   (A) Average and range of intramural attenuation obtained in six subjects with two types of earphones (TDH 39 and an insert earphone, ER-3) (from Killion, M. C., Wilber, L. A., and Gudmundsen, G. I. (1985). Insert earphones for more interaural attenuation. *Hear. Instrum.* 36:34–36). (B) External noise attenuation of four different earphones often used in audiometry (from Killion, M. C., and Berger, E. H. (1987). Noise attenuating earphone for audiometric testing. *J. Acoust. Soc. Am.* (Suppl. 1) 81:S5).

transfer to the cochlear fluid. By increasing the sound transmission selectively to the oval window of the cochlea, the middle ear creates a difference in the force that acts on the two windows of the cochlea.

MIDDLE EAR AS AN IMPEDANCE TRANSFORMER

Theoretical considerations show that the transmission of sound to the oval window would be improved by 36 dB if the middle ear acted as an ideal impedance transformer with the correct transformer ratio. However, the transformer ratio of the human middle ear is not quite optimal and that causes some of the sound to be reflected at the tympanic membrane.

Transfer of Sound to the Cochlear Fluid

The specific impedance of air is 42 cgs units and that of water 1.54×10^5 cgs units (41.5 dynes second/cm^3 and 144,000 dynes second/cm^3), thus a ratio of approximately 1:4,000. Transmission of sound to the oval window will therefore be optimal if the middle ear has a transformer ratio that is equal to the square root of 4,000 (which is 63). This assumes that the input impedance to the cochlea is equal to that of water. In fact it is less. Studies in the cat show that the input impedance of the cochlea is lower at low frequencies than at high frequencies. In the middle frequency range the impedance of the cochlea is about that of seawater [79, 118, 119]. Lynch [79] and Rosowski [118] calculated the overall effectiveness of transferring sound from a free field to the cochlear fluid for the cat (Fig. 2.8). Before that, Siebert [130] suggested that the important matter regarding the impedance match between sound in air and motion of the cochlear fluid is how much acoustic power is transferred to the cochlea.

The impedance transformer action of the middle ear is mainly accomplished by the ratio between the surface area of the tympanic membrane and the area of the stapes footplate, but the lever ratio of the middle ear bones also contri-

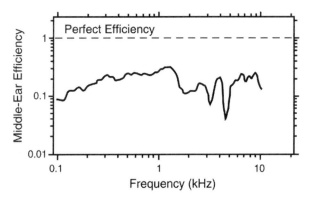

FIGURE 2.8   The efficiency of the cat's middle ear, showing the fraction of sound power entering the middle ear and being delivered to the cochlea (from Rosowski, J. J. (1991). The effects of external- and middle-ear filtering on auditory threshold and noise-induced hearing loss. *J. Acoust. Soc. Am.* 90:124–135).

butes. The ratio of the area of the tympanic membrane to that of the stapes is frequency dependent because it is the effective area of the tympanic membrane and not its geometrical (anatomical) area that determines the transformer ratio.[3] The middle ear has mass and stiffness that make its transmission properties frequency dependent. Its efficiency as an impedance transformer thus becomes a function of frequency. Stiffness impedes the motion mostly at low frequencies and mass impedes motion mostly at high frequencies. The friction in the middle ear causes loss of energy that is independent of frequency. The lever ratio may be frequency dependent because the mode of vibration of the ossicular chain is different at different frequencies. Because the sound transmission through the middle ear is frequency dependent, the transformer ratio also becomes frequency dependent and it is thus an oversimplification to express the transformer action as a single number. The transformer ratio of the middle ear must be described by a function of frequency, i.e., its frequency transfer function).[4]

Estimates of the gain of the middle ear by different investigators vary and there are systematic differences between results obtained in humans and in animals. The total efficiency of the human middle ear is about 10 dB less than ideal for frequencies up to about 200 Hz and its highest efficiency is attained around 1000 Hz, where it is about 3 dB below that of an ideal impedance transformer [117]. Above 1500 Hz the efficiency (in percentage of energy transferred to the cochlea) varies between 20% and 4% (Fig. 2.8), corresponding to losses of 5 and 25 times (7 and 14 dB), respectively.

Measurement of Sound Transmission through the Middle Ear

Direct measurements of the sound transmission through the middle ear have been performed both in anesthetized animals and in human cadaver ears. The small structures of the middle ear and the necessity of measuring very small vibration amplitudes require a high degree of technical skill on the part of the experimenter and advanced technical equipment in order to measure accurately the sound-conducting properties of the middle ear without disturbing its func-

---

[3] The effective area of the tympanic membrane is a rigid, weightless piston that transfers sound in the same way as the membrane.

[4] The frequency transfer function of a transmission system is the ratio between the output and the input plotted as a function of the frequency of a sinusoidal input signal. Such a plot is not a complete description of the transmission properties of a system unless the phase angle between the output signal and the input signal as a function of the frequency is included. This is known as a Bode plot. Nevertheless, often only the amplitude function is shown, often expressed in logarithmic measures (such as decibels, dB).

tion. The ratio between the vibration of the cochlear fluid and the sound pressure of the tympanic membrane has been measured in some studies and in other studies the vibration amplitude of the cochlear fluid has been measured. The middle ear in humans is different from that in animals usually used in auditory experiments; thus, it is important to distinguish between results obtained in human cadavers and in live animals. How to "translate" the results of such different experiments into estimates of sound transmission in live humans is discussed below.

Many of the results of experiments on the function of the middle ear that were obtained almost 60 years ago are still valid. Subsequent studies have confirmed many of these results and other studies have extended our knowledge about the function of the middle ear. More accurate determinations of the gain in transmission of sound to vibration of the cochlear fluid of the middle ear have been obtained more recently using modern techniques.

TRANSFER FUNCTION OF THE MIDDLE EAR

Some of the earliest studies of the frequency transfer function of the middle ear were done in human cadaver ears by von Békésy, the results of which are shown in Fig. 2.9. One of the first animal studies that quantitatively measured the gain of the cat's middle ear in transferring sound to the cochlea was published by Wever, Lawrence, and Smith [150] (Fig. 2.10A).

Animal Studies

The gain of the cat's middle ear is frequency dependent and it is largest in the frequency range between 500 and 10,000 Hz, where it is about 35 dB.

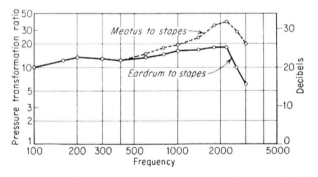

FIGURE 2.9   Ratios of sound pressure between that measured in the ear canal and that measured at the stapes (dashed lines) and between the tympanic membrane and the stapes (from Békésy, von G. (1941). Article 38. *In* G. v. Békésy (1960), "Experiments in Hearing." McGraw-Hill, New York).

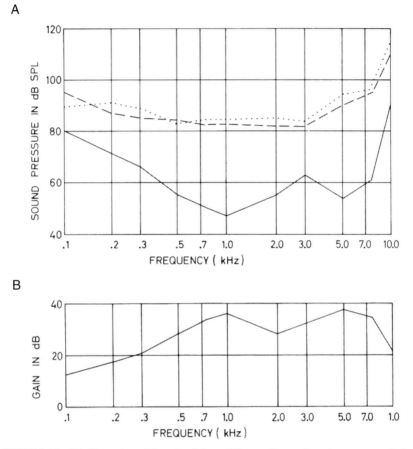

FIGURE 2.10 (A) Illustration of the gain of the middle ear of a cat. Sound pressure needed to produce a CM of an amplitude of 10 μV is shown with the middle ear intact and the sound conducted to the tympanic membrane (solid lines) and after removal of the middle ear and the sound conducted to the oval window (dashes) and round window (dots) separately using a closed sound delivery system (from [150]). (B) Difference between the dotted-dashed curves and the solid curve in A (from [150]).

The transfer function of the middle ear is usually determined using pure tones of different frequencies and measuring the sound pressure at the tympanic membrane that is required to produce cochlear microphonic (CM) potentials of a certain amplitude.[5] Usually the sound pressure that evokes a 10-μV CM response is

[5] The CM is generated in the cochlea and its amplitude is closely related to the volume velocity of the cochlear fluid. The CM in response to pure tones is a sinusoidal waveform the amplitude of which increases with the increase in the sound pressure of the sound that elicits the CM. Recording of the CM is often used to determine changes in sound transmission of the middle ear. The generation of the cochlear microphonic potential (CM) is discussed in detail in Chapter 4.

determined in the frequency range of interest (e.g., from 100 Hz to 10 kHz). To determine the gain of the middle ear such measurements are first done while the middle ear is intact and then repeated after the middle ear is removed surgically. The sound is led directly to the oval window (dashes in Fig. 2.10A) or to the round window (dots in Fig. 2.10A) using a speculum that was attached to the bone of the cochlea. This arrangement ensures that sound only reaches one of the two cochlear windows at a time. When the sound is conducted directly to either the round or the oval window a much higher sound level is needed to obtain a 10-μV CM potential than when conducted via the normal route with the middle ear intact. The difference between the solid curve in Fig. 2.10A and the dotted or the dashed curves are measures of the gain in sound conduction to the cochlea provided by the cat's middle ear (Fig. 2.10B).

In the experiments described above sound was led to only one of the two windows of the cochlea at a time. If sound is led to the middle ear cavity, a different situation arises because sound will then reach both the oval window and the round window with about the same intensity. (Hearing loss due to middle ear damage is discussed in Chapter 13.)

More recently, the transfer function of the middle ear has been studied in anesthetized cats by measuring the vibration amplitude of the stapes using microscopic techniques with stroboscopic illumination [45] or by using a capacitive probe to measure the vibration of the round window [88] (Fig. 2.11). The vibration of the round window represents the motion of the fluid in the cochlea because the cochlear walls are assumed to be rigid and it is a measure of the excursions or the "pumping" action of the footplate of the stapes.

The vibration amplitude of the manubrium of the malleus in a cat for a constant sound pressure at the tympanic membrane (Fig. 2.12) is nearly constant up to 1200 Hz, above which it decreases at a rate of approximately 12 dB/octave up to 4000 Hz. [The frequency of maximal vibration amplitude (approximately 1200 Hz) is often called the resonance frequency although that notion would indicate that the middle ear was a simple combination of mass, elasticity, and friction (see the insert in Fig. 2.12). This is an oversimplification and it is discussed later in this chapter]. Above 4000 Hz the vibration amplitude again increases and the reason for that is not known. The vibration amplitude of the round window follows that of the incus for low frequencies and becomes smaller than that of the incus for frequencies above 600 Hz, probably as a result of the elasticity in the incudo-stapedial joint. (Elasticity in a joint that conveys mechanical vibration reduces transfer of vibrations of high frequencies.) The middle ear transfers sounds of very high frequencies effectively despite the elasticity in the incudo-stapedial joint. The reason for that is probably that the incudo-stapedial joint is much less flexible in a direction of the pistonlike motion of the stapes than in the direction perpendicular to the pistonlike motion. (The flexibility of the incudostapedial joint in that direction is evident from the fact that contractions of the stapedius muscle can cause large movements that tilt the stapes.)

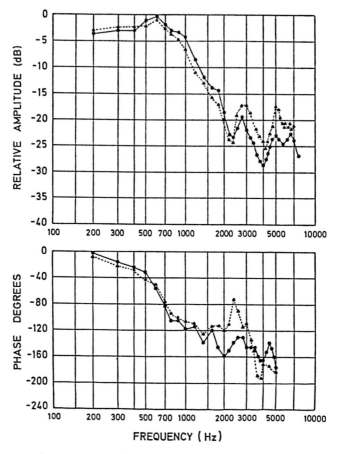

FIGURE 2.11 Vibration amplitude of the round window (circles and solid lines) and the incus (triangles and dashed lines) of the ear of a cat for constant sound pressure at the tympanic membrane. The vibration amplitude was measured using a capacitive probe (from [88]).

In the cat, the ratio between the pressure in the cochlear fluid and the sound pressure at the tympanic membrane corresponds to a gain in pressure that is slightly below 30 dB in the frequency range between 500 and 10 kHz, above and below which the gain is lower [104]. More recently, Merchant et al. [83] arrived at gain values of about 20 dB between 250 and 500 Hz with a maximum of 25 dB at 1 kHz, above which the gain decreases at a rate of 6 dB/octave. These results by different investigators thus show a gain of the middle ear that is in the range of 25 to 30 dB.

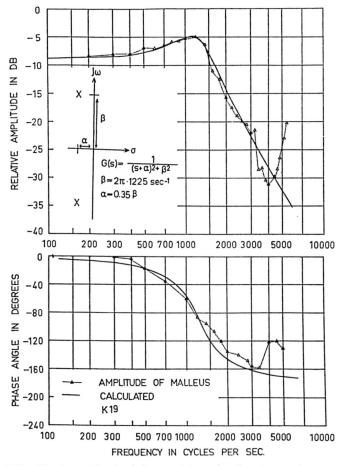

FIGURE 2.12 Vibration amplitude of the manubrium of malleus in a cat for constant sound pressure at the tympanic membrane. (Insert) A simple circuit that contains one mass component, one stiffness component, and a friction component (from Møller, A. R. (1963). Transfer function of the middle ear. *J. Acoust. Soc. Am.* **35**:1526–1534).

Transfer Function of the Human Middle Ear

Despite the development of equipment, such as laser interferometry, that can measure very small displacements, equipment and experimental methods have yet to be developed for measurements of the stapes displacement in living humans. Measurements of the transfer function of the human middle ear are therefore limited to studies in cadavers. The ratio between the vibration amplitude of the ossicles (umbo and the stapes) in human cadaver ears and the sound pressure close to the tympanic membrane (Fig. 2.13) [46, 70] reveals similar transfer functions as those obtained in animals. The displacement amplitude of the ossicles is nearly constant for low frequencies up to the resonance frequency of the middle ear (about 900 Hz). These results are similar to those obtained by von Békésy and published in 1941 [7], thus more than 50 years earlier than the data shown in Fig. 2.9.[6] The similarity between these results and those obtained using modern techniques is remarkable in the light of the technical difficulties associated with such measurements at the time that von Békésy did these studies. The findings by Kurukawa and Goode [70] also showed a considerable individual variation, attributed mainly to variations in the function of the tympanic membrane.

The irregularities in the transfer function of the middle ear seen in Figs. 2.8–2.13 suggest that the function of the middle ear is more complex than a combination of a few elements of mass and stiffness. Several models of the middle ear were developed during the past 3 or 4 decades to account for such complexity [87, 118, 164].

IMPULSE RESPONSE OF THE HUMAN MIDDLE EAR[7]

Using a capacitive probe for measuring the vibrations of different parts of the middle ear, von Békésy measured the response of the manubrium of the malleus to a sharp impulse (Fig. 2.14A). He then transformed these data to obtain the frequency transfer function of sound to the vibration of the manubrium of the malleus (Fig. 2.14A) by using a mechanical device for computing the spectrum of the vibration in response to sound impulse. These results were published 1937. The equipment that von Békésy used was of his own design and his work often involved ingenious inventions and experimental methods that were far ahead of his time. Instead of obtaining the impulse response directly, other investigators [92] have estimated the impulse response of the cat's middle ear (Fig. 2.14B) by computing the inverse

---

[6] All results reported by von Bekesy were taken from the book *Experiments in Hearing* (G. von Bekesy, 1960, McGraw–Hill, New York). This book contains translations of original articles by von Bekesy, published in German. When referenced in this book, the date (year) of the original publication is used along with the reference to the 1960 book. The reason is to give proper credit to the work of von Bekesy and to emphasize when the work was first published. For practical purposes, references to the 1960 book are also given because that book is more readily available than the original articles.

[7] The impulse response of a transmission system such as the middle ear is, by definition, the response to an infinitely short impulse. In practice the impulse response is obtained by applying a short impulse to the system that is tested. There is a mathematical relationship between the impulse response and the frequency transfer function known as the Fourier transform. This mathematical operation can convert an impulse response into a transfer function and the inverse Fourier transform can convert a transfer function into an impulse response.

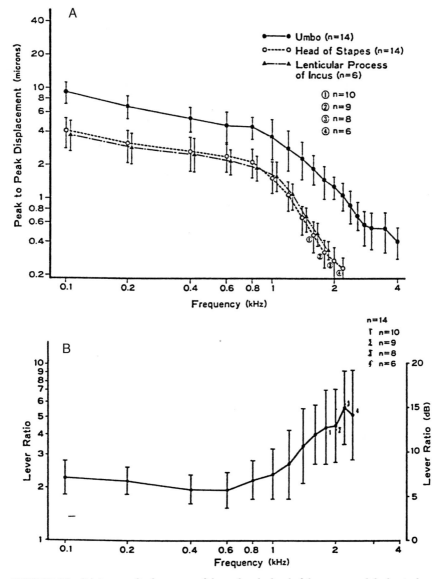

FIGURE 2.13 (A) Average displacements of the umbo, the head of the stapes, and the lenticular process of the incus. (B) The lever ratio at 124 dB SPL at the tympanic membrane in 14 temporal bones. Vertical bars indicate 1 standard deviation (from [46] with permission).

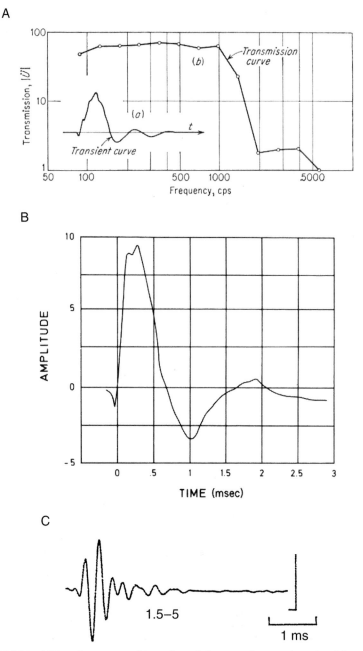

FIGURE 2.14 (A) Impulse response of the malleus of a human cadaver ear (insert) and the transfer function computed from the impulse response (solid line and circles) obtained in experiments, the results of which were published 1937 (from Békésy, von G. (1937). Article 28. *In* G. v. Békésy (1960), "Experiments in Hearing." McGraw-Hill, New York). (B) Impulse response calculated from the transfer function of the middle ear of a cat shown in Fig. 2.11 (from [92]). (C) Impulse response of the umbo obtained in a human (from Svane-Knudsen, V., and Michelsen, A. C. (1985). The impulse response vibration of the human ear drum. *In* "Lecture Notes in Biomathematics" (Vol. 64, pp. 21–27). Springer-Verlag, Berlin).

Fourier transform of the frequency transfer functions such as those seen in Fig. 2.11. These calculations show the displacement of the cochlear fluid in a cat's ear as it would be in response to a brief sound impulse.

More recently, direct measurements of the impulse response of the umbo in awake human volunteers were obtained by Svane-Knudsen and Michelsen [138] by applying an acoustic impulse to the ear and using laser Doppler shift (laser Doppler vibrometer, LDV) to measure the displacement of the umbo (Fig. 2.14C). Goode et al. [42] used a similar method using commercially available LDV equipment to measure the vibration amplitude of the umbo in human volunteers. Although such measurements do not reflect the transmission properties of the middle ear but rather reflect the ability of the tympanic membrane to transform sound into vibration of the manubrium of the malleus, this method might become a useful clinical method for testing the function of the middle ear.

LINEARITY OF THE MIDDLE EAR

The assumption that the middle ear functions as a linear system was supported by the experimental work by Guinan and Peake [45], who found that the stapes (in the cat), at frequencies below 2 kHz, moves in proportion to the sound pressure at the tympanic membrane up to 130 dB SPL and even higher (140–150 dB) for frequencies above 2 kHz.

A transmission system must fulfill several criteria in order to be regarded as a linear system. The output must increase proportionally to the input and if two different input signals (such as two tones with different frequencies) are applied to the input of a linear system the output must be the sum of the output in response to each of the two signals when applied independently. This is known as the superposition criterion of a linear system. The output of a linear system to which one or more sinusoidal signals (e.g., tones) are applied only contains energy at the same two frequencies as the input. The transmission properties of a linear system can be determined by using different kinds of input signals in connection with mathematical operations on the results.

The properties of a nonlinear system cannot be described in a universal way. The transfer function of a nonlinear system may depend on the level of the input, and spectral components other than those present at the input may appear at the output. Which components are generated depends on the type of nonlinearity. The properties obtained by using one kind of test signal may be different from the properties obtained using other test signals. Nonlinear systems are therefore much more difficult to describe than linear systems. This is one of the reasons why most human-made systems, such as electronic filters and amplifiers, are designed to be linear systems within their normal working range of input amplitudes. This is in contrast to most biologic systems that are nonlinear, at least over a large part of their working range. Only recently has it been recognized that linear systems are not always optimal transmission and control systems. The middle ear, being a linear system, is an exception, but the cochlea is a nonlinear spectrum analyzer and the transduction of sounds into a neural code is highly nonlinear.

ACOUSTIC IMPEDANCE OF THE EAR

The ear's acoustic impedance is a measure of the resistance of the tympanic membrane to being set in motion by sound. Studies of the ear's acoustic impedance can provide important insight into how the middle ear functions, including the role of the different parts of the middle ear in transferring sound into vibration of the cochlear fluid and for studies of pathologies of the middle ear. Measurement of the acoustic impedance of the ear has not only played an important role in studies of the function of the middle ear but it is now used routinely in diagnosing disorders of the middle ear. Measurement of changes in the ear's acoustic impedance is used to record the contractions of the middle ear muscles to study the function of the acoustic middle ear reflex for research purposes as well as in routine otoneurologic diagnosis.

Basic Concepts of Impedance

Mechanical and acoustic systems are often described by their electrical analog circuits because many people are more familiar with electrical circuits than with acoustic and mechanical systems. Explaining the impedance concept of the ear using an electrical analogy may therefore be appropriate. Per definition, the impedance of an electrical system is the resistance with which an applied voltage induces an electrical current in the circuit. In the simplest of all systems, a single resistor, the impedance is the voltage that is needed to set up a unit current; thus, using Ohm's law and knowing the voltage (V) and the current (I) makes it possible to determine the resistance (R): $R = V/I$. When a circuit contains other elements such as capacitors and inductors the impedance can no longer be described by a single number. Instead the impedance (Z) becomes a complex quantity that requires two numbers to be described. A complex quantity can be described by its real and its imaginary components (R and jX, in Fig. 2.15, where j denotes an imaginary quantity). A complex quantity can also be described by an absolute value (length of a vector) and the phase angle (of the vector) (Fig. 2.15). The impedance of a resistor is independent of the frequency at which it is tested, but the impedance of a capacitor decreases as a function of the frequency at which it is measured and that of an inductor increases as a function of frequency. The impedance of a capacitor and an inductor, which are common components of electrical circuits, are pure imaginary values that have opposite signs. Since the impedance of a capacitor and an inductor have opposite signs and the impedance of a capacitor decreases as a function of the frequency and that of an inductor increases, the impedance of a circuit that contains a capacitor and an inductor will be zero at a certain frequency (Fig. 2.15). That frequency is known as the resonance frequency.

Mechanical Impedance

Electrical circuits and mechanical systems are analogous in many ways. Thus current in an electrical circuit, corresponds to vibration velocity in a mechanical system and electrical voltage corresponds to mechanical force. The mechanical impedance is therefore the ratio between force (F) and velocity (V). The different components that make up a mechanical system also have analogies in an electrical

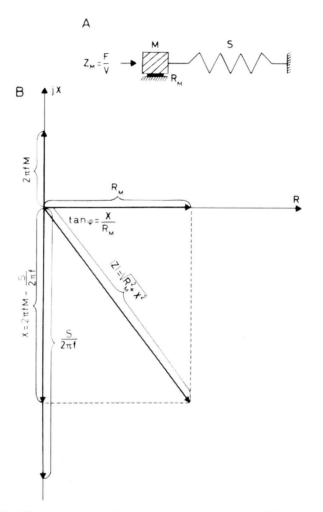

FIGURE 2.15 (A) A simple mechanical system consisting of a mass (M), elasticity (S), and friction (R). (B) Relationship between the different elements of the impedance ($Z = R + jX$) and the frequency (f) of the mechanical system in A (from [89]). (C) The mechanical system in A equipped with a rigid piston to form an acoustic system. (D) Electrical analog of the mechanical system in A (from [89]).

circuit. Thus mechanical friction corresponds to an electrical resistance, mass (or inertia) corresponds to inductance, and a spring (elasticity) corresponds to capacitance. The impedance of the friction is the real component of the impedance and the combined impedance of the mass and the elasticity is the imaginary component.

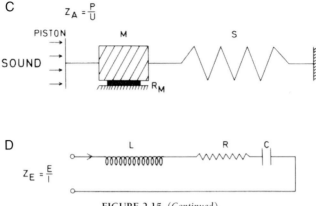

FIGURE 2.15 (*Continued*)

Acoustic Impedance

In an acoustic system volume velocity corresponds to electrical current and sound pressure corresponds to voltage. The acoustic impedance is thus the ratio between sound pressure and volume velocity. Friction in an acoustic system corresponds to electrical resistance and a volume of air corresponds to a capacitor in an electrical circuit. A narrow passage such as that of a narrow tube corresponds to an inductor in an electrical circuit. In studies of the ear, it is the mechanical impedance of the ear transformed to acoustic impedance by the tympanic membrane that is of interest. A mechanical system such as the middle ear is converted into an acoustic system by a piston or a membrane such as the tympanic membrane that converts sound pressure into mechanical force. If the tympanic membrane acted as an ideal piston the mechanical impedance would be the acoustic impedance divided by the surface area of the piston, assuming that appropriate units of measure were used to describe the acoustic and mechanical impedance. How the acoustic impedance of the ear reflects the mechanical properties of the middle ear may be understood by considering a simplified mechanical model of the middle ear system equipped with a piston (Fig. 2.15C) [89].

Admittance

The admittance (Y) is the inverse of the impedance. It is also known as the compliance, because it is a measure of how easily a current is induced in an electrical system or how easily a mechanical system is set into vibration by an external force. In an electrical circuit, the admittance is the current divided by the voltage. In a mechanical system, the impedance is the velocity divided by the force and in an acoustic system, the admittance is the volume velocity divided by the sound pressure. The admittance may be a complex quantity with a real component (G) and an imaginary component (jB).

MEASUREMENT OF THE EAR'S ACOUSTIC IMPEDANCE

The ear's acoustic impedance has been measured in both animals and humans for studies of the function of the middle ear and for studies of pathologies of the middle ear. Despite considerable efforts during many years, measurements of the absolute value of the ear's acoustic impedance never became a useful clinical diagnostic tool. Instead, measurements of changes in the ear's acoustic impedance came into general clinical use for the purpose of measuring the air pressure in the middle ear cavities (tympanometry) and for recording the response of the acoustic middle ear reflex (see page 66).

History of Measurements of the Ear's Acoustic Impedance in Humans

The acoustic impedance is the ratio between sound pressure at the tympanic membrane and the resulting volume velocity in front of the tympanic membrane, but that cannot be directly measured. All methods used so far have measured the acoustic impedance more or less indirectly. The earliest reported measurements of the ear's acoustic impedance were made to gain information about the acoustic properties of the ear in order to improve the design of earphones [40, 145]. Zwislocki [161] measured the ear's acoustic impedance using psychoacoustic methods. The Schuster bridge was used for studies of the ear's acoustic impedance by Metz [84], who was probably the first to publish a study of comparison of the acoustic impedance of normal and pathologic ears. The Schuster bridge, described 1934 [126], is an acoustic equivalent to the Wheatstone bridge for measurement of electrical impedance. The Schuster Bridge, however, was a complex mechanical device that was difficult to use and required the subject's head to be fixated and the measurements were very time consuming.

Zwislocki [161] was probably the first to describe the use of a practical electroacoustic measuring device for direct measurements of the ear's acoustic impedance. He used the measurements of the acoustic impedance he obtained for developing electrical and mathematical models of the middle ear [162]. Subsequently other methods were described for measurements of the ear's acoustic impedance in humans as well as in animals. However, extensive calibrations and computational processing of the obtained measurements were necessary in order to obtain the estimates of the ear's acoustic impedance. These measurements were made before electronic (digital) computers were readily available and such calculations were therefore time-consuming tasks. A modification of the Schuster acoustic bridge was described by Zwislocki, but that instrument never came into general clinical use. It was instead the development of electroacoustic measuring devices to record changes in the ear's acoustic impedance by Terkildsen and Nielsen [142] that became useful clinical tools. These devices use a high impedance sound source connected to the sealed ear canal together with a microphone for measurements of the sound pressure in the ear canal.

ACOUSTIC IMPEDANCE OF THE HUMAN EAR

The acoustic impedance of the human ear can be expressed either as its absolute value and phase angle or as a real and an imaginary component as a function of the frequency. It has been shown that the resistive (real) component varies very little as a function of the frequency while the imaginary (reactive) component is high at low frequencies and decreases with increasing frequency up to about 1000 Hz, indicating that the ear's impedance is dominated by the stiffness of the middle ear. Both the real and the imaginary components vary considerably between individuals (Fig. 2.16) [87]. This occurs even between individuals with normal hearing and no history of middle ear diseases. The variations in the im-

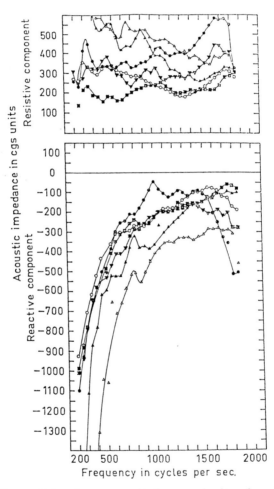

FIGURE 2.16   The acoustic impedance measured in the ear canal and transformed to the estimated plane of the tympanic membrane in six individuals with no known ear disorders (from Møller, A. R. (1961). Network model of the middle ear. *J. Acoust. Soc. Am.* **33**:168–176).

pedance obtained in different individuals are a result of permanent individual differences as evidenced by the fact that measurements of the acoustic impedance in the same individual show a high degree of reproducibility (Fig. 2.17) [86]. This finding also excludes that technical problems was a cause of the variability between individuals. It is this variability that has prevented the use of measurements of the ear's acoustic impedance for diagnosis of middle ear disorders.

> This variation between individuals has several causes. When the tympanic membrane in humans was covered with a thin layer of collodion, the individual variations in the acoustic impedance became smaller and the small irregularities in the curves of the acoustic impedance decreased, indicating that the individual variation and the irregularities in the impedance function result from the tympanic membrane (Fig. 2.18). The irregular pattern of the acoustic impedance of the human ear (Fig. 2.16) is likely to be caused by the properties of a triangular-shaped portion of the tympanic membrane known as the pars flaccida membrana tympani. This part is relatively loose and its vibrations are not transferred to the manubrium of the malleus as effectively as vibrations of other parts of the membrane. Similar irregularities are not present in the ear's acoustic impedance of animals such as the cat, probably because the cat's tympanic membrane does not have as large a pars flaccida. The fact that the irregularities in the acoustic impedance of the human

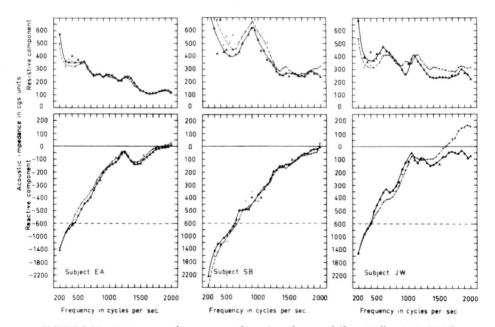

FIGURE 2.17   Acoustic impedance measured at a 2-week interval (from Møller, A. R. (1960). Improved technique for detailed measurements of the middle ear impedance. *J. Acoust. Soc. Am.* **32**:250–257).

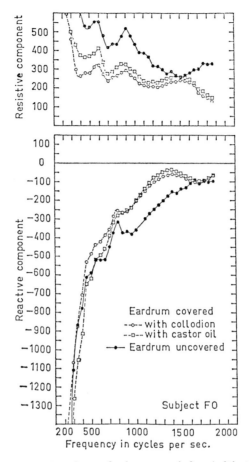

FIGURE 2.18 The acoustic impedance of a human ear before (solid circles) and after the tympanic membrane was covered with a thin layer of collodion (open circles) and castor oil (open squares) (from Møller, A. R. (1961). Network model of the middle ear. *J. Acoust. Soc. Am.* **33**:168–176).

ear disappear when the tympanic membrane is covered with collodion or castor oil (Fig. 2.18) supports the hypothesis that parts of the tympanic membrane is loose [87]. Collodion and castor oil probably add stiffness to the tympanic membrane.

There are other factors that may contribute to the variations in the ear's acoustic impedance. Electroacoustic equipment (and the Schuster Bridge) does not measure the acoustic impedance in the plane of the tympanic membrane but at a certain distance in front of the tympanic membrane. That requires subtraction of the effect of the air volume between the tip of the measuring device and the tympanic membrane and the inner impedance of the measuring device itself. Determination of that volume has considerable uncertainties and that contribute to the variability in the measured impedance. It has been suggested that the acoustic admittance (inverse of the impedance) be used instead of the impedance for determining the resonance frequencies of the middle ear because it is less affected by the inaccuracy in determining the volume of the air that is between the measuring device and the tympanic membrane [86].

CONTRIBUTIONS OF INDIVIDUAL PARTS OF THE MIDDLE EAR TO ITS IMPEDANCE

Studies of the contribution of the different parts of the middle ear to its impedance have been done in animal experiments where the middle ear can be altered experimentally. The possibilities of manipulating the human middle ear are naturally much more limited than what is the case in animals but the use of pathologies for such studies can provide useful information about the function of the middle ear. The effect of pathologic changes such as the immobilization of the ossicular chain as it occurs in patients with otosclerosis has been used in development of electrical and mathematical models of the human middle ear [162]. However, the normal individual variability of the function of the middle ear is an obstacle in using results such as those obtained in patients with stapes immobilization (otosclerosis). Measurements of the ear's acoustic impedance provide an important link between the human ear and the ear of animals that can make data from experiments in animals more useful in explaining the function of the human middle ear.

Functional Properties of the Tympanic Membrane

Measurements of the ear's acoustic impedance indicate that the tympanic membrane does not function as a rigid piston but has a more complex pattern of motion. The properties of the tympanic membrane can be determined by measuring its own acoustic impedance; that is, the ear's impedance when the manubrium is prevented from vibrating. This can be done in animals by comparing the ear's acoustic impedance before and after the malleus is glued to the wall of the middle ear cavity. When the malleus is immobilized the vibrations of the tympanic membrane are not transferred to a motion of the malleus and the measured acoustic impedance is that of the tympanic membrane itself. In the cat the acoustic impedance of the tympanic membrane with the malleus immobilized is very high for frequencies below 3 kHz (Fig. 2.19) [90] indicating that it functions in a similar way as a rigid piston for those frequencies. These results do not provide information regarding whether the equivalent area of this "piston" is different for different frequencies.

Comparing the ear's acoustic impedance with the vibration velocity of the malleus for constant sound pressure at the tympanic membrane can provide

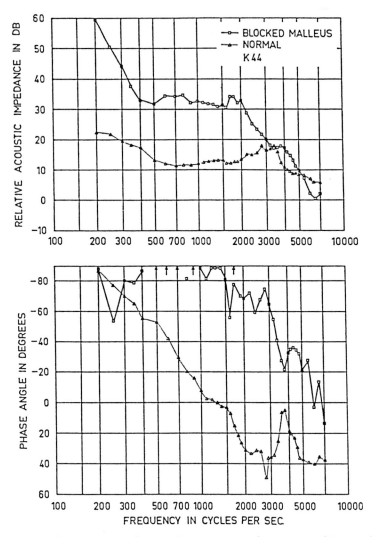

FIGURE 2.19 The acoustic impedance at the tympanic membrane measured in a cat before (dashed lines and triangles) and after immobilization of the ossicular chain (solid lines and squares)(from [90] with permission).

information about the ability of the tympanic membrane to convert sound into vibration of the manubrium of malleus (Fig. 2.20).

If the tympanic membrane functions in the same way as an (ideal) piston, the mechanical force that acts on the manubrium of the malleus is proportional to the

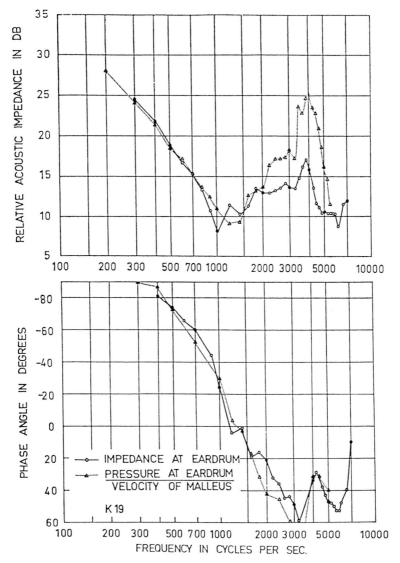

FIGURE 2.20 Comparison of the acoustic impedance at the tympanic membrane with the inverse velocity of the malleus for constant sound pressure at the tympanic membrane in a cat. The impedance is given in decibels relative to 100 cgs units and the inverse vibration velocity is given in arbitrary decibel values. The inverse vibration velocity is expressed in arbitrary units and the two curves were made to superimpose at low frequencies (from Møller, A. R. (1963). Transfer function of the middle ear. *J. Acoust. Soc. Am.* **35**:1526–1534).

sound pressure at the tympanic membrane. The ratio between the vibration velocity of the malleus and the sound pressure will then be equivalent to the velocity of the manubrium divided by the force that acts on the membrane thus the inverse impedance (i.e., admittance). This means that measurement of the vibration velocity of the malleus (for constant sound pressure) is a measure of the ability of the tympanic membrane to convert sound into vibration of the malleus, thus a measure of the function of the tympanic membrane. The velocity of the vibration is the first derivative of the amplitude and the velocity for sinusoidal vibrations (at constant sound pressure level) can therefore be computed from the vibration amplitude by multiplying it with the frequency, which is the same as adding 6 dB/octave to the amplitude when the amplitude is expressed in decibels.

The two curves in Fig. 2.20 showing the acoustic impedance and the inverse velocity of the malleus in the cat are parallel for low frequencies but deviate above 2 kHz, indicating that the tympanic membrane functions similarly to a rigid piston for frequencies only up to about 2 kHz. For frequencies above 2 kHz, the effective area of the tympanic membrane changes with the frequency. The results of experiments obtained in the cat may not be directly applicable to the human ear because the tympanic membrane in humans has a more complex pattern of vibration and it may be less stiff than that of the cat. Studies of the human tympanic membrane done in cadaver ears [61] showed that the tympanic membrane has a smaller effective area at high frequencies than it has at lower frequencies.

Contribution from the Cochlea to the Ear's Acoustic Impedance

Experiments in cats and rabbits show that severing the connection between the incus and the stapes (the incudo-stapedial joint) reduces the resistive component of the ear's acoustic impedance below 4 kHz to very small values (Fig. 2.21) [90] suggesting that the real component (friction) of the ear's acoustic impedance is mainly contributed by the cochlea. Elimination of the friction component of the middle ear makes the resonance of the middle ear more pronounced. Below 4 kHz the reactive (imaginary) component of the ear's acoustic impedance was only little altered by disconnecting the cochlea, indicating that the cochlea contributes little elasticity and mass to the middle ear. The effect on the ear's acoustic impedance from interrupting the incudo-stapedial join is more complex for frequencies above 4 kHz than below (Fig. 2.21), as has been observed by other investigators [143].

Contribution from the Middle Ear Cavities to the Ear's Acoustic Impedance

Animal experiments have shown that the reactive component of the ear's acoustic impedance decreases after opening of the middle ear cavity for frequen-

A

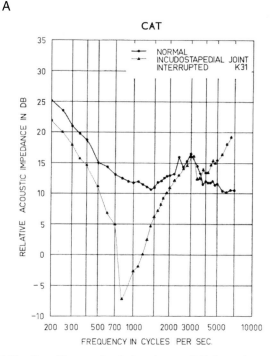

FIGURE 2.21 (A) The effect of interrupting the incudo-stapedial joint on the acoustic impedance of the ear of a cat. Absolute value of the impedance (given in decibels relative to 100 cgs units). (B) The effect of interrupting the incudo-stapedial joint on the acoustic impedance of the ear of a cat. The same data as in A with the real and the imaginary parts of the impedance shown separately (from [90] with permission).

cies below 3 kHz [90]. This is because the middle ear cavities add stiffness to the middle ear. The cat has a bony septum separating the middle ear cavity into two compartments that communicate via a small hole in the septum. Comparison of the acoustic impedance of the cat's ear before and after removal of that septum confirm that the hole, together with the cavities, acts as a Helmholz resonator, which makes the effect of the middle ear cavities in the cat different from that in other animals such as the rabbit. The reactive component of the acoustic impedance of the cat's ear changes rapidly as the frequency changes around 4 kHz because of this resonator. Removing the bony septum of the middle ear makes the middle ear cavity act as a simple stiffness component similar to that in the rabbit, in which a single middle ear cavity adds stiffness to the middle ear [90]. The middle ear cavity in the human is different from that of these animals in that it is much larger and contains many air cells.

B

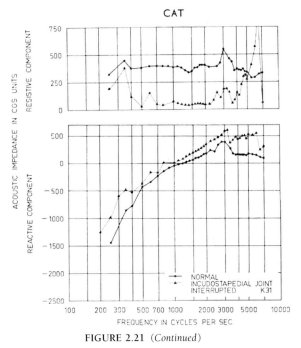

FIGURE 2.21 (*Continued*)

Effect of Pressure Difference between the Two Sides of the Tympanic Membrane

The air pressure in the middle ear cavity is normally kept close to the ambient pressure by the occasional opening of the Eustachian tube that connects the middle ear cavity with the pharynx. If the air pressure is not the same on both sides of the tympanic membrane, the sound conduction to the cochlea becomes impaired and the ear's acoustic impedance deviates from its normal value. The decrease in sound conduction is greatest for low frequencies.

Some of the earliest published studies of the effect of a difference in the static pressure in the ear canal and the middle ear cavity on the transfer function of the human (cadaver) middle ear were done by von Békésy (1929), who showed that the effect from a pressure difference between the two sides of the tympanic membrane on the sound transmission through the middle ear is largest at low frequencies (Fig. 2.22A). These results are in close agreement with the results of measurements done in live subjects using psychoacoustic methods (loudness balance) (Fig. 2.22B). More recently, other investigators confirmed that the ear's

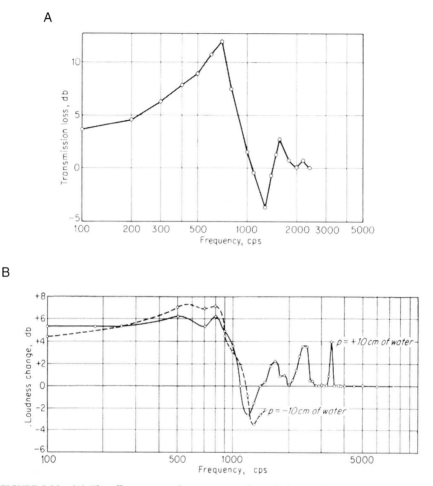

FIGURE 2.22 (A) The effect on sound transmission through the middle ear from static air pressure of 10 cm H$_2$O measured in a human cadaver ear. The attenuation is given in positive decibel values. (B) The effect on sound transmission through the middle ear from static air pressure of 10 cm H$_2$O measured by loudness matching. The attenuation is given in positive decibel values (from Békésy, von G. (1941). Article 38. *In* G. v. Békésy (1960), "Experiments in Hearing." McGraw-Hill, New York).

acoustic impedance is lowest when the pressure in the middle ear cavity is the same as the pressure in the ear canal, corresponding to equal pressure on both sides of the tympanic membrane and that the ear's acoustic impedance increases both when the pressure is increased and when it is decreased (Fig. 2.23) [90, 149]. The decrease in sound transmission to the cochlea when the air pressure

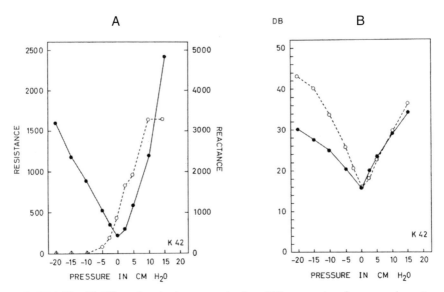

FIGURE 2.23   (A) Effect of static air pressure in the middle ear cavity of a cat on the ear's acoustic impedance at 1000 Hz (resistive and reactive components shown separately). (B) Effect on static air pressure in the middle ear cavity of cat. Comparison between the change in the ear's acoustic admittance and change in its transmission. The admittance is given in decibels relative to 100 cgs units and the transmission is given in arbitrary decibel values (from [90] with permission).

in the middle ear cavities differed from that of the ear canal was more pronounced at low frequencies than at high frequencies and it was larger for a negative pressure in the middle ear cavity (corresponding to a positive pressure in the ear canal) than for positive pressure of the same value (Fig. 2.24) [90]. Examination of the reactive component of the ear's acoustic impedance revealed that both positive pressure and negative pressure in the middle ear cavity increase the stiffness of the middle ear. Negative pressure in the middle ear cavity, in addition, reduces the resistive component of the ear's acoustic impedance (Fig. 2.23A). Positive pressure in the middle ear cavity causes the acoustic admittance to decrease by approximately the same amount as the decrease in the transmission of sound through the middle ear, indicating that the effect of positive pressure in the middle ear cavity is what makes the middle ear more stiff (Fig. 2.23B). It is seen that negative pressure in the middle ear cavity causes a larger decrease in transmission than the same amount of positive pressure. This may be explained by the fact that negative pressure in the middle ear cavity reduces the resistive component of the ear's acoustic impedance (Fig. 2.23) and that the resistive component of the ear's acoustic impedance mainly originates in the cochlea (Fig. 2.21). That negative pressure in the middle ear cavity reduces the resistive com-

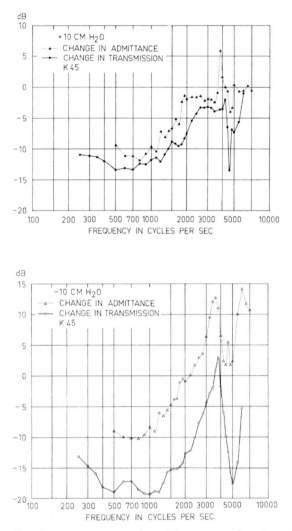

FIGURE 2.24 Effect of negative pressure in the middle ear cavity of a cat's ear (corresponding to a positive pressure in the ear canal) of 10 cm H_2O on the acoustic admittance (inverse of the impedance) (dashed lines) and the transmission through the middle ear (solid lines). The change in the ear's transmission was obtained by measuring the cochlear microphonics (CM) recorded from the round window of the cochlea and adjusting the sound pressure of the tone so that the CM was identical to the CM recorded when the air pressure was equal on both sides of the tympanic membrane. Both the change in transmission and in the admittance (inverse of the impedance) are shown in logarithmic measures, thus 20 times the logarithm of the ratio between the admittance or transmission before and after the change in the middle ear pressure. A positive decibel value shows an increase in transmission and admittance (decrease in impedance) and a negative decibel value shows a decrease in transmission and admittance (from [90] with permission).

ponent of the ear's acoustic impedance thus indicates that negative pressure causes the cochlea to become decoupled from the middle ear in a similar way as interruption of the incudo-stapedial joint. This likely explains why a negative pressure causes a larger decrease in the transmission of sound to the cochlea than a positive pressure.

Measurement of changes in the ear's acoustic impedance when the air pressure in the sealed ear canal is varied has found widespread clinical usage as a diagnostic tool because it makes it possible to determine pressure in the middle ear cavity. It is known as tympanometry and it also provides information about the function of the middle ear in general. The clinical use of tympanometry is discussed in Chapter 13. Usually the acoustic impedance is measured at a single frequency but the variation in the ear's impedance as a result of air pressure in the ear canal is different for different frequencies (Fig. 2.25). Some investigators have made use of that fact to gain more diagnostic information from tympanometry [22].

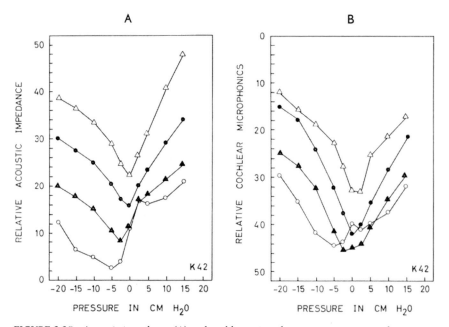

FIGURE 2.25   Acoustic impedance (A) and cochlear microphonics at constant sound pressure at the tympanic membrane (B) as a function of air pressure in the middle ear cavity of a cat for different frequencies: open triangles: 500 Hz; solid circles: 1000 Hz; solid triangles: 2000 Hz; open circles: 3000 Hz (from [90] with permission).

EFFECT OF CONTRACTION OF THE MIDDLE EAR MUSCLES ON THE FUNCTION OF THE MIDDLE EAR

Contraction of the tensor tympani muscle pulls the manubrium of malleus inward, and the stapedius muscle pulls the stapes in a direction that is perpendicular to the pistonlike motion of the stapes in response to sound and causes a sliding movement in the incudo-stapedial joint, parallel to its surface. (The middle ear muscles normally contract as an acoustic reflex, see Chapter 12.) When the tensor tympani muscle contracts, the tympanic membrane moves inward, the sound transmission through the middle ear decreases, and the ear's acoustic impedance changes (Fig. 2.26) [91]. Contraction of the stapedius muscle also changes the sound transmission through the middle ear and the ear's acoustic impedance, but it causes little or no movement of the tympanic membrane. When both muscles are brought to contraction simultaneously, the movement of the tympanic membrane is smaller than it is when the tensor tympani is brought to contract alone (Fig. 2.26), but the change in transmission and the ear's acoustic impedance are larger than when these muscles are

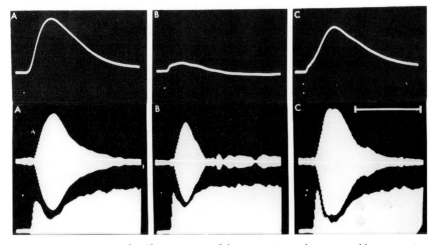

FIGURE 2.26   Upper graphs: The movement of the tympanic membrane caused by contraction of (A) the tensor tympani muscle, (B) the stapedius muscle, and (C) both muscles together recorded by measuring the change in the air pressure in the sealed ear canal. The tensor tympani muscle and the stapedial muscle are contracted independently by electrical stimulation. Middle graphs: Change in the acoustic impedance or the ear measured at 800 Hz. Lower graphs: Change in the CM recorded from the round window for an 800-Hz stimulation (from [91]).

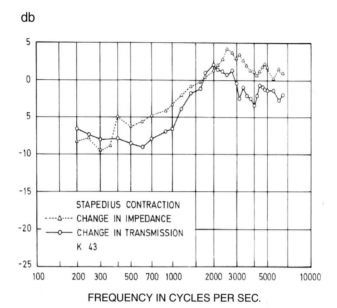

FIGURE 2.27 Change in sound transmission through the middle ear in a cat as a result of contraction of the stapedius muscle (solid lines and circles) together with the concomitant change in the ear's acoustic admittance (dashed lines and triangles) (from [90] with permission).

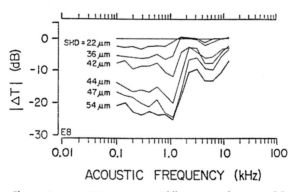

FIGURE 2.28 Change in transmission in one middle ear as a function of frequency for six different sound intensities (expressed in stapes displacement in micrometers) (from Pang, X.-D., and Peake, W.-T. (1986). How do contractions of the stapedius muscle alter the acoustic properties of the ear? *In* "Lecture Notes in Biomathematics." Peripheral Auditory Mechanisms. Springer-Verlag, Berlin).

brought to contract one at a time. Thus, contraction of the stapedius muscle impedes the motion of the tympanic membrane induced by contraction of the tensor tympani muscle. (The use of changes in the ear's acoustic impedance to record contractions of the middle ear muscles is discussed in Chapter 12.)

> The displacement of the tympanic membrane by contraction of the tensor tympani muscle can be recorded by measuring the change in the air pressure in the sealed ear canal as demonstrated in Fig. 2.26. Kato [59] was probably the first to report on recordings of contractions of the middle ear muscles by the displacement of the tympanic membrane by measuring changes in the air pressure in the sealed external ear canal in animal experiments. At about the same time Mangold [80] used a similar method in humans and elicited a contraction of the middle ear muscles by presenting a loud sound to the opposite ear. Similar methods were used by subsequent investigators for studies of the acoustic middle ear reflex [82, 141]. It was probably Hallpike [48] who first showed experimental evidence that contraction of the middle ear muscles caused a change in the sound transmission through the middle ear. Later, several investigators [39, 131, 152, 155] used recordings of the cochlear microphonic potential from the round window of the cat and observed the change in this potential when the middle ear muscles were brought to contract in response to a loud sound presented to the opposite ear.

While it is confirmed in many studies that contraction of the stapedius muscle decreases sound transmission to the cochlea, the function of the tensor tympani muscle is less well understood. It has been suggested that contractions of the tensor tympani improve air exchange in the tympanic cavity because it contracts during swallowing when the eustachian tube is opening. If not replaced, the air is assumed to get a low content of oxygen because of the oxygen absorption on the mucosal surface in the middle ear cavity.

The sound-conducting properties of the middle ear decrease over a large range of frequencies when the stapedius muscle contracts. The change in transmission is largest at low frequencies and it is negligible above the resonance frequency of the middle ear (about 1500 Hz), indicating that contraction of the stapedius muscle adds stiffness to the middle ear. The attenuation is approximately 8 dB in the cat for frequencies below 1000 Hz (Fig. 2.27) [90]. The change in sound transmission through the middle ear is slightly greater than the change in the ear's admittance, indicating that contraction of the stapedius muscle may cause a decoupling of the cochlea in addition to adding stiffness to the middle ear. Comparisons of the change in the acoustic impedance and the change in the transmission properties of the middle ear also support the hypothesis that contraction of the stapedius muscle causes some kind of "decoupling" between the middle ear and the cochlear fluid. The sound transmission through the middle ear decreases gradually as a function of the stapes displacement (Fig. 2.28) [107]. The attenuation is largest in the low-frequency range but during strong contractions sound transmission is also reduced in the high-frequency range.

Physiology of the Cochlea

ABSTRACT

1. The cochlea separates sounds according to their frequency (spectrum) so that different spectral components of sounds activate different populations of auditory nerve fibers.
2. Sensory transduction occurs in inner hair cells.
3. Outer hair cells are active elements that act as "motors" to reduce the influence of friction on the motion of the basilar membrane. This action of the outer hair cells increases the vibration amplitude of the basilar membrane (by about 50 dB) and increases its frequency selectivity.
4. The role of outer hair cells in increasing the frequency selectivity of the basilar membrane is greatest at low sound intensities. Therefore the frequency selectivity decreases with increasing stimulus intensity and the location of the maximal response shifts toward the base of the cochlea.
5. The nonlinear action of the cochlea also provides amplitude compression of sounds before initiation of nerve impulses in the auditory nerve. Without that, it is not possible to code sounds in the auditory nerve in the large range of intensities that are covered by hearing.

71

6. The cochlea can generate different kinds of sounds. These sounds are conducted "backward" by the middle ear, setting the tympanic membrane in motion and thereby generating sounds that can be recorded by a sensitive microphone placed in the ear canal. This is known as otoacoustic emission (OAE).

7. There are several kinds of OAE:
   A. Transient evoked otoacoustic emission (TEOAE) is elicited by a transient sound and generated by reflection of the traveling wave on the basilar membrane.
   B. Spontaneous otoacoustic emission (SOAE) is a sustained sound that is generated without any sound being applied to the ear.
   C. Distortion product otoacoustic emission (DPOAE) is a measure of nonlinear distortion in the cochlea. Distortion product otoacoustic emission is elicited by applying two tones to the ear and measuring the amplitude of the difference tone (usually the $2f_2-f_1$ tone).

8. The olivocochlear efferents influence the function of outer hair cells and by that they affect the OAE.

INTRODUCTION

Sensory cells in the cochlea transform sound into a code of nerve impulses in the auditory nerve, thus conveying information to the brain about sounds that reach the ear within the audible range. The cochlea separates the information into sounds according to their spectrum (frequency) so that different populations of hair cells become activated by sounds within different parts of the audible spectrum range. Besides that, the cochlea compresses the amplitude scale of sounds to accommodate the large dynamic range of natural sounds.

Interplay between theoretical and experimental work has been extremely useful in unraveling the intricate functions of the cochlea both with regard to the frequency analysis in the cochlea and with regard to the sensory transduction. The more knowledge that is accumulated about the function of the cochlea the more it becomes evident that the cochlea is a far more complex system than envisioned by the early investigators. Many features not included in the earlier hypotheses have been added as a result of the extensive experimental work regarding the function of the cochlea that has been done during the past 2 decades.

An example of how new information has totally revised the conception of the function of the cochlea was the discovery that the two groups of sensory cells, inner and outer hair cells, have fundamentally different functions. The outer hair cells act as "motors" that improve the ear's sensitivity and sharpen its frequency selectivity for weak sounds by compensating for the loss of energy in the cochlea. The inner hair cells convert the vibration of the basilar membrane into a neural code in the individual fibers of the auditory nerve.

The range of sound intensities that the ear is capable of handling is enormous and compression of the amplitude before the sound is coded in the discharge pattern of cochlear nerve fibers seems essential in order to code such a large range of sound intensities. The neural transduction in the cochlea, together with the nonlinearity of the motion of the basilar membrane, provides amplitude compression that makes it possible for the auditory system to process sound over a range of approximately 100 dB.

It has been questioned whether the frequency selectivity of the cochlea is indeed the basis for our ability to detect changes in the frequency of a pure tone the size of only a few hertz. The results of recent studies have also cast doubt about the role of spectral analysis in the ear as the basis for discrimination of complex sounds such as speech sounds. Recent studies instead seem to emphasize the role of temporal coding of such features and it is now believed that the main role of frequency selectivity of the basilar membrane is to divide sounds into different spectral bands before the information is processed by the auditory nervous system. The mammalian ear can process a 10-ocatave sound spectrum and that would not be possible without separating the spectrum into suitably sized pieces so that the temporal information in different frequency bands can be coded independently in the discharge pattern of different populations of auditory nerve fibers (discussed in Chapters 6 and 7).

FREQUENCY SELECTIVITY OF THE BASILAR MEMBRANE

Sound analysis in the cochlea is normally equated with spectral analysis that is ascribed to the interplay between the dynamic properties of the basilar membrane and that of the surrounding fluid. Helmholtz [53] was the first to formulate and prove that the ear performs spectral analysis of sounds. Before that, Ohm [106] suggested that the ear could separate a sound into its frequency components. These earlier hypotheses were inspired by the fact that any complex waveform (such as natural sounds) can be divided into a series of sinusoidal waveforms. Fourier analysis is the mathematical technique of separating a complex waveform into sine waves. Helmholtz suggested that the basilar membrane performed such spectral analysis and he believed that was accomplished because the basilar membrane functioned as a series of resonators that were tuned to different frequencies covering the audible range, a function similar to that of the strings of a piano.[1]

[1] Spectrum and frequency are used synonymously in this Chapter.

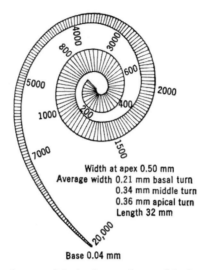

Width at apex 0.50 mm
Average width 0.21 mm basal turn
0.34 mm middle turn
0.36 mm apical turn
Length 32 mm

Base 0.04 mm

FIGURE 3.1 Schematic drawing of the basilar membrane of the human cochlea showing that the width of the basilar membrane increases from the base of the cochlea to its apex. High frequencies are processed in the basal end of the cochlea and lower frequencies toward the apex (from Stuhlman, O. (1943). "An Introduction to Biophysics." Wiley, New York).

While it thus had been hypothesized for many years that the cochlea is involved in frequency analysis of sounds it was the fundamental research by von Békésy that brought the proof. He showed experimental evidence that a tone of a certain frequency caused the highest vibration amplitude of a certain point along the basilar membrane. This means that each point along the basilar membrane is tuned to a certain frequency and a frequency scale can be laid out along the cochlea with high frequencies located at the base and low frequencies at the apex (Fig. 3.1). von Békésy also confirmed that sounds set up a traveling wave motion along the basilar membrane.[2] The stiffness of the basilar membrane decreases from the base of the cochlea to its apex and von Békésy concluded that the traveling wave motion is a result of this change in stiffness.

Other investigators had suggested earlier that there were other kinds of wave motion along the basilar membrane. Ewald's hypothesis that sounds give rise to standing waves on the basilar membrane dates back to 1898 [34]. von Békésy convincingly demonstrated that the traveling wave motion is the basis for the

[2] Georg von Békésy did his fundamental work on the function of the ear between 1928 and 1956. His early work was published in German. All his work has been translated into English and published in a book [5].

frequency selectivity and not resonance of the basilar membrane, as proposed by Helmholz [53].

During the time when our understanding of the function of the cochlea steadily increased, theoretical work by investigators like Ranke [113] and Zwislocki [160] was important in guiding experimentalists by asking relevant questions to be tested experimentally. Experimental studies of the vibration of the basilar membrane that could confirm the various hypotheses about the function of the basilar membrane as a spectrum analyzer have been hampered by the extremely small vibrational amplitude of the basilar membrane. Until the early 1970s data about the frequency selectivity of the basilar membrane were obtained in studies done in human cadaver ears by a single investigator (see [6]). This work was mostly done in the 1930s when limitations in technology made it necessary to use extremely high sound levels in these studies. The results obtained showed that the basilar membrane was broadly tuned. The results obtained at these extreme high sound intensities were taken to represent auditory frequency selectivity in the entire range of hearing because it was assumed that the basilar membrane was a linear structure that allowed such extrapolations of the experimental findings. (Readers who are interested in details about the development of hypotheses and experimental studies of the cochlea as a frequency analyzer are referred to [24], [25], [151], and [153]).

TRAVELING WAVE MOTION

Sounds set the cochlear fluid into motion which in turn sets the basilar membrane into motion. The mechanical properties of the basilar membrane (stiffness and mass) and how they vary along the membrane determine which kind of wave motion a sound gives rise to. The traveling wave motion on the basilar membrane is a result of the fact that the stiffness of the basilar membrane decreases from the region of the cochlear windows toward the cochlear apex. The transfer of energy from the cochlear fluid to the stiff basal part of the basilar membrane is more efficient than energy transfer to other (less stiff) parts of the basilar membrane. The energy that is transferred to the basal portion of the basilar membrane propagates as a traveling wave toward the cochlear apex (Fig. 3.2). As the wave travels along the basilar membrane toward less stiff parts of the basilar membrane, the propagation velocity of the wave decreases (Fig. 3.2) and consequently the wavelength of the motion decreases. (The wavelength is the distance between two identical points of the wave that travels along the basilar membrane.) When the wave motion slows, energy piles up, causing the vibration amplitude to increase [75]. The increase in amplitude is counteracted by frictional losses of energy. When the wavelength of the traveling wave becomes small, these losses increase rapidly and the wave propagation eventually comes to a halt extinguishing the traveling wave. The location on the basilar membrane where that occurs is a function of the frequency of the sound and that is the basis for the separation of sounds according to their frequency (spectrum).

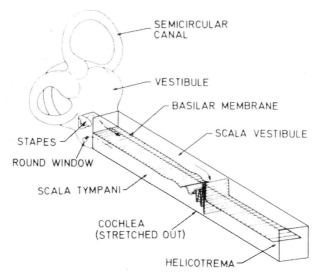

FIGURE 3.2 Schematic illustration of the traveling wave motion along the basilar membrane. The cochlea is shown schematically as a straight tube (from Zweig, G., Lipes, R., and Pierce, J. R. (1976). The cochlear compromise. *J. Acoust. Soc. Am.* **59**:975–982).

BASILAR MEMBRANE FREQUENCY TUNING IS NONLINEAR

The frequency tuning of the basilar membrane as it was known from von Békésy's experiments (Fig. 3.3) was much too broad to explain psychoacoustic data on frequency discrimination in the auditory system. This resulted in many

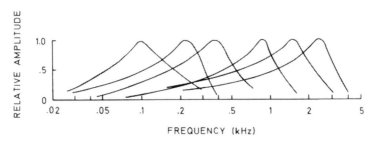

FIGURE 3.3 The vibration amplitude of the basilar membrane at different locations along the membrane of the cochlea in a human cadaver, shown as a function of frequency (from Békésy, von G. (1942). Article 42. *In* G. v. Békésy (1960), "Experiments in Hearing." McGraw-Hill, New York).

attempts to explain how the broad frequency tuning of the basilar membrane could be sharpened. When technology in the beginning of the 1970s had advanced to a level that made it possible to measure the vibration of the basilar membrane for sounds in the upper physiologic intensity range (90–70 dB SPL) [115], it became evident that the frequency selectivity of the basilar membrane was greater at low sound levels than at high levels (Fig. 3.4) [115]. The vibration of the basilar membrane is thus nonlinear and the amplitude of the basilar membrane vibration is compressed relative to the sound stimulus.

Frequency Tuning of Basilar Membrane Depends on Metabolic Energy

The frequency selectivity of the basilar membrane deteriorates after death [161] (Fig. 3.5) and that is an indication that metabolic energy is necessary to maintain

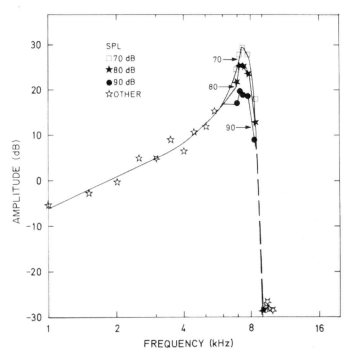

FIGURE 3.4   Amplitude of vibration of a single point on the basilar membrane of an anesthetized squirrel monkey for constant displacement amplitude of the malleus using the Mössbauer technique. The three solid curves show the vibration at three different sound intensities. The curves were shifted vertically so that they would coincide if the basilar membrane functioned in the way of a linear system (from Rhode, W. S. (1971). Observations of the vibration of the basilar membrane in squirrel monkeys using the Mossbauer technique. *J. Acoust. Soc. Am.* 49:1218–1231).

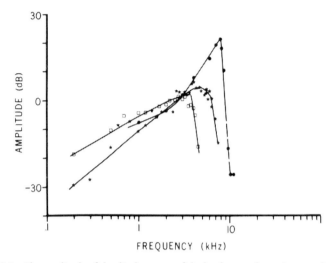

FIGURE 3.5   The amplitude of the displacement of the basilar membrane in a monkey obtained in a similar way as the results shown in Fig 3 4 The top curve shows the results when the monkey was alive (anesthetized), and the two other curves show results obtained 1 hour after the death of the monkey and 7 hours after its death (from [116]).

the high frequency selectivity of the basilar membrane. This, together with the high sound levels used by von Békésy in his studies of the frequency selectivity, explains why the tuning of the basilar membrane he obtained was so broad in the human cadaver ears. (The sound levels used in those studies were probably in the range of 145 dB SPL.) Figure 3.5 also shows that the frequency a point of the basilar membrane is tuned to shifts toward lower frequencies after death.

That the sharp tuning of auditory nerve fibers deteriorates after oxygen deprivation was shown by Evans as early as 1975 [33] (Fig. 3.6). Thus an indication that the frequency tuning of auditory nerve fibers depends on metabolic energy. These results were later confirmed by other investigators [14]. At the time these results were published (1975) it was believed that the high degree of frequency selectivity of single auditory nerve fibers was the result of a sharpening (by a "second filter") of the basilar membrane, which at that time was believed to be broadly tuned. The results of Rhode's study that showed that the frequency selectivity of the basilar membrane deteriorates after death, support that metabolic energy is necessary to maintain the sharp frequency tuning of the basilar membrane. This dependence on metabolic energy was explained much later when it was discovered that the outer hair cells are active elements that make the tuning of the basilar membrane nonlinear and that require metabolic energy [15].

The fact the frequency analysis in the cochlea depends on the intensity of sounds has resulted in doubts about the functional importance of cochlear

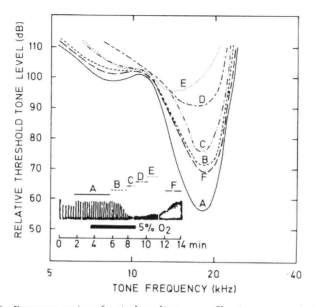

FIGURE 3.6 Frequency tuning of a single auditory nerve fiber in a guinea pig to show the changes of cochlear tuning after exposure to anoxia. The different curves were obtained with normal oxygenation (A) and the other curves (B–E) were obtained in succession while oxygenation was decreased gradually and then increased (F) The insert shows the amplitude of the compound action potential recorded from the round window of the cochlea. Reprinted from Evans, E. F. (1975). Normal and abnormal functioning of the cochlear nerve. *Symp. Zool. Soc. Lond.* 37:133–165, with permission from Cambridge University Press.

frequency selectivity for auditory frequency discrimination, matters that are discussed in more detail in Chapters 6 and 7.

ROLE OF THE OUTER HAIR CELLS IN BASILAR MEMBRANE MOTION

The discovery in 1985 by Brownell and co-workers [16] that outer hair cells act as "motors" that compensate for frictional losses of energy in the cochlea at low sound intensities brought our understanding of the functioning of the cochlea a large step forward. It changed the concept of how the cochlea functions in a fundamental way and it explained several different kinds of experimental results. Not so long ago it was believed that the role of both outer and inner hair cells was exclusively to transduce the vibration of the basilar membrane into a neural code. Now it is known that the ability of the basilar membrane to separate sounds based on their spectrum is intensity dependent because the outer hair cells contribute active and nonlinear elements to basilar membrane motion. For low sound intensities the outer hair cells act as "motors" that compensate for the energy losses in the basilar membrane and that increases the sensitivity and the frequency selectivity of the ear. That also explains why metabolic energy is necessary to maintain the normal sensitivity and frequency selectivity of the ear and it explains the differences between frequency tuning of the basilar membrane in cadavers and in living animals (Fig. 3.5) [116]. It also explains why metabolic insult induced by oxygen deprivation causes the threshold of auditory nerve fibers to increase and the tuning to become wider (Fig. 3.6) [33] (see also Chapters 6 and 14).

Motility of Hair Cells

The role of outer hair cells in cochlear mechanics is related to their motility [2, 16]. This means that outer hair cells can contract and expand. Two kinds of motility of outer hair cells have been demonstrated: One is a fast change in length of a maximum of about 5%, and the other is a slow change that can be much larger. The fast change can follow sound frequencies, is believed to sharpen the cochlear frequency selectivity, and it probably also causes amplitude compression. The slow change occurs over seconds and it may (slowly) change the sensitivity of the ear. These changes in length, fast and slow, can be elicited in different ways, one being by sound that reaches the ear. The fast change, first observed in outer hair cells that were isolated from the cochlea (of guinea pigs), can be elicited by passing electrical current through the hair cell [2, 16]. Slow changes in the length of hair cells occur in response to changes in the concentration of potassium ions in the surrounding fluid.

It is believed that the fast motility of outer hair cells can be elicited by receptor currents evoked by sound stimulation. Thus, sound stimulation that causes motion of the outer hair cells generates receptor currents that can cause (further) motion of the outer hair cells. This is assumed to be the basis for the positive feedback that explains how the outer hair cells become "motors" that amplify the vibration of the basilar membrane. If too strong, this positive feedback may cause self-sustained vibrations that result in generation of sound that can be measured in the ear canal (otoacoustic emission, OAE). Some investigators believe that hair cells may be able to generate their own intrinsic oscillations and that would explain some forms of OAE (see p. 84). The role of the slow change in length of the outer hair cells is less well understood.

Implications of the Active Cochlea for Hearing

The most obvious implication of the active process mediated by the outer hair cells is increased sensitivity and greater acuity of frequency analysis for weak sounds. The active process of the cochlea is assumed to account for about

50 dB of the ear's sensitivity, and total loss of the function of outer hair cells causes hearing loss of about that amount. This is why injuries that affect mainly outer hair cells, such as noise-induced hearing losses, cause elevation of the hearing threshold by 40–50 dB at the most (see Chapter 14).

Other Implications of Cochlear Nonlinearity

The nonlinearlity of the cochlea also causes the location of the maximal amplitude of the deflection of the basilar membrane in response to stimulation with pure tones to shift as a function of the intensity of the sound. This shift was demonstrated in studies of the cochlea by Honrubia and Ward in 1968 [54] and later in studies of the tuning of single auditory nerve fibers [95] and in recordings from cochlear hair cells [167]. That the location of maximal deflection of the basilar membrane is not only a function of the frequency (spectrum) of sounds but also depends on their intensity has implications for theories about the physiologic basis for discrimination of frequency. This is discussed in Chapter 7.

Olivocochlear Efferents Can Affect the Mechanical Properties of the Basilar Membrane

It has been known for a long time that neural activity in the olivo-cochlear efferent fibers controls the excitability of hair cells [36, 43, 44, 146, 154]. The fact that efferent neural activity controls the function of outer hair cells implies that efferent neural activity can change the mechanical properties of the basilar membrane. That function is mediated by the medial olivocochlear, the fibers of which terminate directly on the outer hair cells (see Chapter 1). Activity in these fibers releases transmitter substances which affects the mechanical properties of the outer hair cells and, in turn, the mechanical properties of the basilar membrane.

Earlier studies of the function of the olivo-cochlear bundle have been done by electrically stimulating the olivocochlear bundle where it comes close to the surface of the floor of the fourth ventricle, which can only be done in animal experiments. The fact that the medial olivo-cochlear efferent fibers that terminate on outer hair cells can be activated by sound stimulation of the opposite ear makes it possible to study the function of outer hair cells on the sensitivity of the ear and on basilar membrane tuning in humans (and in animals) [147, 148]. Studies have demonstrated that such activation of the efferent system by contralateral sound stimulation can change basilar membrane tuning and otoacoustic emission [21, 103, 112] (See Chapter 14).

AMPLITUDE COMPRESSION

Another important feature of the active cochlea is the nonlinear conversion of sound into vibration of the basilar membrane. It has been estimated that a 10-dB increase of the sound at the tympanic membrane results in an increase of only 2.5 dB of the vibration of the basilar membrane [19]. This amplitude compression takes place before the transduction into a neural code and functions similarly to the automatic gain control that is often incorporated in human-made communication systems. These matters are discussed in more detail in Chapters 14 and 16 in connection with the pathophysiology of hyperacusis.

ROLE OF THE TECTORIAL MEMBRANE FOR BASILAR MEMBRANE TUNING

Zwislocki [165] speculated that the tectorial membrane could possibly play an important role in creating the frequency selectivity of the basilar membrane. He suggested that the tectorial membrane together with the hairs (stereocilia) of the hair cells form a mass-stiffness resonator that is coupled to the basilar membrane and contributes to its frequency selectivity properties [1, 160, 168]. Zwislocki and Kletsky [169] showed that this hypothesis was plausible by using a mechanical model composed of a steel reed with a mass on top. When the reed was set into up and down vibrations, the mass would exhibit lateral movements when the frequency of the vibrations was equal to the resonance frequency of the reed–mass combination. Later, these theoretical and model studies were confirmed in animal experiments [168] showing that the mechanical properties of the tectorial membrane and the stereocilia of the outer hair cells contribute to the frequency selectivity of the basilar membrane by forming resonators along the basilar membrane. These resonators are tuned to different frequencies at different locations along the basilar membrane because the mass of the tectorial membrane varies along the membrane and the length and thereby the stiffness of the hairs also vary along the basilar membrane. These resonators, together with the traveling wave motion, are the bases for the frequency selectivity of the cochlea.

> It is interesting that resonators are again introduced as a contributor to the frequency selectivity but in a different way than Helmholtz's resonance theory because it is the tectorial membrane and hairs of hair cells and not the fibers of the basilar membrane that form the resonators. Also, the tectorial membrane is assumed to act in conjunction with the traveling wave motion in creating the frequency selectivity of the cochlea and therefore the basilar membrane motion is not the only mechanism of frequency selectivity.

COCHLEA AS A GENERATOR OF SOUND

It was reported anecdotally many years ago that the ears of some animals (dogs) sometimes emitted sounds that could be heard by an observer but it was not until Kemp [60] published his study on cochlear echoes that sound generation by the cochlea was described scientifically. Since then, many papers have been published on the subject and it has become evident that the cochlea can generate several different kinds of sounds, now commonly known as otoacoustic emission (OAE). The ability of the cochlea to generate sound is closely associated with the action of outer hair cells as active elements.

COCHLEAR ECHOES

When a transient sound is presented to the ear, a reflected sound (echo) can be recorded in the ear canal. That reflected sound, first known as the cochlear echo (or Kemp echo, after the person who described it, Kemp [60]), occurs with a latency of 5–15 ms. The cochlear echo, or transient evoked otoacoustic emission (TEOAE) (Fig. 3.7), is assumed to be caused by reflection of the

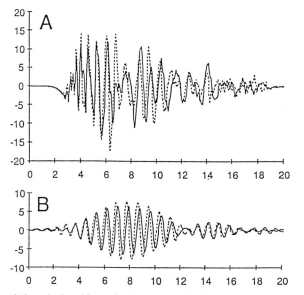

FIGURE 3.7   Click-evoked cochlear echo (transient evoked otoacoustic emission, TEOAE) recorded in standing position (solid lines) and in Trendelenburg position (supine with head lowered). (A) Unfiltered response; (B) filtered response (from Büki, B., Avan, P., and Ribari, O. (1996). The effect of body position on transient otoacoustic emission. *In* A. Ernst, R. Marchbanks, and M. Samii (Eds.), "Intracranial and Intralabyrinthine Fluids" (pp. 175–181). Springer-Verlag, Berlin).

traveling wave at some point along the basilar membrane. It is not entirely clear why that happens but any inhomogeneity in the basilar membrane may cause reflection of the traveling wave. Normally the traveling wave propagates smoothly without any reflection along the basilar membrane and all energy is dissipated before the wave reaches the cochlear apex. A slight inhomogeneity at a certain location along the basilar membrane, however, causes some of the energy to be reflected and that energy travels in the opposite direction, i.e., toward the base of the cochlea. When it reaches the basal region of the cochlea it sets the cochlear fluid into motion and that causes the stapes to vibrate. That vibration is conducted (backward) by the ossicular chain so that the tympanic membrane generates sound in the ear canal. The reflected wave is assumed to be amplified by the cochlear amplifier, which is how the active process in the cochlear becomes involved in generation of TEOAE [129].

Since the hair cells contribute to the mechanical properties of the basilar membrane, loss of hair cells or injury to hair cells at a certain location along the basilar membrane could cause such discontinuities. The reflected sound from a transient sound such as a click appears in the ear canal with a certain delay after presentation of the sound. The delay is the travel time on the basilar membrane to the location of the inhomogeneity and back again to the base of the cochlea. The main component of the reflected sound in response to broadband clicks is usually an oscillation with a narrow spectrum (Fig. 3.7), indicating that it originates from a narrow segment of the basilar membrane. The frequency of the oscillation is different for different individuals but stable over many years in a certain individual.

> Detailed analysis of the TEOAE reveals the frequency contents of the TEOAE are related to the stimulus that elicited the TEOAE in a complex way and the TEOAE contains energy at frequencies not represented in the stimulus. Normally, broadband clicks are used to elicit TEOAE, but when the spectrum of the stimulus clicks are limited, e.g., by high-pass filtering, it emerges that the TEOAE contain frequency components outside (below) the range of the stimulus sounds [158]. There is reason to believe that the amplitude of the different frequency components of the TEOAE reflect the physiologic condition of the areas of the cochlea that are tuned to these frequencies. Intermodulation distortion between the components of the stimulus may generate some of the spectral components of the TEOAE.

OTOACOUSTIC EMISSIONS THAT DEPEND ON ACTIVE PROCESSES

The TEOAE could be explained without involvement of the active properties of outer hair cells by reflection of the wave motion on the basilar membrane as was done by Kemp [60]. However, the fact that the TEOAE is largest for low-stimulus intensities and that its amplitude grows in a nonlinear fashion when the stimulus intensity is increased indicates that the cochlear echo is caused by active processes in the cochlea. That the TEOAE is generated by active processes in the cochlea was further supported by Wilson [156], who showed that in some individuals, the TEOAE did not die away with time but persisted for long periods, indicating that a supply of energy in the cochlea is necessary to generate at least some forms of TEOAE. Otoacoustic emission may also be elicited by a single pure tone. That kind of OAE is known as single-frequency otoacoustic emissions (SFOAE).

DISTORTION PRODUCT OTOACOUSTIC EMISSIONS (DPOAE)

When two tones are presented simultaneously a series of combination tones can be recorded by a microphone in the ear canal. The most prominent of these distortion products is the tone with the frequency $2f_1-f_2$, where f_1 and f_2 are frequences of the two tones that are presented (cubic distortion) (Fig. 3.8). These combination tones can be affected by stimulation of the olivocochlear

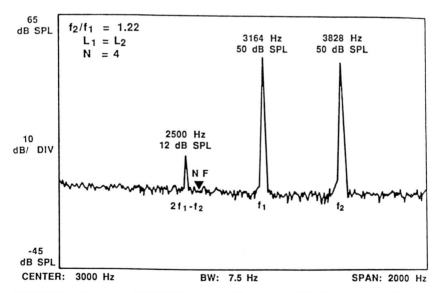

FIGURE 3.8   Illustration of DPOAE from a normal human ear, elicited by two tones of the same intensity (50 dB SPL), with frequencies of 3.16 and 3.82 kHz. The $2f_1-f_2$ component (2.5 kHz) has an intensity of 12 dB SPL (from Lonsbury-Martin, B. L., and Martin, G. K. (1990). The clinical utility of distortion-product otoacoustic emissions. *Ear Hear.* 11:144–154).

(efferent) bundle [112], which confirms that outer hair cells are actively involved in generation of distortion products. This relationship with the function of outer hair cells is the basis for the clinical use of recordings of DPOAE to test the function of outer hair cells. Recording the DPOAE is now an important clinical method that provides information about the function of the cochlea and its frequency selectivity.

Spontaneous Otoacoustic Emission

Vibrations of the outer hair cells can occur spontaneously. Self-oscillation of the outer hair cells generates continuous sounds by the cochlea without any external sound eliciting it, which is known as spontaneous otoacoustic emission (SOAE). The SOAE has the character of a pure tone, which indicates that the SOAE may be produced by a narrow segment of the basilar membrane where the outer hair cells oscillate.

All kinds of otoacoustic emissions are normally very weak sounds and sensitive microphones and recording equipment are required to record them. In rare cases, however, SOAE can be heard by an observer. It is interesting to note that administration of aspirin can abolish otoacoustic emission.

EPOCHS OF RESEARCH IN COCHLEAR MECHANICS

Dallos [24] offered a perspective on the development of theories of cochlear function and experimental work by dividing the eras of research on the function of the cochlea as a spectrum analyzer into three epochs. The first period was characterized by Helmholtz's theories [53], where a series of lightly damped mechanically tuned resonant elements were located along the basilar membrane (reviewed by Wever [53]). The second epoch, from late 1940s to early 1970s, was dominated by von Békésy's experimental demonstration of the spectral analysis in the cochlea as being the result of a traveling wave motion along the basilar membrane (Fig. 3.2). The third epoch, starting in the 1970s, includes the present time. That period is dominated by the finding that the traveling wave motion is boosted by active processes of the outer hair cells that inject energy into the system, compensating for frictional losses. The result is greater sensitivity of the ear and sharper tuning of the basilar membrane for sounds of low intensities.

Recognition of the role of a resonant system consisting of the tectorial membrane and the hairs of hair cells by Zwislocki and co-workers [168] may be regarded as a fourth epoch in our understanding of cochlear micromechanics (see [24]). That resonator, together with the traveling wave properties of the basilar membrane, is the basis for the frequency selectivity of each small segment of the basilar membrane. The frequency selectivity of the cochlea depends on the sound intensity because the properties of outer hair cells depends on the sound intensity.

In summary, as we understand it now, cochlear frequency selectivity is a result of a combination of at least three different mechanisms: (1) the traveling wave motion, (2) the resonance of the tectorial membrane and its attachment (mainly the stereocilia of the outer hair cells), and (3) the active function of the outer hair cells that inject energy into the motion of the basilar membrane. The traveling wave properties of the basilar membrane can be described by a linear system, which means that they are independent of the sound intensity. The active process of the outer hair cells is nonlinear, which is what causes the frequency selectivity of the basilar membrane to depend on the sound intensity. The resonator formed by the stereocilia and the tectorial membrane is also nonlinear because it involves the outer hair cells.

SENSORY TRANSDUCTION IN THE COCHLEA

Sensory transduction in the cochlea has been the subject of many studies and probably more is known about the sensory transduction in the ear than that in any other sensory system. Early research efforts were directed toward understanding how the two groups of hair cells, the inner and the outer hair cells, converted the vibrations of the basilar membrane into a neural code in single auditory nerve fibers. The fact that so few nerve fibers terminated on the outer hair cells was puzzling before it was understood that the outer and inner hair cells have fundamentally different functions. It is now assumed that only inner hair cells participate in the transduction of the motion of the basilar membrane into a neural code in the auditory nerve.

It was shown many years ago that the stereocilia of hair cells in the lateral line organ of fish are sensitive to bending [37]. Bending of the stereocilia of such hair cells in the direction toward the kinocilium depolarizes the hair cells and that causes the release of a chemical neurotransmitter substance, which causes excitation of the auditory nerve fibers that terminate on the hair cells. Deflections in the opposite direction hyperpolarize the cells (Fig. 3.9A) [37, 56]. Cochlear hair cells are assumed to function in a similar way but they do not have kinocilia. A basal body is located at the place of the kinocilium. Studies of mammalian hair cells have later confirmed these studies of hair cells of the lateral line organ [123] (Figs. 3.9B and 3.9C). Deflection of the hairs toward the location of the basal body has been regarded to be excitatory. This is also the direction toward the tallest row of stereocilia. However, more

A

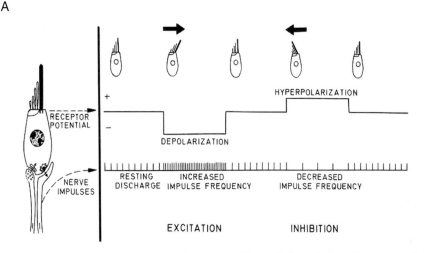

FIGURE 3.9 (A) Schematic illustration of excitation of hair cells from the lateral line organ of a fish. Intracellular potentials are affected by bending of the stereocilia of hair cells in the lateral line organ of fish hairs (from Flock, A. (1965). Transducing mechanisms in lateral line canal organ receptors. *Cold Spring Harbor Symp. Quant. Biol.* **30**:133–146). (B) Waveform of the voltage recorded intracellularly (receptor potentials) from inner hair cells in a guinea pig at different sound intensities. (C) Voltage of the response seen in B (from Russel, I. J., and Sellick, P. M. (1983). Low frequency characteristic of intracellularly recorded receptor potentials in guinea-pig cochlear hair cells. *J. Physiol.* **338**:179–206).

recent studies show that the question regarding which direction of stereocilia displacement is excitatory is more complex than earlier assumed.

Displacement of the cilia opens specific ionic channels that are located at or near the tips of the stereocilia [55]. The inflow of ions results in the release of a neurotransmitter, which is how sensory transduction in hair cells occurs. The connections between inner hair cells and auditory nerve fibers are in many ways similar to other synapses but its ability to transmit timing information may be better than other synapses. It is not yet known what the neurotransmitter is but glutamate has been suspected.

Which Phases of a Sound Excite Hair Cells (Rarefaction or Condensation)?

If deflection of the stereocilia of inner hair cells toward the basal body is excitatory, motion of the basilar membrane toward the scala vestibuli would be excitatory. At a first approximation, that corresponds to an outward movement of the stapes and thus that the rarefaction phase of sounds is excitatory.

B

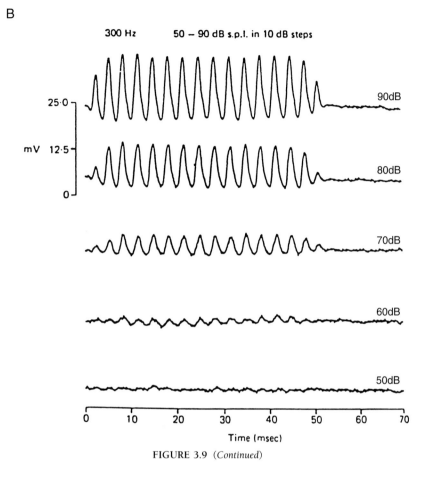

FIGURE 3.9 *(Continued)*

This concept was supported by early recordings of the compound action potential from the round window of the cochlea in response to click stimulation [108]. However, these results were challenged by later studies [69, 132, 170], which revealed a more complex relationship between the acoustical waveform of a sound and the excitation of cochlear hair cells. Inner hair cells were shown to respond to the motion of the basilar membrane in both directions (up and down) or only in one direction. Nerve fibers that were tuned to high frequencies respond when the basilar membrane was deflected toward the scala tympani while neurons that were tuned to low-frequency sounds respond to motion of the basilar membrane either toward scala vestibuli or towards scala tympani [122].

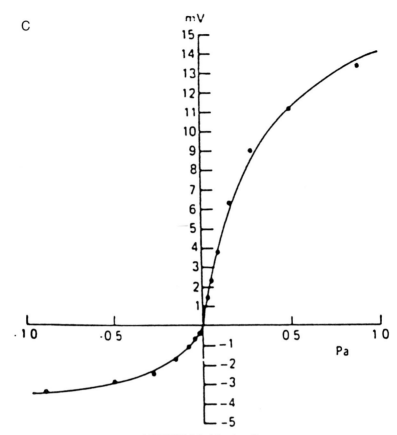

FIGURE 3.9 (*Continued*)

ARE HAIR CELLS SENSITIVE TO VELOCITY OR DISPLACEMENT OF THE BASILAR MEMBRANE?

In early studies of the function of hair cells as mechanotransducers, when low-frequency tones (e.g., 300 Hz) were used as stimuli, it appeared that hair cells responded to the amplitude of the displacement of the basilar membrane. More recently, when tones of higher frequencies have been used, it has become evident that inner hair cells may be excited by either the amplitude of the basilar membrane displacement or by the velocity of the basilar membrane motion [132, 170]. These results, which were obtained in studies of the response of single auditory nerve fibers, can be understood by considering how the force that acts on the stereocilia of the hair cells is generated. This force can be generated either by the shearing motion between the reticular lamina and the tectorial membrane or by the flow of endolymph that is caused by displacement of the tectorial membrane. The visco-elastic coupling between the tectorial membrane and the stereocilia makes hair cells sensitive to the velocity of the motion of the basilar membrane. This is because a visco-elastic coupling transmits the change in displacement, thus the velocity. Since the velocity of the motion increases proportionally with the frequency, low-frequency motion of the basilar membrane is less effective than high-frequency motion in displacing the stereocilia.

Ruggero [122] confirmed that excitation in the cochlea is complex and concluded that it can best be described in the following way: Auditory nerve fibers tuned to low frequencies respond to low-intensity sounds when the basilar membrane moves at its highest speed in the direction toward the scala vestibuli, whereas nerve fibers that are tuned to high frequencies respond to low-frequency sounds of moderate intensity (60–80 dB SPL) when the basilar membrane moves at its highest velocity toward scala tympani and when it is maximally displaced toward the scala tympani. Recent studies of the stimulation of the inner hair cells by basilar membrane motion have shown further evidence that the motion of the stereocilia is more complex than that of the basilar membrane and that the deflection of the stereocilia of inner hair cells, i.e. the basis for excitation of auditory nerve fibers, is not a direct function of the motion of the basilar membrane [166]. Mountain and Cody [102] recorded the receptor potentials from the inner hair cells of guinea pigs and compared these recordings with published data on basilar membrane vibration. Their results showed evidence that excitation of the inner hair cells is partly a result of motion of the outer hair cells and is modified by the visco-elastic coupling between the basilar membrane motion and the deflection of the stereocilia of inner hair cells. The relationship between these two different kinds of mechanical stimulation depend on the stimulus intensity. At low sound intensities the outer hair cell component dominates, in the middle range both contribute equally, and at high stimulus intensities it is the motion of the basilar membrane that controls the deflection of the hairs of the inner hair cells.

The conversion of motion of the basilar membrane into deflection of the stereocilia of inner hair cells is thus complex and different modes of the deflection of the basilar membrane can excite hair cells depending on the stimulus intensity.

EFFERENT CONTROL OF INNER HAIR CELLS

It has been shown that electrical stimulation of the olivocochlear efferent bundle decreases sound-evoked activity in single auditory nerve fibers [36,

154]. Lateral efferent fibers terminate on auditory (afferent) nerve fibers where these leave the inner hair cells (see Chapter 1), which explains how stimulation of these efferent fibers can modulate (decrease) the excitability of the afferent fibers. Activity in the efferent nerve fibers releases a transmitter substance that alters the excitability of afferent nerve fibers where they leave the inner hair cells. It is not known in detail what the transmitter substances are. Evidence has been presented that acetylcholine plays an important role, but other known neurotransmitters may also be involved.

AUTONOMIC CONTROL OF THE COCHLEA

There is anatomical evidence of considerable adrenergic innervation of the sensory cells of the cochlea (Chapter 1) [27, 28, 134], but there is little evidence concerning the function of this adrenergic innervation. It has been shown that norepinephrine is present in the cochlea but it does not seem to be liberated from sympathetic terminals [114]. Electrical stimulation of the stellate ganglion that gives rise to the adrenergic innervation of the cochlea has little effect on the click-evoked AP recorded from the round window of the cochlea [74, 109]. Electrical stimulation of the stellate ganglion, however, also affects the blood flow in the cochlea [73] and that may be at least partly responsible for the (small) changes in the cochlear potentials. Sympathectomy of the ear reduces temporary threshold shift caused by exposure to loud sounds [11] and it has been suggested that the sympathetic innervation of the cochlea may mediate protection against noise-induced (permanent) hearing loss. Sympathectomy is an effective treatment of certain forms of tinnitus and it has therefore been suggested that the sympathetic nervous system may modulate (increase) the sensitivity of cochlear hair cells (see Chapter 16).

NONAUDITORY ASPECTS OF THE COCHLEA

ELECTRICAL POTENTIALS IN THE COCHLEA

The endocochlear potential (EP) is present as a potential difference between the perilymphatic and endolymphatic fluid spaces. The EP was studied by von Békésy [9], who found that the endolymphatic space was about +80 mV relative to the tissue surrounding the cochlea and that the potential in the scala vestibuli was about 5 mV (Fig. 3.10). Thus the difference between the electrical potential in the perilymphatic and the endolymphatic space is approximately 75 mV. The EP is higher (80–120 mV) near the base of the cochlea than what it was in higher turns where it was 50–80 mV. These studies were

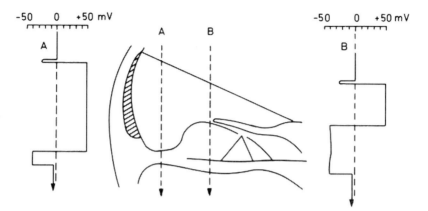

FIGURE 3.10   Electrical potentials recorded in different locations of the cochlea (from Békésy, von G. (1941). Article 38. *In* G. v. Békésy (1960), "Experiments in Hearing." McGraw-Hill, New York).

done in the guinea pig but similar values were found in the cat. The EP is generated by the stria vascularis and it serves as a"battery" that seems to be important for the normal functioning of the cochlear hair cells. (Sound evoked potentials are discussed in detail in Chapter 4.)

Autoregulation of Blood Flow to the Cochlea

In order for the cochlea to function properly the blood supply to the cochlea must be kept within relatively narrow limits and arterial pulsation must not activate the hair cells. Pulsation of the flow in the cochlea could excite hair cells, which would result in constantly hearing one's own pulse. Autoregulation of cerebral blood flow is an important mechanism for maintaining constant perfusion of the brain independent of fluctuations in systemic blood pressure. In the brain, autoregulation is maintained by controlling the width (lumen) of arterioles. It is not known if there is a similar regulation of cochlear blood flow. The labyrinthine artery consists of many small arterioles which could be the anatomical basis for such autoregulation. However, the fact that cochlear flood flow is affected by catecholamines (such as epinephrine) speaks against autoregulation.

One way the vascular changes in the cochlea are minimized is via branching of the labyrinthine artery. The labyrinthine artery consists of many parallel arteries, which reduces blood pulsation because narrow vessels act as low-pass filters and thus help to supply the cochlea with a smooth flow of blood.

CHAPTER 4

Electrical Potentials in the Cochlea

ABSTRACT

1. Three different sound-evoked potentials can be recorded from the cochlea: the cochlear microphonics (CM), the summating potential (SP), and the action potential (AP).
2. The endolymphatic potential (EP) is a steady potential that is not evoked by sound.
3. The CM follows the waveform of a sound.
4. The amplitude of the CM increases with increasing stimulus intensity in a linear fashion up to a certain intensity above which it reaches a plateau. Further increase in sound intensity results in a decrease in the amplitude of the CM.
5. The CM recorded from the round window is mainly generated by outer hair cells in the basal portion of the cochlea.
6. The SP is the most variable of the cochlear potentials, but it may depend on the pressure in the cochlea in a systematic way.
7. The SP is generated by cochlear hair cells.
8. The AP is the compound action potentials of the auditory nerve.

9. Recorded from the round window of the cochlea in animals, the AP has two negative peaks (N_1 and N_2).
10. The N_1 of the AP is generated in the most peripheral portion of the auditory nerve and the N_2 is mainly generated by the cochlear nucleus.
11. The latency of the AP decreases with increasing stimulus intensity and its amplitude increases.
12. The latency of the AP is shorter in response to high-frequency sounds than to low-frequency sounds of the same intensity.
13. Cochlear potentials recorded from the human ear are known as the electrocochleogram (ECoG). It comprises all the cochlear potentials (SP, CM, and AP).

INTRODUCTION

Recordings of sound-evoked potentials from the cochlea in animal experiments have played an important role in understanding the function of the cochlea. Auditory-evoked potentials recorded from electrodes placed in the human ear (electrocochleogram, ECoG) are used as a diagnostic tool in assessing pathologies of the ear.

Three distinctly different kinds of sound-evoked potentials can be identified in recordings from the cochlea, i.e., the cochlear microphonics (CM), the summating potential (SP), and the action potential (AP). The CM and the AP were discovered first and the SP was identified later. The CM and the SP are generated by cochlear hair cells, while the AP is generated by the auditory nerve. When recorded from the round window of the cochlea in animals, potentials generated in the cochlear nucleus contribute to the AP. The endolymphatic potential (EP) is a steady potential that is not evoked by sound. The EP is generated by the ionic differences between the different compartments of the cochlea.

In animals, sound-evoked cochlear potentials are commonly recorded from an electrode placed on or near the round window of the cochlea or by electrodes placed inside the cochlea. In humans evoked potentials from the cochlea, known as electrocochleographic (ECoG) potentials, are recorded by electrodes placed on the cochlear capsule or in the ear canal near the tympanic membrane.

RECORDINGS FROM THE ROUND WINDOW

All three sound-evoked potentials (CM, AP, and SP) can be recorded simultaneously from an electrode placed at the round window when an appropriate sound stimulus is used (Fig. 4.1). The typical AP response, recorded in a small animal consists of two negative peaks, N_1 and N_2. The N_1 appears approximately 1.5 ms after the onset of a high-frequency tone-burst stimulus. The CM, in response to pure tones, appears as a sinusoidal oscillation that is present throughout the duration of a tone-burst stimulus. The SP occurs as a deflection of the baseline during the tone burst. The CM and SP occur without any noticeable latency.

Depending on the stimulus, each one of these potentials may dominate the record. The response to a transient sound, such as a click, will be dominated by the AP response. The response to a low-frequency tone will be dominated by the CM. When a high-frequency tone is used as stimulus, the low-pass filter that is a part of commonly used physiologic amplifiers may attenuate the recorded CM so that it is not visible in the record and only the SP and the AP will appear (Fig. 4.2). Since the AP only appears at the onset of a high-frequency tone burst (and to some degree at the offset of a tone), the potentials that occur after a few milliseconds will almost entirely be the SP.

The polarity of the CM reverses when the polarity of the sound is reversed (e.g., from a rarefaction to a condensation click), while the polarities of the SP and the AP are independent of the click polarity. The latency of the AP

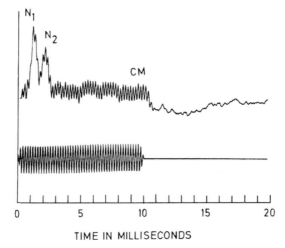

FIGURE 4.1 Response recorded from the round window of the cochlea of a rat to 5-ms long bursts of a 5-kHz tone with a rapid onset showing CM and AP (N_1 and N_2). The SP is represented by the baseline shift during the tone burst. The sound is shown below. A negative potential is shown as an upward deflection (from [96]).

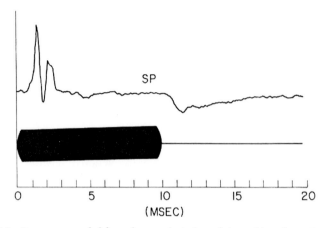

FIGURE 4.2   Response recorded from the round window of the cochlea of a rat in response to a 20-kHz tone burst to show AP and SP (from Møller, A. R. (1983a). "Auditory Physiology." Academic Press, New York (pp. 320).

may be slightly different when elicited by rarefaction clicks compared with condensation clicks.

COCHLEAR MICROPHONICS (CM)

The cochlear microphonic (CM) was first recorded in the 1930s by placing an electrode in contact with the round window of an experimental animal. When a person spoke into the animal's ear and the amplified CM was passed on to a loudspeaker, an observer could hear the speech sounds as if a microphone had been connected to the amplifier. This began the era of auditory physiology and it gave the impression that the ear functioned in a way similar to a microphone. This is also when the term "cochlear microphonics" began to be used. But, this hypothesis regarding the function of the ear proved later to be too simplistic and the CM is less important for hearing than earlier believed. Between 1930 and 1960 much research effort was devoted to studying the CM recorded from the round window in animals. The relationship between the amplitude of the CM and the sound intensity was studied in detail as was the harmonic distortion of the CM elicited by pure tones. It was found that the amplitude of the cochlear microphonic potentials recorded from the round window of a cat is a linear function of the input sound intensity up to a certain intensity where the amplitude no longer increases as the sound level increases [151]. Nevertheless, recordings of the CM became important in studies of the function of the cochlea and studies of CM have probably produced more journal articles than studies of any other single phenomenon of hearing.

Where Is Cochlear Microphonics (CM) Generated?

The CM recorded from an electrode placed at the round window of the cochlea is generated mostly by outer hair cells that are located in the basal portion of the cochlea [23]. The CM recorded in that way has an initial positive deflection in response to a rarefaction click [62, 64]. The CM recorded from an electrode at the round window is the sum of potentials generated by a large population of hair cells. A pair of electrodes placed inside the cochlea will record from a small portion of the sensory epithelium and such recordings are more useful in studies of the function of the cochlea. Such recordings show frequency selectivity in accordance with the tuning of the basilar membrane.

> The guinea pig has been used frequently for such studies because its cochlea protrudes into the middle ear space, which makes it possible to gain access to all the turns of the cochlea. The technique of recording from pairs of fine wires placed inside the cochlear capsule near a specific location of the basilar membrane was introduced early in the history of cochlear electrophysiology [139]. These investigators connected two such electrodes to the two inputs of a differential amplifier. Later that technique was used in many investigations to study the difference in the electrical potentials in the scala media, scala tympani, and scala vestibuli. For a review, see [23].

SUMMATING POTENTIALS (SP)

The SP is, as the name indicates, a summation of sound-evoked potentials. The SP appears as a slow potential that follows the envelope of a sound, and it can therefore be readily demonstrated when tone bursts are used as stimuli and can be recorded by placing an electrode on the round window. But the best way to record the SP is by placing recording electrodes inside the cochlear capsule. Recorded in that way, the SP is dependent on many factors and it is highly sensitive to impairment of the function of the cochlea. The amplitude and perhaps the polarity of the SP is affected by the pressure in the scala media or rather by the distension of the Reissners's membrane and presumed downward deflection of the basilar membrane. The SP can be identified in recordings from humans (it is a part of the ECoG) and it has found some use in diagnosis of disorders associated with distension of the Reissners's membrane (Meniere's disease, see Chapter 14).

The SP may be regarded as a distortion product, and it is a result of the basilar membrane not being deflected the same amount in both directions. The amplitude of the SP is thus a measure of the asymmetry of the motion of the basilar membrane. The SP has contributions from both inner and hair cells, but it is not totally clear which population of hair cells is the main generator of the SP. The relative contributions from the two groups of hair calls depend on the sound level used to elicit the SP and how it is recorded.

> In studies made of the vestibular nerve section it was shown that the SP changes when the olivocochlear bundle is severed (the olivocochlear bundle travels together with the vestibular nerve, see Chapter 5). Interruption of the normal efferent neural activity to the hair cells is assumed to alter the function of outer hair cells and these findings thus lend support to the assumption that outer hair cells contribute to the SP.

ACTION POTENTIAL (AP)

The AP has two distinct components that appear as negative peaks (N_1 and N_2) when recorded from the round window of the cochlea in animals such as the cat, guinea pig, or rat.[1] The N_1 component of the AP is the compound action potential (CAP) of the auditory nerve.

The AP has been recorded in small animals from electrodes placed at the round window and inside the cochlear capsule as well as from electrodes placed on the surface of the cochlear capsule. The waveform is similar when recorded from these different locations but the amplitude is highest when recorded from inside the cochlea. The N_1 of the AP is the electrical activity generated in the auditory nerve, probably at the location where impulse activity in the fibers of the auditory nerve is initiated (the first node of Ranvier). It therefore reflects the neural transduction in the cochlea. The AP is the most important of the cochlear potentials for studying the normal functioning and pathologies of the cochlea. The basic properties of the AP were studied in the 1960s and 1970s [108, 140].

Since the N_1 of the AP is the summation of neural discharges of many nerve fibers, its amplitude is highest when many nerve fibers discharge simultaneously. Transient sounds, such as clicks, cause many nerve fibers to discharge within a narrow time interval and the AP is therefore best recorded in response to transient sounds. Since the activation of hair cells in the (high-frequency) basal portion of the basilar membrane has the highest degree of synchronization, the AP has its largest amplitude when elicited by clicks that contain much energy at high frequencies or by high-frequency tone bursts with a fast onset.

Since the N_1 component of the AP is a result of auditory nerve fibers firing, it depends on the function of the hair cells as well as the synaptic transmission

---

[1] This abbreviation for action potentials was first used by researchers of cochlear electrophysiology many years ago and has been used ever since. Although the initial negative peak (N_1) is the same as the compound action potentials (CAP) of a nerve, we use "AP" for the compound action potentials recorded from the cochlea.

between the hair cells and the auditory nerve fibers. The N_2 component of the AP recorded in small animals is generated mostly in the cochlear nucleus [97] and it therefore depends on the integrity of the auditory nerve.

Latency of the AP

The latency of the cochlear AP depends on the intensity and the spectrum of the stimulus sound [96], as illustrated by the responses to band-pass-filtered clicks (Fig. 4.3). At a given stimulus intensity the latency of the N_1 (and the N_2) is shortest when the energy of the stimulus clicks is concentrated to high frequencies. The latency of the AP changes more as a function of the stimulus intensity when the stimuli have their energy in the low-frequency range than when the energy of the click stimuli is located in the high-frequency range.

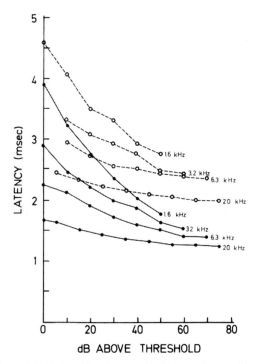

FIGURE 4.3   Latency of the N_1 (solid lines) and N_2 (dashed lines) peaks of the response recorded from the round window of a rat to 1/3-octave band-pass-filtered clicks. The center frequencies of the band-pass filter setting used are given by legend numbers (from Møller, A. R. (1983a). "Auditory Physiology." Academic Press, New York (pp. 320).

The decrease in latency as a function of increasing stimulus intensity has several causes. A nerve fiber discharges when the excitatory postsynaptic potential (EPSP) has exceeded a certain threshold value. The higher the stimulus intensity, the steeper the rise of the EPSP and it thus takes a shorter time to reach the threshold of firing when the stimulus intensity is high [93]. Another cause is related to the fact that the cochlea is nonlinear (see Chapter 3). This causes the maximal deflection of the basilar membrane to shift toward the base of the cochlea when the sound intensity is increased [98]. A shift of the maximal displacement of the basilar membrane toward the base of the cochlea results in a decrease of the travel time of the displacement of the basilar membrane to reach maximal deflection. This contributes to the decrease in the latency with increasing stimulus intensity of the recorded AP [98].

Amplitude of the AP

The amplitude of the N_1 peak of the AP increases with increasing stimulus intensity. The commonly used stimuli for studies of the AP are click sounds that are generated by applying 100-μs rectangular waves to an earphone. The amplitude of the N_1 of the AP in response to such stimuli raises more slowly from threshold to approximately 50 dB above the threshold than it does above 50 dB above the threshold. A curve that shows the amplitude of the N_1 peak as a function of the stimulus intensity thus has two different segments with different slopes (Fig. 4.4). These findings were based on recordings from the round window of the cochlea in experimental animals using stimuli generated by applying 100-μs rectangular waves to standard earphones. Only the response to rarefaction clicks has a clear two-segment relationship between amplitude and stimulus intensity (Fig. 4.4) and the amplitude of the N_1 in the AP response to condensation clicks increases monotonically as a function of the stimulus intensity in response to stimulation with clicks or short tone or noise bursts. This two-segment increase in amplitude was earlier regarded as a sign of two different excitatory mechanisms in the cochlea, each operating in a different intensity range [108]. However, it was shown later that the shape of the stimulus response curves of the N_1 peak is related to the spectrum of the stimulus sounds and these peculiarities are absent when clicks with a wider spectrum are used as stimuli (Fig. 4.5). Thus, the dependence on the click polarity and the two-segment amplitude function is only present in response to low-pass-filtered clicks. When the spectrum of the stimulus clicks was varied, by varying the duration of the rectangular wave applied to the sound transducer, only stimuli generated by 100-μs rectangular waves caused two-

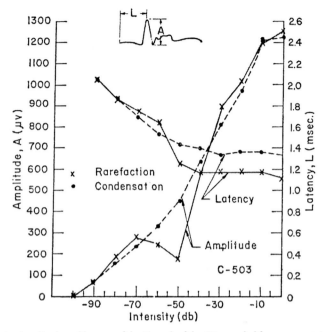

FIGURE 4.4 Amplitude and latency of the N_1 peak of the AP recorded from concentric electrodes in the internal meatus (eighth nerve) of a cat to condensation and rarefaction clicks. The intensity is given in decibels with an arbitrary reference (from Peake, W. T., and Kiang, N. Y.-S. (1962). Cochlear responses to condensation and rarefaction clicks. *Biophys. J.* 2:23–34).

segment stimulus–response curves (Fig. 4.5).[2] The choice of stimuli based on 100-μs rectangular waves was an unfortunate one but they are still the most commonly used stimuli for research and clinical studies of auditory evoked potentials (see Chapters 10 and 11).

> Studies of the response to low-pass-filtered clicks confirm that the difference between the response to condensation and rarefaction clicks is more pronounced when the spectrum of the stimulus sound is limited to the low-frequency range (Fig. 4.7). When low-pass-filtered clicks are added to broadband clicks, the response

[2] The spectrum of click sounds generated by a sound transducer with a broad frequency range depends mainly on the duration of the rectangular impulse that is used to drive the sound transducer. The spectrum of a rectangular wave falls off toward high frequencies and dips at a frequency of $1/T$, where T is the duration of the rectangular wave (Fig. 4.6). Thus, the spectrum of a rectangular wave the duration of which is 100 μs has a dip (null) at 10 kHz. The sound spectrum of a click generated by applying a 100 μs rectangular wave to a transducer will therefore have a dip at 10 kHz independent on the characteristics of the transducer.

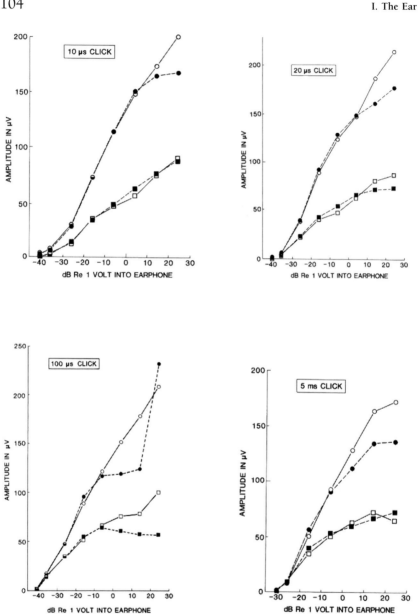

FIGURE 4.5 Amplitude and latency of the N_1 peak of the AP recorded from the round window of the cochlea of a rat to click sound generated by applying rectangular waves of different duration to the sound transducer. Circles, N_1; squares, N_2; open symbols and solid lines, rarefaction clicks; solid symbols and dashed lines, condensation clicks. The duration of the clicks is given by legends (from Nilsson, R., and Borg, E. (1983). Noise-induced hearing loss in shipyard workers with unilateral conduction hearing loss. *Scand. Audiol.* **12**:135).

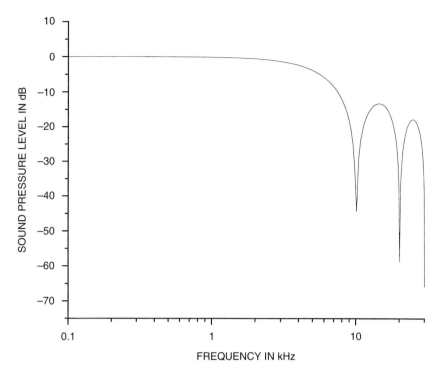

FIGURE 4.6   Spectrum of a rectangular wave with a duration of 100 μs.

to condensation and rarefaction clicks are different only for high-intensity stimulus sounds. Thus adding low-frequency components to broadband clicks reduces the amplitude of the N_1 of the AP response to clicks of high intensity (Fig. 4.7) [99] for condensation clicks, but it increases the amplitude of the N_1 peak of the AP response to rarefaction clicks. It is thus the low-frequency component of a transient sound that causes the irregular (two-segment) increase in the amplitude of the N_1 peak of the AP when recorded from the round window in response to transient sounds. The amplitude of the N_2 peak is smaller than that of the N_1 of the AP and it increases in a similar way as that of the N_1, although the dependence on the spectrum of the sound stimuli is less pronounced than what it is for the N_1 peak.

When low-frequency sounds (e.g., low-pass-filtered clicks) are used as stimuli and recordings are made from the round window the CM may interfere with the AP because the low-frequency sounds have a longer duration than the broadband click sounds. The AP and the CM can be studied separately by using masking with broadband noise. Broadband noise masks the AP, leaving the CM unchanged. Appropriate masking can therefore remove the AP and make it possible to study the CM in isolation. When such a "clean" CM response is obtained and then subtracted from the unmasked response containing both CM and AP, a "clean" AP

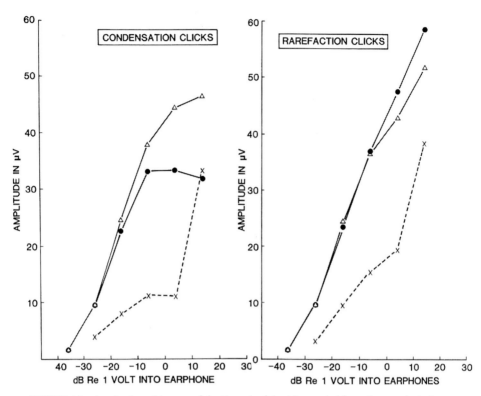

FIGURE 4.7 Amplitude and latency of the N_1 peak of the AP recorded from the round window of the cochlea of a rat to click sounds generated by applying 20-μs rectangular waves to the ear (triangles and solid lines) and by low-pass-filtered 10-μs rectangular waves (2300 Hz cutoff) (dashed lines and crosses). The solid lines and circles show the response obtained when these two sounds were presented simultaneously (from Nilsson, R., and Borg, E. (1983). Noise-induced hearing loss in shipyard workers with unilateral conduction hearing loss. *Scand. Audiol.* **12**:135).

response without contamination by the CM is the result (Fig. 4.8) [99]. This is a better method for eliminating the CM response than the commonly used method of reversing the polarity of every other stimuli (alternating condensation and rarefaction clicks) because the AP may be different in response to condensation and rarefaction clicks.

Where Is the N_1 of the AP Generated?

The N_1 peak of the AP has a slightly shorter latency than the responses recorded from single auditory nerve fibers (Fig. 4.9), which supports the hypothesis that the N_1 peak is generated by the distal part of the auditory nerve.

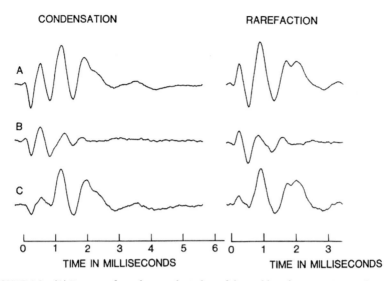

FIGURE 4.8 (A) Response from the round window of the cochlea of a rat to 2300-Hz low-pass-filtered clicks. (B) Response to the same stimulus to which broadband noise is added to mask the AP response, showing the "clean" CM. (C) The difference between the masked (B) and the unmasked responses (A) to show the "clean" AP (from Møller, A. R. (1986). Effect of click spectrum and polarity on round window N_1N_2 response in the rat. *Audiology* **25**:29–43).

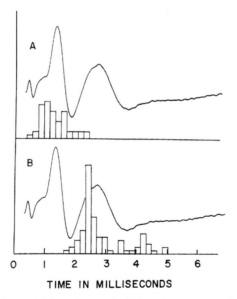

FIGURE 4.9 The AP recorded from the round window of a rat compared with the distribution of latencies of single nerve fibers of the auditory nerve (A) and the distribution of latencies of single nerve cells of the cochlear nucleus (B) (from [97]).

Relationship between the AP and Discharges of Single Auditory Nerve Fibers

Kiang and co-workers [63] simultaneously recorded responses from the round window of the cochlea in small animals and discharges from a single auditory nerve fiber. These investigators averaged the response from the round window with the signal averager triggered by the recorded discharges from an auditory nerve fiber. After a sufficient number of responses were averaged, the record from the round window of the cochlea revealed an AP response that resulted from the discharges of the single auditory nerve fiber from which recording was made. Kiang and co-workers identified a component of the averaged round window response that occurred 0.25–0.5 ms before the discharge they recorded from the single auditory nerve fiber (Fig. 4.10) and which triggered the averager. They called this the N_0 and interpreted their results to support the assumption that the AP recorded from the round window is a summation of discharges of the most peripheral portion of the auditory nerve. The recorded potentials had very small amplitudes because they were generated by only one single nerve fiber. They used either spontaneous activity of the auditory nerve fiber or activity elicited by stimulation with pure tones with a frequency equal to the fiber's characteristic frequency.

Where Is the N_2 of the AP Generated?

There are two different hypotheses regarding the generation of the N_2 peak. One hypothesis postulates that the N_2 component of the AP is generated by

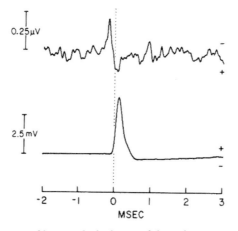

FIGURE 4.10   Illustration of how single discharges of the auditory nerve contribute to the AP recorded from the round window in a cat. The record was obtained by averaging the response from the round window with the averager triggered by single discharges of a fiber of the auditory nerve. No stimuli were applied to the ear. The record represents 12,600 discharges. The characteristic frequency of the fiber was 21 kHz and its spontaneous rate was 30 discharges per second (from Kiang, N. Y.-S., Moxon, E. C., and Kahn, A. R. (1976). The relationship of gross potentials recorded from the cochlea to single unit activity in the auditory nerve. In R. J. Ruben, C. Elberling, and G. Salomon (Eds.), Electrocochleography (pp. 95–115). Univ. Park Press, Baltimore).

a second firing of auditory nerve fibers, while the first firing contributes to the N_1 peak. The other hypothesis states that the N_2 peak is generated in the cochlear nucleus and passively conducted to the recording site. This hypothesis seems plausible in view of the small distance between the cochlea and the cochlear nucleus in (small) animals and is further supported by comparing the AP response with a histogram of the distribution of latencies of single auditory nerve fibers in the rat and cells in the cochlear nucleus in the same animal species (Fig. 4.9). The latency of the N_2 peak is similar to that of the peak in the histogram of latencies of responses from single cells in the cochlear nucleus. This supports the hypothesis that postulates the N_2 peak is generated by the cochlear nucleus and passively conducted to the recording site at the round window. Removal of the cochlear nucleus in the rat eliminated the second peak of the AP recorded from the round window (Fig. 4.11) [97] and further supports the hypothesis that the N_2 peak of the AP recorded from the round window is generated by the cochlear nucleus.

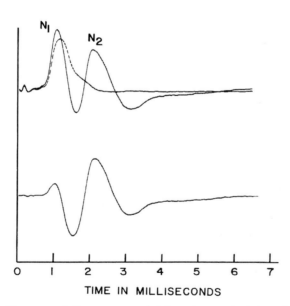

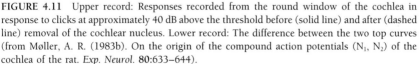

FIGURE 4.11 Upper record: Responses recorded from the round window of the cochlea in response to clicks at approximately 40 dB above the threshold before (solid line) and after (dashed line) removal of the cochlear nucleus. Lower record: The difference between the two top curves (from Møller, A. R. (1983b). On the origin of the compound action potentials (N_1, N_2) of the cochlea of the rat. *Exp. Neurol.* 80:633–644).

Use of the AP to Determine Frequency Tuning of the Auditory Nerve

Cochlear frequency selectivity (frequency tuning curves) can be studied from recordings of the AP in animals [25] by using (weak) tone bursts to elicit the AP (test tone) and another tone to mask the response. These two tone bursts may either overlap each other (simultaneous masking) or the masking tone may be presented a brief time before the test sound that elicits the AP response (forward masking). The test sound is set to the frequency at which tuning is to be determined and the frequency of the masking tone is varied in a frequency range above and below the test frequency while its amplitude is adjusted so that it reduces the amplitude of the evoked AP by a certain amount (20 or 30%). The intensity of the masking tone is called the masking tone level. The AP tuning curves are similar to frequency tuning curves obtained using recordings from single auditory nerve fibers (Fig. 4.12) [50].

ELECTROCOCHLEOGRAPHIC (ECoG) POTENTIALS

It was probably Ruben and his colleagues [121] who first recorded sound-evoked potentials from the human cochlea. Auditory-evoked potentials recorded from the cochlear capsule or the round window in humans came into use much later as a clinical test and became known as electrocochleographic (ECoG) potentials. The ECoG potentials are not fundamentally different from sound-evoked potentials (CM, SP, and AP) that can be recorded from the round window of the cochlea in animals (Fig. 4.13) and the ECoG consists of the CM, SP, and AP [30, 32, 47, 136]. When the ECoG was first introduced, it was common to use tone bursts or band-pass-filtered clicks to elicit the response, but, more recently, broadband clicks have been the most commonly used. Band-pass-filtered clicks have well-defined spectra and such sounds are often generated by applying short impulses to 1/3-octave band-pass filters [136]. The clinical value of the ECoG using tones of different frequencies or band-pass-filtered clicks over broadband clicks was never convincingly proven and tones and filtered clicks did not gain widespread clinical use, probably because of the greater complexity in generating such stimuli compared to broadband clicks.

The neural component of the ECoG recorded from the human ear is mainly generated by the distal end of the auditory nerve in the cochlea and the contributions from the cochlear nucleus is negligible because of the longer distance between the cochlea and the cochlear nucleus. (The auditory nerve in humans is much longer than it is in the small animals usually used in studies of the responses from the cochlea (see Chapter 5).)

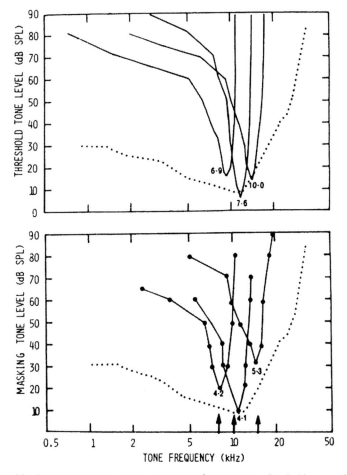

FIGURE 4.12   Frequency tuning curves. Upper graph: Frequency threshold curves for three normal guinea pig cochlear nerve fibers with characteristic frequencies of 9.4, 11.6, and 14 kHz. Lower graph: The AP tuning curves obtained from the same animal. The test frequencies were 8, 10, and 15 kHz and simultaneous masking was used (from Harrison, R. V., Aran, J. M., and Erre, J. P. (1981). AP tuning curves from normal and pathological human and guinea pig cochlea. *J. Acoust. Soc. Am.* **69**:1374–1385).

Recording the ECoG from the surface of the cochlear capsule (otic capsule) involves piercing the tympanic membrane with the recording electrode; thus it is an invasive approach (transtympanic ECoG) that requires certain skills and also involves risk of infection. To avoid this, investigators [20, 47, 100, 101, 124] developed extratympanic methods for recording ECoG using an

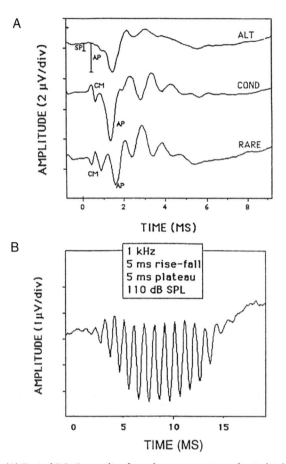

FIGURE 4.13   (A) Typical ECoG recording from the promontorium of an individual with normal hearing in response to click stimuli (indicated by an arrow). Note that negativity is shown as a downward deflection. (B) ECoG response to tone bursts (1 kHz) (from Harrison, R. V., Aran, J. M., and Erre, J. P. (1981). AP tuning curves from normal and pathological human and guinea pig cochlea. *J. Acoust. Soc. Am.* **69**:1374–1385).

electrode placed deep in the ear canal or at the tympanic membrane (Fig. 4.14) [157]. Such recordings yield potentials with similar waveforms but the amplitude is lower than those recorded from the surface of the otic capsule (promontorium).

Recordings of ECoG are used clinically. In one application, the SP component is used in diagnosis of cochlear hydrops (see Chapter 14).

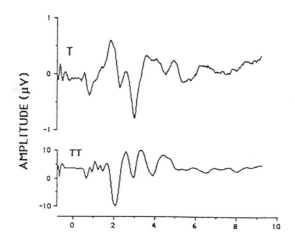

TIME (MS)

FIGURE 4.14 Extratympanic ECoG recorded from the surface of the ear canal (T) compared with transtympanic ECoG (TT) (from Harrison, R. V., Aran, J. M., and Erre, J. P. (1981). AP tuning curves from normal and pathological human and guinea pig cochlea. *J. Acoust. Soc. Am.* **69**:1374–1385).

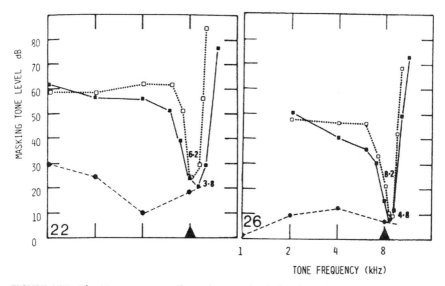

FIGURE 4.15 The AP tuning curves obtained in an individual with nearly normal hearing using simultaneous masking (solid lines) and using forward masking (dotted lines) (from Harrison, R. V., Aran, J. M., and Erre, J. P. (1981). AP tuning curves from normal and pathological human and guinea pig cochlea. *J. Acoust. Soc. Am.* **69**:1374–1385).

USE OF THE ECoG TO DETERMINE FREQUENCY TUNING OF THE AUDITORY NERVE IN HUMANS

Frequency tuning curves can be obtained in humans using recordings of ECoG potentials using the same masking technique as described above [31, 50]. This is useful because it can be obtained in individuals with normal hearing as well as in individuals with hearing loss and it can thus be used to study the effect of cochlear injury on frequency selectivity. As in animal experiments, studies in humans have used either forward masking or simultaneous masking. In humans, forward masking yields slightly narrower tuning curves than simultaneous masking (Fig. 4.15).

Section I References

1. Allen, J. B. (1980). Cochlear micromechanics—A physical model of transduction. *J. Acoust. Soc. Am.* **68**:1660–1670.
2. Ashmore, J. F. (1981). A fast motile response in guinea pig outer hair cells: The cellular basis of the cochlear amplifier. *J. Physiol.* **388**:323–347.
3. Axelsson, A. and Ryan, A. F. (1988). Circulation of the inner ear. I. Comparative study of vascular anatomy in the mammalian cochlea. *In* A. F. Jahn, and J. Santos-Sacchi. (Eds.), "Physiology of the Ear" (pp. 295–315). Raven, New York.
4. Ballantyne, S. (1928). Effect of diffraction around the microphone in sound measurements. *Physiol. Rev.* **32**:988–992.
5. Békésy, von G. (1929). Article 3. *In* G. v. Békésy (1960), "Experiments in Hearing." McGraw-Hill, New York.
6. Békésy, von G. (1937). Article 28. *In* G. v. Békésy (1960), "Experiments in Hearing." McGraw-Hill, New York.
7. Békésy, von G. (1941). Article 38. *In* G. v. Békésy (1960), "Experiments in Hearing." McGraw-Hill, New York.
8. Békésy, von G. (1942). Article 42. *In* G. v. Békésy (1960), "Experiments in Hearing." McGraw-Hill, New York.
9. Békésy, von G. (1952). Article 62. *In* G. v. Békésy (1960), "Experiments in Hearing." McGraw-Hill, New York.
10. Blauert, J. (1982). Binaural localization. *Scand. Audiol. Suppl.* **15**:7–26.
11. Borg, E. (1982). Auditory thresholds in rats of different age and strain: A behavioral and electrophysiological study. *Hear. Res.* **8**:101–105.
12. Breschet, G. (1836). Recherches anatomiques et physiologiques sur l'organe de l'ouie et sur l'audition, dans l'homme et les animaux vertebres. *Memoires, Academie de Medecine, Paris* **5**:229–524.
13. Brodel, M. (1946). "Three Unpublished Drawings of the Anatomy of the Human Ear." W. B. Saunders, Philadelphia.
14. Brown, M. C., Nuttall, A. L., Masta, R. I. and Lawrence, M. (1983). Cochlear inner hair cells: Effects of transient asphyxia on intracellular potentials. *Hear. Res.* **9**:131–144.
15. Brownell, W. E. (1983). Observation on the motile response in isolated hair cells. *In* W. R. Webster and L. M. Aiken (Eds.), "Mechanisms of Hearing" (pp. 5–10). Monash Univ. Press, Melbourne.
16. Brownell, W. E., Bader, C. R., Bertrand, D. and de Ribaupierre, Y. (1985). Evoked mechanical responses of isolated cochlear hair cells. *Science* **227**:194–196.

17. Büki, B., Avan, P. and Ribari, O. (1996). The effect of body position on transient otoacoustic emission. *In* A. Ernst, R. Marchbanks, and M. Samii (Eds.), "Intracranial and Intralabyrinthine Fluids" (pp. 175–181). Springer-Verlag, Berlin.

18. Carlborg, B., Konradsson, K. S. and Farmer, J. C. (1996). Pressure relation between labyrinthine and intracranial fluids: Experimental study in cats. *In* A. Ernst, R. Marchbanks, and M. Samii (Eds.), "Intracranial and Intralabyrinthine Fluids" (pp. 63–72). Springer-Verlag, Berlin.

19. Chatterjee, M. and Zwislocki, J. J. (1998). Cochlear mechanisms of frequency and intensity coding. II. Dynamic range and the code for loudness. *Hear. Res.* **124**:170–181.

20. Coats, A. C. (1974). On electrocochleographic electrode design. *J. Acoust. Soc. Am.* **56**:708–711.

21. Collet, L., Kemp, D. T., Veuillet, E., Duclaux, R, Moulin, A., and Morgon, A. (1990). Effect of contralateral auditory stimuli on active cochlear micro-mechanical properties in human subjects. *Hear. Res.* **43**:251–262.

22. Colletti, V. (1977). Multifrequency tympanometry. *Audiology* **16**:278–287.

23. Dallos, P. (1973). "The Auditory Periphery: Biophysics and Physiology." Academic Press, New York.

24. Dallos, P. (1992). The active cochlea. *J. Neurosci.* **12**:4575–4585.

25. Dallos, P., and Cheatham, M. A. (1976). Compound action potential (AP) tuning curves. *J. Acoust. Am.* **59**:591–597.

26. Davis, H., Benson R. W., Covel, W. P., Fernandez, C., Goldstein, R., Katsuki, Y., Legouix, J. P., McAuliffe, D. R., and Tasaki, I. (1953). Acoustic trauma in guinea pig. *J. Acoust. Soc. Am.* **25**:1180–1189.

27. Densert, O. (1974). Adrenergic innervation in the rabbit cochlea. *Acta Otolaryngol. (Stockh).* **78**:345–346.

28. Densert, O., and Flock, A. (1974). An electron microscope study of adrenergic innervation of cochlea. *Acta Otolaryngol. (Stockh.)* **77**:185–197.

29. Durrant J. D., Lovrinic, J. H. (1984) "Bases of Hearing Science." Williams & Wilkins, Baltimore.

30. Eggermont, J. J. (1976). Electrocochleography. *In* W. D. Keidel and W. D. Neff (Eds.), "Handbook of Sensory Physiology" (Vol. 3, Chapter 15, pp. 625–705). Springer-Verlag, New York.

31. Eggermont, J. J. (1977). Compound action potential tuning curves in normal and pathological human ears. *J. Acoust. Soc. Am.* **62**:1247–1251.

32. Eggermont, J. J., Spoor, A., and Odenthal, D. W. (1976). Frequency specificity of toneburst electrocochleography. *In* R. J. Ruben, C. Elberling, and G. Salomon (Eds.), "Electrocochleography" (pp. 215–246). Univ. Park Press, Baltimore.

33. Evans, E. F. (1975). Normal and abnormal functioning of the cochlear nerve. *Symp. Zool. Soc. Lond.* **37**:133–165.

34. Ewald, J. R. (1898). Ueber eine neue Hoertheorie. *Wien. Klin. Wochenschr.* **11**:721.

35. Feddersen, W. E., Sandel, T. T., Teas, D. C., Jeffress, L. A. (1957) Localization of high frequency tones. *J. Acoust. Soc. Am.* **29**:988–991.

36. Fex, J. (1962). Auditory activity in centrifugal and centripetal fibers in cat. *Acta. Phsyiol. Scand.* **55** (suppl 189):1–68.

37. Flock, A. (1965). Transducing mechanisms in lateral line canal organ receptors. *Cold Spring Harbor Symp. Quant. Biol.* **30**:133–146.

38. Fullerton, B. C., Levine, R. A., Hosford-Dunn, H. L., and Kiang, N. Y. S. (1987). Comparison of cat and human brain stem auditory evoked potentials. *Hear. Res.* **66**:547–570.

39. Galambos, R., and Rupert, A. L. (1959). Action of the middle-ear muscles in normal cats. *J. Acoust. Soc. Am.* **31**:349–355.

40. Geffcken, W. (1934). Untersuchungen uber akustische Schwellenwerte. *Poggendorff's Ann. Phys. Chem.* **19**:829–848.
41. Gelfand, S. A. (1997). "Essentials of Audiology." Thieme, New York. 1997.
42. Goode, R. L., Ball, G., and Nishihara, S. (1993). Measurement of umbo vibration in human subjects—Method and possible clinical applications. *Am. J. Otol.* **3**:247–251.
43. Guinan, J. J., Jr, and Gifford, M. L. (1988a). Effects of electrical stimulation of efferent olivocochlear neurons on cat auditory-nerve fibers. I. Rate-level functions. *Hear. Res.* **33**:97–114.
44. Guinan, J. J., Jr, and Gifford, M. L. (1988b). Effects of electrical stimulation of efferent olivocochlear neurons on cat auditory-nerve fibers. II. Spontaneous rate. *Hear. Res.* **33**:115–128.
45. Guinan, J. J., and Peake, W. T. (1967). Middle-ear characteristics of anesthetized cats. *J. Acoust. Soc. Am.* **41**:1237–1261.
46. Gyo, K., Aritomo, H., and Goode, R. L. (1987). Measurement of the ossicular vibration ratio in human temporal bones by use of a video measuring system. *Acta Otolaryng.* **103**:87–95.
47. Hall, J. W. (1992). "Handbook of Auditory Evoked Responses." Allyn and Bacon, Boston.
48. Hallpike, C. S. (1935). On the function of the tympanic muscles. *J. Laryng.* **50**:362–369.
49. Harrison, R. V. (1988) "The Biology of Hearing and Deafness." Charles C. Thomas, Springfield, IL.
50. Harrison, R. V., Aran, J. M., and Erre J. P. (1981). AP tuning curves from normal and pathological human and guinea pig cochlea. *J. Acoust. Soc. Am.* **69**:1374–1385.
51. Harrison, R. V., and Hunter-Duvar, I. M. (1988). An anatomical tour of the cochlea. *In* A. F. Jahn, and J. Santos-Sacchi (Eds.), "Physiology of the Ear" (pp. 159–171). Raven, New York.
52. Hawkins, J. E. (1988). Auditory physiologic history: A surface view. *In* A. F. Jahn, and J. Santos-Sacchi (Eds.), "Physiology of the Ear" (pp. 1–28). Raven, New York.
53. Helmholtz, H. L. F. (1863). "Die Lehre von den Tonempfindungen als physiologische Grundlage fur die Theorie der Musik." Vieweg-Verlag, Brunswick, Germany.
54. Honrubia, V., and Ward, P. H. (1968). Longitudinal distribution of cochlear microphonics inside the cochlear duct (guinea pig). *J. Acoust. Soc. Am.* **44**:951–958.
55. Hudspeth, A. J. (1989). How the ear's works work. *Nature* **341**:397–404.
56. Hudspeth, A. J, and Corey, D. P. (1977). Sensitivity, polarity, and conductance change in the response of vertebrate hair cells to controlled mechanical stimuli. *Proc. Natl. Acad. Sci. USA* **74**:2407–2411.
57. Hughes, G. B. (Ed.) (1985). "Textbook of Otology." Thieme-Stratton, New York.
58. Jepsen, O. (1955). "Studies of the Acoustic Stapedius Reflex in Man." Universitetsforlaget, Aarhus.
59. Kato, T. (1913). Zur Physiologie der Binnenmuskelen des Ohres. *Pflugers Arch. Ges. Physiol.* **150**:569–625.
60. Kemp, D. T. (1978). Stimulated acoustic emissions from within the human auditory system. *J. Acoust. Soc. Am.* **64**:1386–1391.
61. Khanna, S. M., and Tonndorf, J. (1972). Tympanic membrane vibrations in cats studied by time-averaged holography. *J. Acoust. Soc. Am.* **51**:1904–1920.
62. Kiang, N. Y.-S., Moxon, E. C., and Levine, R. A. (1970). Auditory-nerve activity in cats with normal and abnormal cochleas. *In:* G. E. W. Wolstenholme and J. Knight, (Eds.). Sensorineural Hearing Loss (pp. 241–268). CIBA Foundation, J. A. Churchill, London.
63. Kiang, N. Y.-S., Moxon, E. C., and Kahn, A. R. (1976). The relationship of gross potentials recorded from the cochlea to single unit activity in the auditory nerve. *In* R. J. Ruben, C. Elberling, and G. Salomon (Eds.), Electrocochleography (pp. 95–115). Univ. Park Press, Baltimore.

64. Kiang, N. Y.-S., and Peake, W. T. (1960). Components of electrical responses recorded from the cochlea. *Ann. Otol. Rhinol. and Laryng.* **69**:448–458.

65. Kiang, N. Y.-S., and Peake, W. T. (1962). Cochlear responses to condensation and rarefaction clicks. *Biophys. J.* **2**:23–34.

66. Killion, M. C., and Berger, E. H. (1987). Noise attenuating earphone for audiometric testing. *J. Acoust. Soc. Am.* (Suppl. 1) **81**:S5.

67. Killion, M. C., and Villchur, E. (1989). Comments on "Earphones in Audiometry" [Zwislocki *et al.* (1988). *J. Acoust. Soc. Am.* **83**:1688–1689]. *J. Acoust. Am.* **85**:1775–1778.

68. Killion, M. C., Wilber, L. A., and Gudmundsen, G. I. (1985). Insert earphones for more interaural attenuation. *Hear. Instrum.* **36**, 34–36.

69. Konishi, T. and Nielsen, D. W. (1973). The temporal relationship between motion of the basilar membrane and initiation of nerve impulses in the auditory nerve fibers. *J. Acoust. Soc. Am.* **53**:325.

70. Kurokawa, H. and Goode, R. L. (1995). Sound pressure gain produced by the human middle ear. *Otolaryngol. Head Neck Surg.* **113**:349–355.

71. Lang, J. (1981). Facial and vestibulocochlear nerve, topographic anatomy and variations. *In* M. Samii and P. J. Jannetta (Eds.), "The Cranial Nerves" (pp. 363–377). Springer-Verlag, New York.

72. Lang, J. (1983). "Clinical Anatomy of the Head." Springer-Verlag, New York.

73. Laurikainen, E., Kim, D., Didier, A., Ren, T., Miller, J., Quirk, W., Nuttall, A. (1993). Stellate, ganglion drives sympathetic regulation of cochlear blood flow. *Hear. Res.* **64**:199–204.

74. Lee, A. H., and Møller, A. R. (1985). Effects of sympathetic stimulation on the round window compound action potential in the rat. *Hear. Res.* **19**:127–134.

75. Lighthill, J. (1991). Biomchanics of hearing sensitivity. *J. Sound Vibrat.* **113**:1–13.

76. Lilly, D. J. (1964). Some properties of the acoustic reflex in man. *J. Acoust. Soc. Am.* **36**:2007–2008.

77. Lim, D. J. (1986). Effects of noise and ototoxic drugs at the cellular level in the cochlea: A review. *Am. J. Otolaryngol.* **7**:73–99.

78. Lonsbury-Martin, B. L., and Martin, G. K. (1990). The clinical utility of distortion-product otoacoustic emissions. *Ear Hear.* **11**:144–154.

79. Lynch, T. J., Nedzelnitsky, V., and Peake, W. T. (1982). Input impedance of the cochlea in cat. *J. Acoust. Soc. Am.* **72**:108–130.

80. Mangold, E. (1913). Willkurliche Kontraktionen des Tensor tympani und die graphische Registrierung von Druckschwankungen im ausseren Gehorgang. *Pflugers Arch. Ges. Physiol.* **149**:539–587.

81. Marchbanks, R. J. (1996). Hydromechanical interactions of the intracranial and intralabyrinthine fluids. *In* A. Ernst, R. Marchbanks, and M. Samii (Eds.), "Intracranial and Intralabyrinthine Fluids (pp. 51–61). Springer-Verlag, Berlin.

82. Mendelson, E. (1957). A sensitive method for registration of human intratympanic muscle reflexes. *J. Appl. Physiol.* **11**:499–502.

83. Merchant, S. N., Ravicz, M. E., Puria, S., Voss, S. E., Wittemore, K. R., Peake, W. T., and Rosowski, J. J. (1997). Analysis of middle ear mechanics and application to diseased and reconstructed ears. *Am. J. Otol.* 1 **18**:139–154.

84. Metz, O. (1946). The acoustic impedance measured on normal and pathological ears. Acta Otolaryng. (Stock.) Suppl. 63.

85. Møller, A. R. (1958). Intra-aural muscle contraction in man, examined by measuring acoustic impedance of the ear. *The Laryngoscope* LXVIII (1):48–62.

86. Møller, A. R. (1960). Improved technique for detailed measurements of the middle ear impedance. *J. Acoust. Soc. Am.* **32**:250–257.

87. Møller, A. R. (1961). Network model of the middle ear. *J. Acoust. Soc. Am.* **33**:168–176.

88. Møller, A. R. (1963). Transfer function of the middle ear. *J. Acoust. Soc. Am.* **35**:1526–1534.

89. Møller, A. R. (1964). The acoustic impedance in experimental studies on the middle ear. *Int. Audio.* **3**:123–135.

90. Møller, A. R. (1965a). An experimental study of the acoustic impedance of the middle ear and its transmission properties. *Acta Otolaryngol.* **60**:129–149.

91. Møller, A. R. (1965b). Effect of tympanic muscle activity on movement of the eardrum, acoustic impedance, and cochlear microphonics. *Acta Oto-Laryngol.* **58**:525–534.

92. Møller, A. R. (1972). The middle ear. In J. V. Tobias (Ed.), "Foundation of Modern Auditory Theory," (Vol. II, pp. 133–194). Academic Press, New York.

93. Møller, A. R. (1975a). Latency of unit responses in the cochlear nucleus determined in two different ways. *J. Neurophysiol.* **38**:812–821.

94. Møller, A. R. (1975b). Noise as a health hazard. *Ambio* **4**:6–13.

95. Møller, A. R. (1977). Frequency selectivity of single auditory-nerve fibers in response to broadband noise stimuli. *J. Acoust. Soc. Am.* **62**:135–142.

96. Møller, A. R. (1983a). "Auditory Physiology." Academic Press, New York (pp 320).

97. Møller, A. R. (1983b). On the origin of the compound action potentials (N_1,N_2) of the cochlea of the rat. *Exp. Neurol.* **80**:633–644.

98. Møller, A. R. (1985). Origin of latency shift of cochlear nerve potentials with sound intensity. *Hear. Res.* **17**:177–189.

99. Møller, A. R. (1986). Effect of click spectrum and polarity on round window N_1N_2 response in the rat. *Audiology* **25**:29–43.

100. Montandon, P. B., MeGill, N. D., Kahn, A. R., Peaks, W. T., and Kiang, N. Y.-S. (1975). Recording auditory-nerve potentials as an office procedure. *Ann. Otol. Rhinol. Laryngol.* **84**:2–10.

101. Montandon, P. B., Shepard, N. T., Marr, E. M., Peaks, W. T., and Kiang, N. Y.-S. (1975). Auditory-nerve potentials from ear canals of patients with otologic problems. *Ann. Otol. Rhinol. Laryngol.* **84**:164–173.

102. Mountain, D. C., and Cody, A. R. (1999). Multiple modes of inner hair cell stimulation. *Hear. Res.* **132**:1–14.

103. Mountain, D. C., Geisler, C. D., and Hubbard, A. E. (1980). Stimulation of efferents alters the cochlear microphonic and the sound-induced resistance changes measured in scala media of the guinea pig. *Hear. Res.* **3**:231–240.

104. Nedzelnitski, V. (1980). Sound pressure in the basal turn of the cat cochlea. *J. Acoust. Soc. Am.* **68**:1676–1689.

105. Nordlund, B. (1962). Physical factors in angular localization. *Acta Oto. Laryngol. (Stockh.)* **54**:76–93.

106. Ohm, G. S. (1843). Uber die Definition des Tones nebst daran geknupfter Theorie der Direne und ahnlicher tonbildender Vorrichtungen. *Annalen Phys.* **59**:513–565.

107. Pang, X.-D., and Peake, W.-T. (1986). How do contractions of the stapedius muscle alter the acoustic properties of the ear? In "Lecture Notes in Biomathematics." Peripheral Auditory Mechanisms. Springer-Verlag, Berlin.

108. Peake, W. T., and Kiang, N. Y.-S. (1962). Cochlear responses to condensation and rarefaction clicks. *Biophys. J.* **2**:23–34.

109. Pickles, J. O. (1979). An investigation of sympathetic effects on hearing. *Acta Oto. Laryngol.* **87**:69–71.

110. Pickles, J. O., Comis, S. D., and Osborne, M. P. (1984). Cross-links between stereocilia in the guinea pig organ of Corti, and their possible relation to sensory transduction. *Hear. Res.* **15**:103–112.

111. Pierson, L. L., Gerhardt, K. J., Rodriguez, G. P., and Yanke, R. B. (1994). Relationship between outer ear resonance and permanent noise-induced hearing loss. *Am. J. Otolaryngol.* **15**:37–40.

112. Puel, J. L., and Rebillard, G. (1990). Effect of contralateral sound stimulation on distortion product 2F1-F2: Evidence that the medial efferent system is involved. *J. Acoust. Soc. Am.* **87**:1630–1635.

113. Ranke, O. F. (1950). Hydrodynamik der Schneckenflussigkeit. *Z. Biol.* **103**:409–434.

114. Rarey, K. E., Ross, M. D., and Smith, C. B. (1981). Quantitative evidence for cochlear, non-neuronal norepinephrine. *Hear. Res.* **5**:101–108.

115. Rhode, W. S. (1971). Observations of the vibration of the basilar membrane in squirrel monkeys using the Mossbauer technique. *J. Acoust. Soc. Am.* **49**:1218–1231.

116. Rhode, W. S. (1973). An investigation of post-mortem cochlear mechanics using the Mossbauer effect. *In* A. R. Møller (Ed.), "Basic Mechanisms in Hearing" (pp. 49–63). Academic Press, New York.

117. Rosowski, J. J. (1991). The effects of external- and middle-ear filtering on auditory threshold and noise-induced hearing loss. *J. Acoust. Soc. Am.* **90**:124–135.

118. Rosowski, J. J. (1996). Models of external- and middle-ear function. *In* H. L. Hawkins, T. A. McMullen, A. N. Popper, and R. R. Fay (Eds.), "Auditory Computation" (Chapter 2). Springer-Verlag, New York.

119. Rosowski, J. J., Carney, L. H., Lynch, T. J., III, and Peake, W. T. (1986). The effectiveness of external and middle ears in coupling acoustic power into the cochlea. In J. B. Allen, J. L. Hall, A. Hubbard, S. T. Neely, and A. Tubis (Eds.), Lecture Notes in Biomathematics. "Peripheral Auditory Mechanisms" (pp. 3–12). Springer-Verlag, New York.

120. Ruben, R. J., Hudson, W., and Chiong, A. (1962). Anatomical and physiological effects of chronic section of the eighth cranial nerve in cat. *Acta Otolaryngol.* **55**:473–484.

121. Ruben, R. J., and Walker, A. E. (1963). The VIIIth nerve action potential in Méniére's disease. *Laryngoscope* **11**:1456–1464.

122. Ruggero, M. A. (1992). Responses to sound of the basilar membrane of the mammalian cochlea. *Curr. Op. Neurobiol.* **2**:449–456.

123. Russel, I. J., and Sellick, P. M. (1983). Low frequency characteristic of intracellularly recorded receptor potentials in guinea-pig cochlear hair cells. *J. Physiol.* **338**:179–206.

124. Salomon, G. and Elberling, C. (1971). Cochlear nerve potentials recorded from the ear canal in man. *Acta Otolaryngol. (Stockh.)* **71**:319–325.

125. Santi, P. (1988). Cochlear microanatomy and ultrastructure. *In* A. F. Jahn, and J. Santos-Sacchi (Eds.), "Physiology of the Ear" (pp. 173–199). Raven, New York.

126. Schuster, K. (1934). Eine Metode zum Vergleich akustisher Impedanzen. *Phys. Z.* **35**:408–409.

127. Shaw, E. A. C. (1974). The external ear. *In* W. D. Keidel, and W. D. Neff (Eds.), "Handbook of Sensory Physiology" (Vol. V(1), pp. 450–490). Springer-Verlag, New York.

128. Shaw, E. A. C. (1974). Transformation of sound pressure level from the free field to the eardrum in the horizontal plane. *J. Acoust. Soc. Am.* **56**:1848–1861.

129. Shera, C. A., and Guinan, J. J. (1999). Evoked otoacoustic emissions arise by two fundamentally different mechanisms: A taxonomy for mammalian OAE's. *J. Acoust. Am.* **105**:782–798.

130. Siebert, W. M. (1970). Simple model of the impedance matching properties of the external ear. *In* "Quarterly Progress Report 96: Research Laboratory of Electronics" (pp. 236–242). MIT Press,

131. Simmons, F. B. (1959). Middle ear muscle activity at moderate sound levels. *Ann. Otol. (St. Louis)*, **68**:1126–1143.

132. Sokolich, W. G., Hamernick, R. P., Zwislocki, J. J., Schmiedt, R. A. (1976). Inferred response polarities of cochlear hair cells. *J. Acoust. Soc. Am.* **59**:963–974.

133. Spoendlin, H. (1970). Structural basis of peripheral frequency analysis. *In* R. Plomp and G. F. Smoorenburg (Eds.) "Frequency Analysis and Periodicity Detection in Hearing" (pp. 2–36). A. W. Sijthoff, Leiden, The Netherlands.

134. Spoendlin, H., and Lichtensteiger, W. (1966). The adrenergic innervation of the labyrinth. *Acta. Otolaryngol. (Stockh.)* **61**:423–434.

135. Spoendlin, H, and Schrott, A. (1989). Analysis of the human auditory nerve. *Hear. Res.* **43**, 25–38.

136. Spoor, A. Eggermont, J. J., and Odenthal, D. W. (1976). Comparison of human and animal data concerning adaptation and masking of eighth nerve compound action potentials. *In* R. J. Ruben, C. Elberling, and G. Salomon (Eds.), "Electrocochleography" (pp. 183–198). Univ. Park Press, Baltimore.

137. Stuhlman, O. (1943). "An Introduction to Biophysics." Wiley, New York.

138. Svane-Knudsen, V. and Michelsen, A. C. (1985). The impulse response vibration of the human ear drum. *In* "Lecture Notes in Biomathematics" (Vol. 64, pp. 21–27) Springer-Verlag, Berlin.

139. Tasaki, I., Davis, H. and Legouix, J. P. (1952). The space–time pattern of the cochlear microphonics (guinea pig), recorded by differential electrodes. *J. Acoust. Soc. Am.* **24**: 502–518.

140. Teas, D. C., Eldredge, D. H., and Davis, H. (1962). Cochlear responses to acoustic transients: An interpretation of whole-nerve action potentials. *J. Acoust. Soc. Am.* **32**:1438–1459.

141. Terkildsen, K. (1957). Movements of the eardrum following intraaural muscle reflexes. *Arch. Otolaryngol.* **66**:484–488.

142. Terkildsen, K., and Nielsen, S. S. (1960). An electroacoustic impedance measuring bridge for clinical use. *Arch. Otolaryngol.* **72**:339–346.

143. Tonndorf, J., Khanna, S. M., and Fingerhood, B. (1966). The input impedance of the inner ear in cats. *Ann. Otol. Rhinol. Laryngol.* **75**:752–763.

144. Tos, M. (1985). "Manual of Middle Ear Surgery: Mastoid Surgery and Reconstructive Procedures (Vol. 2). Thieme Medical, Stuttgart.

145. Tröger, J. (1930). Die Schallaufnahme durch das aussere Ohr. *Phys. Z.* **31**:26–47.

146. Warr, W. B., and Guinan, J. J. (1979). Efferent innervation of the organ of Corti: Two separate systems. *Brain Res.* **173**:152–155.

147. Warren, E. H., III, and Liberman, M. C. (1989a). Effects of contralateral sound on auditory-nerve responses. I. Contributions of cochlear efferents. *Hear. Res.* **37**: 89–104.

148. Warren, E. H., III, and Liberman, M. C. (1989b). Effects of contralateral sound on auditory-nerve responses. II. Dependence on stimulus variables. *Hear. Res.* **37**:105–122.

149. Wever, E. G., Lawrence, M., and Smith, K. R. (1948a). The effects of negative air pressure in the middle ear. *Ann. Otol. Rhinol. Laryngol.* **57**:418–428.

150. Wever, E. G., Lawrence, M., and Smith, K. R. (1948b). The middle ear in sound conduction. *Arch. Otolaryngol.* **48**:19–35.

151. Wever, E. G., and Lawrence, M. (1954). "Physiological Acoustics." Princeton Univ. Press, Princeton, NJ.

152. Wever, E. G., and Vernon, J. A., (1955). The effect of the tympanic muscle reflexes upon sound transmission. *Acta Otolaryngol. (Stockh.)* **45**:433–439.

153. Wever, E. G. (1949). "Theory of Hearing". Wiley, New York.

154. Wiederhold, M. L., and Kiang, N. Y.-S. Effects of electric stimulation of the crossed olivocochlear bundle on single auditory-nerve fibers in the cat. *J. Acoust. Am.* **48**:950–965.

155. Wiggers, H. C. (1937). The function of the intraaural muscles. *Am. J. Physiol.* **120**:771–780.

156. Wilson J. P. (1980). Evidence for cochlear origin for acoustic re-emissions, threshold fine-structure and tonal tinnitus. *Hear. Res.* **2**:233–252

157. Winzenburg, S. M., Margolis, R. H., Levine, S. C., Haines, S. J., and Fournier, E. M. (1993). Tympanic and transtympanic electrocochleography in acoustic neuroma and vestibular nerve section surgery. *Am. J. Otol.* **14**:63–69.

158. Yates, G. K., and Withnell, R. H. (1999). The role of intermodulation distortion in transient-evoked otoacoustic emissions. *Hear. Res.* **136**:49–64.

159. Zweig, G., Lipes, R., and Pierce, J. R. (1976). The cochlear compromise. *J. Acoust. Soc. Am.* **59**:975–982.

160. Zwislocki, N. J. (1948). Theorie der Schneckenmechanick. *Act. Oto-Laryngol. Suppl.* **72**:1–76.

161. Zwislocki, J. (1957a). Some measurements of the impedance at the eardrum. *J. Acoust. Soc. Am.* **29**:349–356.

162. Zwislocki, J. J. (1957b). Some impedance measurements on normal and pathological ears. *J. Acoust. Soc. Am.* **29**:1312–1317.

163. Zwislocki, J. J. (1962). Analysis of the middle ear function. I: Input impedance. *J. Acoust. Soc. Am.* **34**:1514–1523.

164. Zwislocki, J. J. (1963). Analysis of the middle-ear function. II. Guinea-pig ear. *J. Acoust. Soc. Am.* **35**:1034–1040.

165. Zwislocki, J. J. (1980). Five decades of research on cochlear mechanics. *J. Acoust. Soc. Am.* **67**:1679–1685.

166. Zwislocki, J. J. (1986). Are nonlinearities observed in firing rates of auditory-nerve afferents reflections of a nonlinear coupling between the tectorial membrane and the organ of Corti? *Hear. Res.* **22**:217–222.

167. Zwislocki, J. J. (1991). What is the cochlear place code for pitch? *Acta Otolarying. (Stockh.)* **111**:256–262.

168. Zwislocki, J., and Cefaratti, L. K. (1989). Tectorial membrane. II. Stiffness measurements *in vivo. Hear. Res.* **42**:211–228.

169. Zwislocki, J. J., and Kletsky, E. J. (1980). Micromechanics in the theory of cochlear mechanics. *Hear. Res.* **2**:505–512.

170. Zwislocki, J. J., and Sokolich, W. G. (1973). Velocity and displacement responses in auditory nerve fibers. *Science* **182**:64–66.

Auditory Nervous System

The auditory nervous system is the most complex of all sensory pathways. Studies of the function of the auditory nervous system have earlier been regarded to be mainly of academic interest but recent studies have revealed that many of the pathologies that earlier were thought to be located to the ear are in fact caused by changes in the auditory nervous system. The complexity of the anatomy of the auditory nervous system is considerable and a thorough understanding of the anatomy and the normal functioning of the auditory nervous system is a prerequisite to understand such pathologies.

ANATOMY

The auditory nervous system consists of ascending and descending systems. Two ascending systems, known as the classical and nonclassical auditory systems, and three descending systems, the corticofugal system, the cortico cochlear system and the olivocochlear system, have been identified. In the ascending system, some investigators have divided the nonclassical system into two separate systems, the diffuse system and the polysensory system.

The classical ascending pathway has three main nuclei with several additional nuclei where some of the ascending neural activity is interrupted. This is the basis for the parallel and hierarchical neural processing that is characteristic of the classical ascending auditory pathway. There are connections between the two sides of the brain at several levels of the classical ascending auditory pathway. The nonclassical pathways are complex structures that connect to a large number of other brain areas, but the anatomy of these pathways is incompletely known. Of the two separate descending auditory pathways, one, the corticofugal system, connects the cerebral auditory cortex with more peripheral nuclei and one, the olivocochlear system, connects pontine nuclei with hair cells of the cochlea.

The anatomy of the auditory nervous system is described in chapter 5. More detailed coverage of the anatomy of the auditory system can be found in Ehret and Romand's *The Central Auditory System* [28] and other listed references.

NEUROPHYSIOLOGY

The physiology of the auditory nervous system has been studied mostly in animals. Most studies have been concerned with the coding of different kinds of sounds in the discharge pattern of auditory nerve fibers and the transformation of that code as the information travels up the neural axis toward the cerebral cortex. Peripheral parts of the ascending auditory pathway have been studied more extensively than central portions and it is the classical, or lemniscal, ascending auditory pathways that have attracted the most interest; only a few studies have been aimed at other parts of the auditory nervous system. Thus, the nonclassical (adjunct or extralemniscal) ascending auditory pathways have not been studied nearly as much as the classical auditory system and little is known about the coding and transformation of information in these systems.

All information that is available to the auditory nervous system is contained in the neural discharge pattern of auditory nerve fibers. This information undergoes an extensive transformation in the nuclei of the classical ascending auditory pathway, which performs hierachial and parallel processing of information. The representation of frequency in the auditory nervous system is the most studied of the different kinds of processing that occur. The representation in the nervous system of different features of natural sounds is the basis for our ability to discriminate a wide variety of sounds. For humans, speech is the most important sound and it would have been natural to ask the question "How does the auditory nervous system discriminate speech sounds?" Nevertheless, that is too complex a question and it is more realistic to ask simpler questions such as "How is frequency discriminated?" Frequency discrimination is prominent and its physiologic basis has been studied extensively because it is assumed to play an important role in the discrimination of natural sounds. In this section, we first discuss representation of frequency in the auditory nervous system as place and temporal codes (Chapter 6). In Chapter 7 we discuss the relative importance of the temporal and place codes of frequency for discrimination of complex sounds.

Frequency, or frequency spectrum, however, is only one feature of complex sounds. The way sounds change is also an important property of natural sounds and changes in frequency and amplitude of sounds are accentuated by the neural processing of the classical ascending auditory nervous system. Changes in frequency (spectrum) and amplitude are prominent characteristics of natural sounds and therefore it may be assumed that the changes in a particular sound are important for distinguishing between different sounds. Chapter 8 covers coding of complex sounds in the auditory nervous system and focuses on processing of sounds, the frequency and amplitude of which change more or less rapidly. The sounds discussed here are similar to important natural sounds such as speech sounds and yet are better defined. Chapter 9 discusses the neurophysiologic basis for directional hearing and how the perception of space may be formed.

Most of our knowledge about the function of the auditory nervous system is based on recordings from single nerve cells and the impulse traffic in single nerve fibers. However, single elements of the nervous system do not function similarly in the nervous system as in human-made systems such as computers and telephone systems. It is not known how the impulse pattern in nerve cells and groups of nerve cells is related to perception and interpretation of sounds and other sensory input. Recordings of evoked potentials represent the working of many nerve cells but such studies have contributed comparatively little to the understanding of the function of the auditory nervous system. Evoked potentials, however, have the advantage that they can be recorded in animals as well as in humans. Evoked potentials are discussed in Chapters 10 and 11.

Comparing the nervous system to a telephone system was common in the beginning of this century. Such a comparison provides not only an oversimplified picture of the nervous system but also an incorrect one because the function of the nervous system is based on the function of large groups of elements with complex and numerous interconnections. Studies of the response pattern of single nerve cells therefore only provides a view of the working of the nervous system through a narrow and distorted window. At present, the nervous system is often compared with computers. While computers are more complex systems than old-fashioned telephone systems and thus offer a closer similarity, the analogy is still a great oversimplification. Just comparing the numbers of elements illustrates how wrong such a comparison is. It has been estimated that the human central nervous system has approximately 30 billion nerve cells. Compare that to a Pentium II processor, the "brain" of modern personal computers, which has approximately 3 million elements. The central nervous system thus has 10,000 times more elements than a modern computer. Many nerve cells have hundreds of inputs, while the elements in a computer are transistors with only one or two inputs. The nervous system has a vast number of interconnections between nerve cells and that has no analogy in modern computers, even the most advanced computer systems. In the nervous system, the existence of a specific element in the central nervous system and its connections are not known; only the probabilities of their existence may be known. Similar probabilistic systems are unknown in human-made systems because their functioning is utterly impossible to fathom. We can essentially only study the functioning of one or perhaps a few neural elements at a time and we know little about the connections of the elements we study. It is unknown which features of the functioning of a single element in the nervous system are important, which adds uncertainty to the interpretation of the results of studies of the nervous system. The plasticity of the nervous system that is so prevalent is unique to the nervous system. Self-organizing computer programs are emerging but are quite simple compared with the plasticity of the central nervous system. This neural plasticity may be involved in disorders of the nervous system, which are discussed in Chapters 15 and 16 in Section IV.

CHAPTER 5

Anatomy of the Auditory Nervous System

ABSTRACT

1. The ascending auditory pathways are more complex than the ascending pathways of other sensory systems such as the somatosensory system or the visual system.
2. Besides the classical ascending pathway (the tonotopic system), two other ascending systems have been identified, namely the polysensory system and the diffuse system. These two systems are also referred to as the nonclassical or adjunct auditory systems.
3. In the classical ascending auditory system, three main relay nuclei are located between the auditory nerve and the primary auditory cerebral cortex: (1) the cochlear nucleus, (2) the central nucleus of the contralateral inferior colliculus, and (3) the lateral portion of the ventral division of the contralateral medial geniculate body. All information is interrupted in these nuclei by synaptic transmission.
4. Some fibers of the classical ascending pathways connect to neurons in other nuclei such as the nuclei of the superior olivary complex and the nucleus of the lateral lemniscus, and some fibers are interrupted in these nuclei.

5. Several fiber tracts connect the two sides of the ascending auditory pathway at several levels. The lowest level of crossover occurs at the cochlear nuclei, but the most prominent connections occur between the nuclei of the superior olivary complex, the inferior colliculus, and the auditory cerebral cortices.

6. The number of nerve fibers that connect the different nuclei increases from the periphery to the cerebral cortex.

7. The nonclassical (adjunct) ascending auditory pathways (the diffuse system and the polysensory system) branch off the classical system at several levels, the most prominent being the central nucleus of the inferior colliculus.

8. The diffuse system projects, via the dorsal portion of the medial geniculate body, to cortical areas other than the primary auditory cortex.

9. The polysensory system receives input from other sensory systems (somatosensory and visual) in addition to auditory input. It projects, via the medial portion of the medial geniculate body, to different nonprimary auditory cortical areas as well as other brain regions, including the limbic system.

10. Three separate descending paths of the auditory system conduct information from central structures to peripheral structures. One system, the olivocochlear system, connects the cochlea with nuclei of the superior olivary complex. The corticofugal system connects the cerebral cortex with several nuclei of the ascending auditory pathway and the corticocochlear system connects the auditory cerebral cortex with the cochlea.

CLASSICAL ASCENDING AUDITORY PATHWAY

The auditory nerve (AN) extends from the organ of Corti to the cochlear nucleus (CN). From the cochlear nucleus, fibers cross over to the opposite side in three fiber tracts that connect to the contralateral inferior colliculus (IC). Fibers from the IC project to the medial geniculate body (MGB). The fibers from the MGB project to the primary auditory cortex (Fig. 5.1).

AUDITORY NERVE

The auditory nerve in humans has approximately 30,000 fibers. The AN is part of the VIIIth cranial nerve, which also includes the (superior and inferior) vestibular nerve. The AN consists of two types of fibers, Type I and Type II. The nerve fibers of the AN are bipolar cells, which have their cell bodies in the spiral ganglion, located in the modiolar region of the cochlea (for details see [193]). The peripheral portions of the Type I fibers of the AN terminate

A

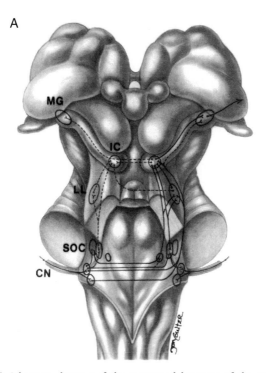

FIGURE 5.1   (A) Schematic drawing of the anatomical locations of the ascending auditory pathway. Abbreviations: AN, auditory nerve; CN, cochlear nucleus; SOC, superior olivary complex; LL, lateral lemniscus; NLL, nucleus of the lateral lemniscus; IC, inferior colliculus; MG, medial geniculate body. (B) Schematic diagram showing of the main nuclei and fiber tracts of the classical ascending auditory system pathway. Abbreviations: CN, cochlear nucleus; LL, lateral lemniscus; IC, inferior colliculus; MGB, medial geniculate body. (C) More detailed drawing of the ascending auditory pathway from the ear to the IC. The pathways that ascend on the ipsilateral side are shown as dashed lines, and those that cross over to the other side are shown as solid lines. Abbreviations: AVCN, anterior ventral cochlear nucleus; PVCN, posterior ventral cochlear nucleus; DCN, dorsal cochlear nucleus; LSO, lateral superior olive; NTB, nucleus of the trapezoidal body; MSO, medial superior olive; SH, stria of Held (intermediate stria); SM, stria of Monakow (dorsal stria); LL, nucleus of the lateral lemniscus; DNLL, dorsal nucleus of the lateral lemniscus; VNLL, ventral nucleus of the lateral lemniscus; IC, inferior colliculus.

on the inner hair cells (Chapter 1) and the central portions terminate on the cells of the cochlear nucleus. The Type I nerve fibers are called the radial fibers and they are thought to carry all the auditory information from the organ of Corti to higher centers of the central nervous system. The average diameter of myelinated (Type I) cochlear nerve fibers in the internal auditory meatus in children is within a narrow size range of about 2.5 μm. The diameter

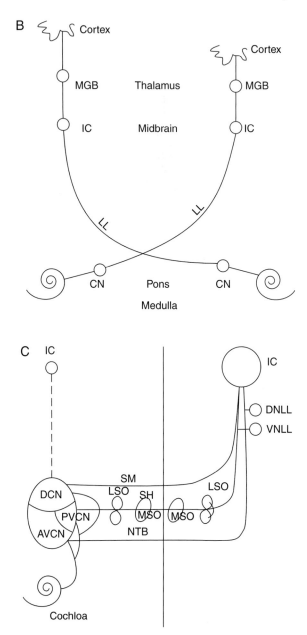

FIGURE 5.1 (*Continued*)

of the myelinated fibers in the osseous spinal lamina is approximately half that of fibers in the internal auditory meatus [210] (Fig. 5.2). At any cross section of the AN, the variations in the diameters of these Type I nerve fibers are small, [210], which implies that the variations in the conduction velocity of different auditory nerve fibers are small. The information that is carried in different auditory nerve fibers will therefore arrive at the cochlear nucleus at very small time differences, ensuring a high degree of temporal coherence of the nerve impulses that arrive at the cochlear nucleus. Evidence has been presented that such coherence is important for discrimination of complex sounds such as speech sounds (discussed in Chapters 6 and 7). The variation in fiber size, and thus in conduction velocity, increases with age [210], which may explain some of the hearing problems that occur with age and which are not directly related to elevation in pure tone threshold (see Chapter 15).

The Type II fibers form the outer spiral fibers and innervate outer hair cells. These fibers constitute only approximately 1% of the total population of nerve fibers in the auditory nerve. Type II nerve fibers project to the cochlear nucleus (mostly the dorsal cochlear nucleus), but their function is unknown. The human spiral ganglion is different from that of the animals most studied, and the cell bodies of the majority of ganglion cells were unmyelinated in humans, but the individual variation is large [169].

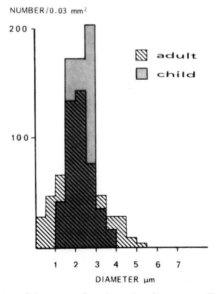

FIGURE 5.2 Distribution of diameters of myelinated auditory nerve fibers in humans. Results obtained in an adult are compared with that found in a child. Reprinted from *Hear. Res.*, Volume 43, Spoendlin, H., and Schrott, A., Analysis of the human auditory nerve, pp. 25–38, copyright 1989, with permission from Elsevier Science.

Most of our knowledge about the morphology of the auditory nerve is based on studies in animals (mostly cat) and it is not completely known to what extend these results are applicable to humans. It is known, however, that the human VIIIth cranial nerve, including the auditory nerve is much longer than it is in animals such as the cat (2.5 cm [97, 98] vs 0.8 cm [45]). One of the reasons the VIIIth nerve is longer in humans than in animals (including monkeys) is because the head of humans is larger, and the subarchnoidal space, is wider. This difference in the size of the subarchnoidal space has not been given much attention, although it may have many implications regarding the development of certain auditory disorders. Like other cranial nerves, the VIIIth nerve is twisted. In its central course, the auditory portion is located caudally with respect to the superior vestibular nerve [98] (Fig. 5.3A) and it

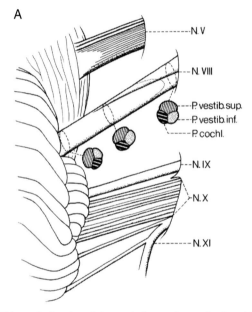

FIGURE 5.3 (A) Schematic drawing of the cerebello pontine angle viewed from the dorsal side with a cross section of the VIIIth cranial nerve showing the different portions of the nerve and how it rotates (from Lang, J. (1981). Facial and vestibulocochlear nerve, topographic anatomy and variations. *In* M. Samii, and P. J. Jannetta (Eds.), "The Cranial Nerves" (pp. 363–377). Springer-Verlag, New York). (B) Drawing of the anatomy of the internal auditory canal as seen from the retro mastoid approach. The posterior wall of the internal auditory meatus has been removed so that it appears as a single canal. Abbreviations: IVN, inferior vestibular nerve; SVN, superior vestibular nerve; FN, facial nerve; VN, (entire) vestibular nerve; CoN, cochlear (auditory) nerve (from Silverstein, H., Norrell, H., Haberkamp, T., and McDaniel, A. B. (1986). The unrecognized rotation of the vestibular and cochlear nerves from the labyrinth to the brain stem: Its implications to surgery of the eighth cranial nerve. *Otolaryngol. Head Neck Surg.* **95**:543–549).

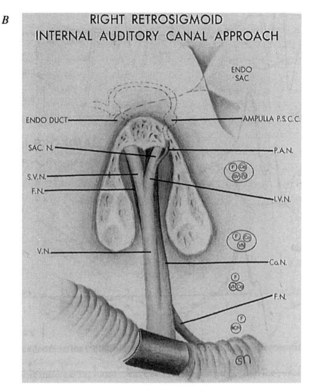

FIGURE 5.3 (*Continued*)

is located dorsally in the internal auditory meatus. In its most peripheral portion the auditory nerve is located ventrally with respect to the vestibular nerve (Fig. 5.3B, [202]).

Each individual myelinated auditory nerve fiber (Type I) is covered with peripheral myelin (Schwann cell) in its peripheral course, up to the point just before it leaves the internal auditory meatus, where the covering changes to central myelin (oligodendrocyte myelin). The most central portion of the auditory nerve (approximately 1 cm) is thus similar to brain tissue. This has practical implications because it implies that the auditory nerve (and, for that matter, the entire VIIIth cranial nerve) is at considerable risk of being injured in surgical manipulations operations in the cerebello pontine angle.

COCHLEAR NUCLEUS

The cochlear nucleus is the first relay nucleus of the ascending auditory pathway. It is located in the lower brainstem at the junction between the medulla and the pons (the pontomedullary junction). The CN has three main divisions, the dorsal cochlear nucleus (DCN), the posterior ventral cochlear nucleus (PVCN), and anterior ventral nucleus (AVCN) (Fig. 5.4). Each auditory nerve fiber connects to all three divisions of the CN [108]. First, each nerve fiber bifurcates and one of the two branches terminates in the AVCN and the other branch bifurcates again before terminating in the cells of the PVCN and the DCN (Fig. 5.4). This represents the initiation of the parallel processing that is abundant in the auditory system.

In small animals, the CN is a prominent structure of the lower brainstem, but in humans, it is comparatively small. In humans, the CN has a rostral-caudal extension of only 3 mm, but a medial-to-lateral extension of approximately 10 mm

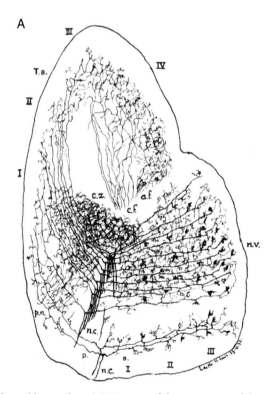

FIGURE 5.4   The cochlear nucleus. (A) Drawings of the connections of the auditory nerve with the cochlear nucleus [108]. (B) Schematic drawing of the cochlear nucleus to show the auditory nerve's connections with the three main divisions and the cochlear nucleus. Abbreviations: DCN, dorsal cochlear nucleus; PVCN, posterior ventral cochlear nucleus; AVCN, anterior ventral cochlear nucleus.

B

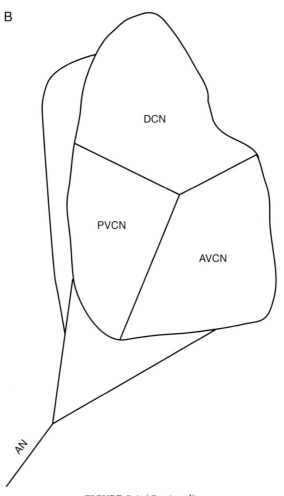

FIGURE 5.4 (*Continued*)

and an extension of 8 mm ventrolaterally [153]. In contrast, the cat CN is more symmetrical, with extensions of approximately 4 mm in all three planes.

The fibers from the CN project to the contralateral inferior colliculus through three fiber tracts, the dorsal stria (stria of Monaco, SM), the intermediate stria (stria of Held, SH), and the ventral stria (trapezoidal body, TB) (Fig. 5.1C). The SM originates in the DCN. The fibers from the PVCN cross in the SH and the output of the AVCN forms the TB. These three striae, after crossing to the opposite side, form the lateral lemniscus (LL), a fiber tract that projects to the central nucleus of the inferior colliculus (ICC).

Some fibers of the output of the AVCN and PVCN do not cross the midline but ascend on the same side to reach the ipsilateral IC (Fig. 5.1C). Thus a few fibers from the PVCN ascend on the same side (uncrossed pathways) to the ventral lateral nucleus of the lateral lemniscus and from there to the (ipsilateral) IC. The ventral cochlear nucleus also sends fibers to the facial motor nucleus and the trigeminal motor nucleus as part of the acoustic middle ear reflex (see Chapter 12).

The two sides' cochlear nuclei are connected [113]. This is the most peripheral connection between the two side's ascending auditory pathway, but its functional importance is unknown. The CN also receives input from the somatosensory system [227], but the functional importance of that is also unknown.

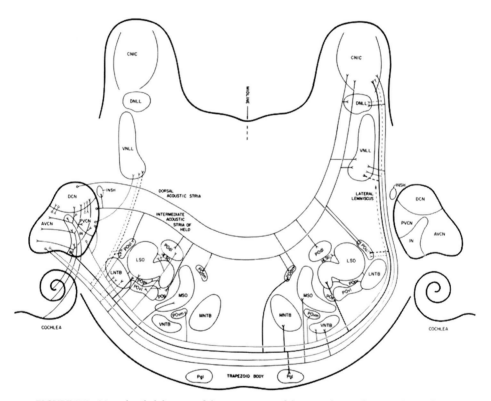

FIGURE 5.5   More detailed drawing of the connections of the ascending auditory pathway (from Kiang, N. Y.-S. (1975). Stimulus representation in the discharge patterns of auditory neurons. In E. L. Eagles (Ed.), "The Nervous System" (pp. 81–96). Raven, New York).

SUPERIOR OLIVARY COMPLEX AND NUCLEI OF THE LATERAL LEMNISCUS

The superior olivary complex (SOC) consists of three main nuclei, the medial superior olivary nucleus (MSO), the lateral superior olivary (LSO) nucleus, and the nucleus of the trapezoidal body (NTB) (Fig. 5.5). Some of the fibers of the three striae (SM, SH, and TB) give off collaterals to some of the nuclei of the SOC and some fibers are interrupted in one of the nuclei of the SOC before forming the lateral lemniscus. Nuclei of the SOC, in particular the MSO, receive input from the CN of both sides.

The SOC is thus the first group of nuclei that integrate information from both ears. The nuclei of the SOC are involved in directional hearing, mainly by comparing arrival times of neural activity from the two ears. The nuclei of the SOC comprise some of the most complicated parts of the ascending auditory pathway and they have the largest variations between different species of mammals. The anatomical arrangements of these nuclei in humans is in many ways different from that of the commonly used experimental animals such as the cat [153].

Lateral Lemniscus and Its Nuclei

The lateral lemniscus is the most prominent fiber tract of the classical ascending auditory pathway (Fig. 5.5). The LL is formed by the three striae that emanate from the CN after they cross the midline. The LL is composed of fibers from all divisions of the contralateral CN. Some fibers of the LL travel uninterrupted all the way from the CN to the IC, while other fibers are interrupted by synapses in different nuclei of the SOC and the nuclei of the lateral lemniscus (NLL). Both the ipsilateral and the contralateral CN contribute to the LL and both the ipsilateral and contralateral lateral SOC contribute to the LL. Contralateral (octopus) cells of the PVCN do not travel directly to the central nucleus of the inferior colliculus (ICC) as other axons do, but instead terminate in the ventral nucleus of the lateral lemniscus (VNLL). Since fibers of the LL extend from different sources, the LL contains both second-, third-, and possibly fourth-order neurons.

Nucleus of the Lateral Lemniscus

The nucleus of the lateral lemniscus (NLL) has two main parts, the dorsal NLL (DNLL) and the ventral NLL (VNLL). Both nuclei receive input from the LL and some fibers in the LL are interrupted in the NLL. The DNLL receives input from both ears and is involved in binaural hearing, while the VNLL mainly receives input from the contralateral ear. Some of the axons that lead away from the DNLL travel to the opposite side as the commissure of Probst and, once there, connect to neurons of the ICC. The ipsilateral CN also has direct connections to the ICC.

Inferior Colliculus

The inferior colliculus is located in the midbrain just caudal to the superior colliculus (SC). The IC is the midbrain relay nucleus where all ascending auditory information is channeled. The IC consists of the central nucleus, (ICC), the external nucleus (ICX) (the ICX is also known as the lateral nucleus, LN), and the cortex of the IC (Fig. 5.6). The ICC is a part of the classical ascending auditory system. The ICC receives its input from the LL and all the fibers of the LL are interrupted by neurons in the IC. The connections between

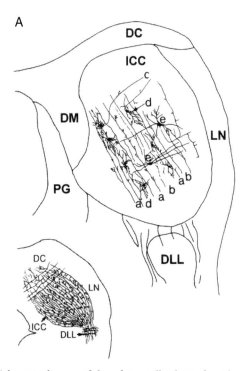

FIGURE 5.6 (A) Schematic drawing of the inferior colliculus in frontal section. Abbreviations: DC, dorsal cortex; DM, dorsomedial nucleus; ICC, central nucleus; LN, lateral nucleus; DLL, dorsal nucleus; PG, periaqeductal gray; a and b, lemniscal axons; c, axons leaving the ICC (from Ehret, G., and Romand, R. (Eds.). (1997). "The Central Auditory Pathway." Oxford Univ. Press, New York). (B) Schematic drawing of connections of the auditory midbrain [3].

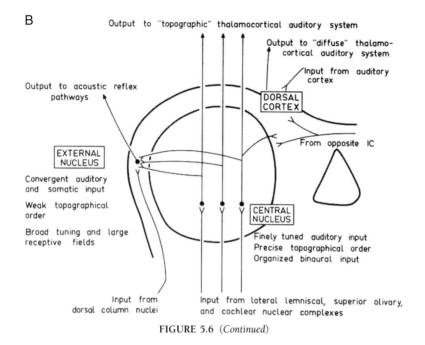

B

Output to "topographic" thalamocortical auditory system

Output to "diffuse" thalamocortical auditory system

Input from auditory
cortex

Output to acoustic reflex
pathways

DORSAL
CORTEX

From opposite IC

EXTERNAL
NUCLEUS

Convergent auditory
and somatic input

Weak topographical
order

Broad tuning and large
receptive fields

CENTRAL
NUCLEUS

Finely tuned auditory input
Precise topographical order
Organized binaural input

Input from
dorsal column nuclei

Input from lateral lemniscal, superior olivary,
and cochlear nuclear complexes

FIGURE 5.6 (Continued)

the two side's ICC are important for directional hearing that is based on the differences in the sound intensity at the two ears. The ICX and the cortex of the IC, which are parts of the nonclassical auditory system receives input from the ICC (Fig. 5.6B). The dorsal cortex of the IC delivers its output to the diffuse thalomocortical auditory system. The external nucleus receives input from the somatosensory system (dorsal column nuclei) and provides auditory output to acoustic reflex pathways (not the acoustic middle ear reflex) (Fig. 5.6B) [3].

The IC has been described and labeled as the auditory reflex center, as it connects to the SC to control eye movements and other motor responses to auditory stimuli. It also connects to other motor centers of the brain.

The main output of the ICC is the brachium of the inferior colliculus (BIC), which leads to the last of the three main auditory nuclei, the medial geniculate body (MGB). However, other parallel pathways also exist. Galambos showed already in 1961 [48] that the auditory cortex could be activated by sound after the BIC was severed. It is interesting that the number of fibers of the BIC is approximately 250,000, thus approximately 10 times that of the fibers of the auditory nerve. This divergence indicates that considerable signal processing takes place between the auditory nerve and the auditory cortex.

Medial Geniculate Body

The medial geniculate body is the thalamic auditory relay nucleus where all fibers that originate in the ICC are interrupted (see Fig. 5.7). The medial geniculate body has three distinct divisions, ventral, dorsal, and medial [155, 156]. The ventral division of the medial geniculate body includes the pars lateralis (LV) and the pars ovoidea (OV). The ventral division receives its input from the central nucleus of the IC and is part of the classical ascending auditory pathway. The LV portion of the MGB probably also receives input from the ipsilateral ear via the ICC. The dorsal and medial divisions of the MGB are parts of the nonclassical auditory system (discussed below) and these divisions receive their input from the cortex of the IC and the ICX. The medial

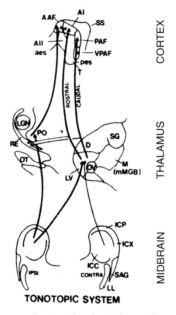

FIGURE 5.7 Schematic diagram showing the classical ascending auditory pathway. Abbreviations: AAF, anterior cortical field; AI, primary auditory cortical field; D, dorsal nucleus of the dorsal division of the MGB; ICC, central nucleus of the inferior colliculus; ICP, pericentral nucleus of the inferior colliculus; ICX, external nucleus of the inferior colliculus; LGN, lateral geniculate nucleus; LL, lateral lemniscus; LV, pars lateralis of the ventral division of the MGB; M, pars magnocellularis of the medial division of the MGB; MGB, medial geniculate body; RE, reticular nucleus of the thalamus; OT, optic tract; OV, pars ovoidea of the ventral division of the MGB; PAF, posterior cortical auditory field; pes, posterior ectosylvian nucleus; VPAF, ventroposterior auditory field; SAG, sagulum; SC, superior colliculus, SG, suprageniculate nucleus of the MGB; SS, suprasylvian sulcus (from Ehret, G., and Romand, R. (Eds.). (1997). "The Central Auditory Pathway." Oxford Univ. Press, New York).

division of the MGB projects to the anterior auditory field (AAF) cortical area (Fig. 5.10). The posterior division of the MGB (PO) may receive input from the ipsilateral ICC.

AUDITORY CEREBRAL CORTEX

Several different regions of the auditory cortex have been identified (Fig. 5.7). The primary auditory cortex (AI) receives its input from the ventral division of the MGB. The primary auditory cortex is not the end station for the auditory information, and fiber tracts from the primary auditory cortex project to other cortical areas. Thus the primary auditory cortex (AI) projects to another region of the auditory cortex, the posterior auditory field (PAF).

In humans, the auditory cortex is located deep in the superior portion of the temporal lobe in the transverse gyrus of Hechel and it is not visible from the surface of the brain. The anatomy of the human auditory cortex is not completely known and different investigators have placed different names on the same parts of the human auditory cortices. The anatomy and the exact anatomical location of the different components of the human auditory cortex vary between individuals.

CONNECTIONS BETWEEN THE TWO SIDES

The most peripheral level where the two sides of the ascending auditory pathway come in contact with each other is the superior olivary complex (Fig. 5.8). The neurons of the SOC receive input from both ears. There are also connections between the two cochlear nuclei [113] but little is known about the anatomy of these connections. Central to the SOC, several large fiber tracts connect the nuclei of the two sides. The commissure of Probst connects the two DNLL with each other and contains fibers from the DNLL on one side that connect to the ICC on the other side. The commissure of the inferior colliculus connects the ICCs of each side and sends fibers from the ICC to the MGB on the other side through the BIC. A larger fiber tract that is a part of the corpus callosum (Fig. 5.8) connects the auditory cerebral cortices on the two sides [28, 164].

DIFFERENCES BETWEEN THE CLASSICAL
AUDITORY PATHWAY IN HUMANS
AND IN ANIMALS

The most obvious difference between the classical auditory nervous system in humans and that of commonly used experimental animals is that the auditory

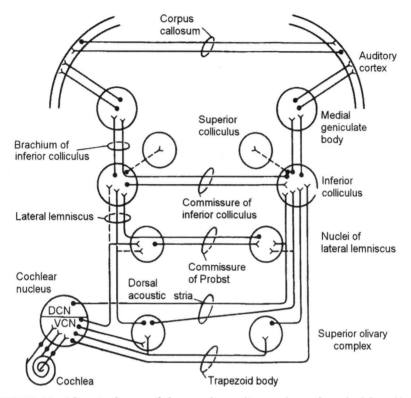

FIGURE 5.8 Schematic diagram of the ascending auditory pathways from the left cochlea, showing the main nuclei and their connections, including the connections between the two sides (from Ehret, G., and Romand, R. (Eds.). (1997). "The Central Auditory Pathway." Oxford Univ. Press, New York).

nerve is much longer in humans than in animals [97]. The fiber tracts of the ascending auditory pathway are also in general longer in humans than in small animals such as the cat [45, 97, 153] which implies that the neural travel time is longer in humans [145, 153] (Fig. 5.9). The most pronounced difference in the nuclei of the ascending auditory pathway is in the superior olivary complex [153]. There are fewer small neurons in the lateral olivary nucleus, the nucleus of the trapezoidal body, and the ventral nucleus of the lateral lemniscus in humans compared with animals such as the cat. There are also fewer small cells in the human cochlear nucleus than in the cat and the dorsal cochlear nucleus is much smaller and less developed in humans compared with the cat or other animals that are used in auditory research. Groups of large neurons are more developed in the human auditory nervous system in

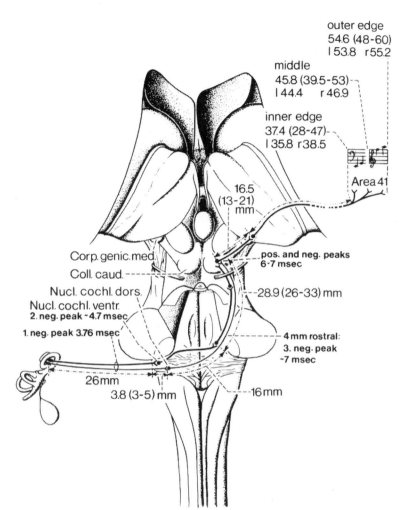

outer edge
54.6 (48-60)
l 53.8 r55.2

middle
45.8 (39.5-53)
l 44.4   r 46.9

inner edge
37.4 (28-47)
l 35.8 r38.5

Area 41

16.5
(13-21)
mm

Corp. genic. med.

Coll. caud.

Nucl. cochl. dors.
Nucl. cochl. ventr.
2. neg. peak ~4.7 msec
1. neg. peak 3.76 msec

pos. and neg. peaks
6-7 msec

28.9 (26-33) mm

4 mm rostral:
3. neg. peak
~7 msec

26mm
3.8 (3-5) mm

16mm

FIGURE 5.9   Length of the main paths of the ascending auditory system in humans (from Lang, J. (1991). Clinical Anatomy of the Posterior Cranial Fossa and Its Foramina. Thieme-Verlag, Stuttgart).

the cochlear nucleus, the medial superior olivary (MSO) nuclei, periolivary nuclei, and the dorsal nucleus of the lateral lemniscus [153]. We can only speculate about the functional importance of these differences between the ascending auditory pathway in humans compared with that of animals that are commonly used in studies of the auditory system. The differences in the length of the auditory nerve and the length of the fiber tracts, however, are important for the interpretation of auditory evoked potentials (BAEP) (see Chapter 11).

NONCLASSICAL ASCENDING AUDITORY

The anatomy of the nonclassical or adjunct ascending auditory system briefly mentioned above. It was described by Graybiel in the early 1970s [60] and later by Aitkin [3], Ehret and Romand [28] and others. Unlike the classical ascending system, it has few distinct nuclei. The nonclassical auditory pathways project to the association cortices, the reticular activating system, and the limbic system. The functional implications of these projections are mostly unknown but they may be involved in specific pathologies as discussed in Chapter 16.

Some authors [28] have shown the classical and the nonclassical ascending auditory pathways in similar ways. The classical auditory pathway (Fig. 5.7) is called the tonotopic system because it has distinct frequency tuning and the neurons are organized anatomically according to the frequency to which they are tuned. These authors have separated the nonclassical pathways in the diffuse system (Fig. 5.10A) and the polysensory systems (Fig. 5.10B). The diffuse system is so named because its neurons are not as clearly tuned and not as clearly organized anatomically as those of the classical ascending pathway. The polysensory auditory system receives input from other sensory systems.

The nonclassical or adjunct system (Fig. 5.10) branches from the classical pathway at the inferior colliculus (Figs. 5.10A and 5.10B) but some studies indicate that it branches off as early as the cochlear nucleus. It is, however, the external nucleus of the inferior colliculus and the cortex of the IC that usually are associated with the adjunct auditory system.

The origin of the diffuse system is the pericentral nucleus (ICP) of the inferior colliculus, (mostly contralateral, but also ipsilateral). The ICP receives its input from the central nucleus of the inferior colliculus (ICC). From there fibers connect to the dorsal nucleus of the dorsal division of the MGB (D in Fig. 5.10A). The dorsal division of the MGB projects to the secondary auditory cortical field (AII, in Fig. 5.10A) rather than the primary auditory cortex, which is the target of the classical pathway. The fibers that leave the dorsal portion of the MGB may also project to the reticular nucleus of the thalamus (RE).

The medial portion of the MGB is part of the polysensory system (Fig. 5.10B), which originates in the contralateral ICC and the ICX on both sides. The ICX and the cortex of the IC receive their input from the LL, but in more complex ways than the ICC. The ICX and ICP also receive input from the

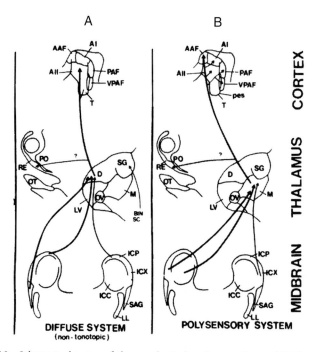

FIGURE 5.10 Schematic drawing of the nonclassical auditory pathway. (A) The nonclassical (diffuse) system. (B) The nonclassical (polysensory) system. Abbreviations: AAF, anterior cortical field; AI, primary auditory cortical field; BIN, nucleus of the brachium of the inferior colliculus; D, dorsal nucleus of the dorsal division of the MGB; IC, inferior colliculus; ICC, central nucleus of the inferior colliculus; ICP, pericentral nucleus of the inferior colliculus; ICX, external nucleus of the inferior colliculus; LGN, lateral geniculate nucleus; LL, lateral lemniscus; LV, pars lateralis of the ventral division of the MGB; M, pars magnocellularis of the medial division of the MGB; MGB, medial geniculate body; RE, reticular nucleus of the thalamus; OT, optic tract; OV, pars ovidea of the ventral division of the MGB; PAF, posterior cortical auditory field; pes, posterior ectosylvian nucleus; VPAF, ventroposterior auditory field; SAG, sagulum; SC, superior colliculus, SG, suprageniculate nucleus of the MGB; SS, suprasylvian sulcus (from Ehret, G., and Romand, R. (Eds.). (1997). "The Central Auditory Pathway." Oxford Univ. Press, New York).

neurons of the ICC. In addition to receiving auditory input, neurons of the ICX also receive input from other sensory systems such as the somatosensory system (the dorsal column nuclei) and from the visual system. The neurons of the medial division of the MGB project to the anterior auditory cortical field, which may send collaterals to the reticular nucleus of the thalamus and connects to nuclei of the limbic system. These neurons receive both inhibitory and excitatory input from the somatosensory system and probably also from the visual system. The cells of the IC also connect to many other parts of the brain whose functions are lesser known.

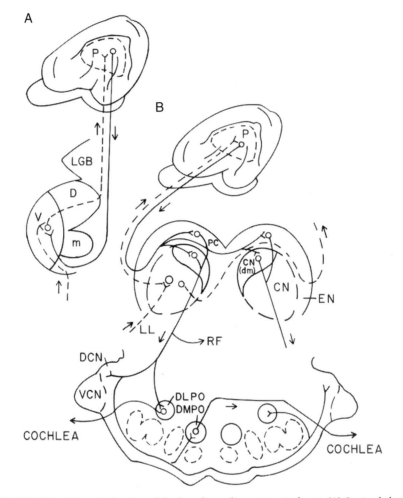

FIGURE 5.11 Schematic drawings of the three descending systems in the cat. (A) Corticothalamic system. (B) Corticocochlear and olivocochlear systems. Abbreviations: P, principle area of the auditory cortex; LGB, lateral geniculate body; D, dorsal division of the medial geniculate body; V, ventral division of the medial geniculate body; M, medial (magnocellular) division of the medial geniculate body; PC, pericentral nucleus of the inferior colliculus; EN, external nucleus of the inferior colliculus; LL, lateral lemniscus; CN dm, dorsal medial part of the central nucleus of the inferior colliculus; DCN, dorsal cochlear nucleus; VCN, ventral cochlear nucleus; DLPO, dorsolateral periolivary nucleus; DMPO, dorsomedial periolivary nucleus; RF, reticular formation (from [65]). (C) Olivocochlear system in the cat. The uncrossed olivocochlear bundle (UCOCB) and the crossed bundle (COCB) are shown (redrawn from Pickles, J. O. (1988). "An Introduction to the Physiology of Hearing" (2nd ed.). Academic Press, London).

The fact that these nonclassical systems also project to the reticular formation and the limbic system may be important in disorders of the auditory system that result from changes brought about by neural plasticity. The IC may thus convey reactions to sound other than auditory sensation and recent results.

Many of these connections between the classical and the nonclassical auditory pathways may normally be dormant in adults but the synaptic efficacy of the connections to these systems is dynamic and may change (neural plasticity) by external or internal neural activity. Such alterations in function may be the cause of certain pathologies such as tinnitus and hyperacusis (Chapter 16).

DESCENDING PATHWAYS

Three different descending (efferent) pathways have been identified. The corticofugal system originates in the auditory cerebral cortex (Fig. 5.11A). The corticocochlear system projects from the auditory cortex to the cochlear nucleus and the cochlea (Fig. 5.11B). Both systems include a crossed and an uncrossed pathway. The olivocochlear system (Fig. 5.11C) projects from SOC to the cochlea. It has two parts, one that projects mainly to the ipsilateral cochlea and whose fibers travel close to the surface of the floor of the fourth ventricle [62]. The other part of the olivocochlear system projects mainly to the contralateral cochlea and the fibers of that system travel deeper into the

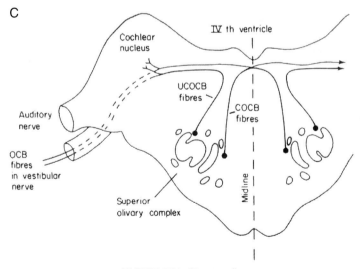

FIGURE 5.11 (*Continued*)

brainstem. The ipsilateral fibers originate in the lateral part of the SOC. The system that mainly projects to the contralateral cochlea originates from the medial part of the SOC. Both systems project to hair cells in the cochlea but the system that originates in the LSO mainly terminates on afferent fibers of inner hair cells, whereas axons of the medial system terminate mainly on outer hair cells. This description refers to the cat, and the olivocochlear system may be different in different species, including humans.

Representation of Frequency in the Auditory System

ABSTRACT

1. Frequency selectivity is a prominent property of the auditory system which can be demonstrated at all anatomical levels. The frequency selectivity of the basilar membrane is assumed to be the originator of the frequency tuning of auditory nerve fibers and cells in the classical ascending auditory pathway.

2. The frequency selectivity of auditory nerve fibers and cells in the nuclei of the classical ascending auditory pathway has been the most extensively studied.

3. The threshold of the responses of an auditory nerve fiber is lowest at one frequency, known as that fiber's characteristic frequency (CF), and a fiber is said to be tuned to that frequency. Different auditory nerve fibers are tuned to different frequencies.

4. A plot of the threshold of an auditory nerve fiber as a function of the frequency of a tone is known as a frequency threshold curve, or tuning curve.

151

5. The tuning curves of cells of the nuclei of the classical ascending auditory pathway have different shapes.

6. Nerve fibers of the auditory nerve and cells of auditory nuclei and those of the auditory cerebral cortex are arranged anatomically according to their characteristic frequency. This is known as tonotopical organization.

7. An auditory nerve fiber's response to one tone can be inhibited by a second tone when that tone is within a certain range of frequencies and intensities (inhibitory tuning curves).

8. Statistical signal analysis of the discharge pattern of single auditory nerve fibers in response to continuous broadband noise reveals great similarity with the tuning of the basilar membrane over a large range of stimulus intensities.

9. The waveform of a tone or of complex sounds is coded in the time pattern of discharges of single auditory nerve fibers.

10. This coding of a sound's time pattern in single auditory nerve fibers is known as "phase locking." Phase locking can be demonstrated experimentally in the auditory nerve for sounds with frequencies at least up to 5000 Hz but may also exist at higher frequencies.

11. Coding of the temporal pattern of sounds can be demonstrated in discharges of some cells in nuclei of the auditory system but the upper frequency limit is lower than it is in the auditory nerve and decreases along the axis of the ascending auditory pathway.

12. It is assumed that temporal information is degraded in synaptic transmission and it has therefore been assumed that temporal information must be decoded before it is degraded by synaptic transmission, but it is not known how and where that may occur.

13. The convergence of input from many nerve fibers on one nerve cell can improve the temporal precision of phase locking by a process similar to that of signal averaging, which may preserve, or even enhance phase locking, but its importance is unknown.

INTRODUCTION

We can discriminate very small changes in the frequency of a tone. In fact even moderately trained ears can detect the difference between a 1000-Hz tone and a 1002-Hz tone (a 2/10 of 1% difference in frequency). The enormous sensitivity of the human auditory system to change in frequency has aroused many investigators' curiosity and much effort has been made to determine the mechanism that is used by the ear and the auditory nervous system to discriminate such subtle differences in the frequency of a tone. Our ability to discrimi-

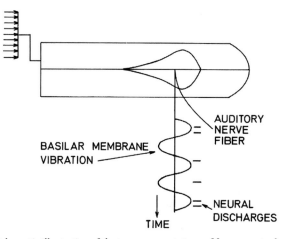

FIGURE 6.1   Schematic illustration of the two representations of frequency in the auditory nerve (from [131]).

nate changes in the spectrum of complex sounds is also prominent and this ability is assumed to be essential for discrimination of speech.[1]

Two hypotheses of neural coding of the frequency of sounds have been presented to explain the physiologic basis for frequency discrimination. One hypothesis, the *place principle,* claims that frequency discrimination is based on the frequency selectivity of the basilar membrane. The other hypothesis, the *temporal principle,* claims that frequency discrimination is based on coding the waveforms (temporal patterns) of sounds in the discharge pattern of auditory neurons. This is known as phase locking (Fig. 6.1). Both the frequency spectrum (power spectrum) and the time pattern of a sound are thus coded in the responses of neurons in the classical auditory nervous system.[2]

> When the spectra of sounds are discussed in this chapter, and in Chapter 7 it refers to the power spectrum. The power spectrum is a measure of the power within a (narrow) band of frequencies as a function of the frequency. The complete description of the spectrum of a sound is a complex quantity that can be described by a vector. The length of which is the amplitude of the spectrum and the angle of which is the phase angle of the spectrum. A spectrum can also be described by its real and imaginary values. These values are functions of frequency. The power

[1] Complex sounds are sounds that have their energy distributed over a large part of the audible frequency range and the amplitude and spectrum of complex sounds varies more or less rapidly over time. Most natural sounds are complex sounds, of which speech is an example.

[2] The statement that certain features of a sound are coded in the discharge pattern of neurons in the auditory system means that these features can be recovered by analyzing the discharge pattern of neurons in the auditory nervous system. We do not know the functional importance of such coding.

spectrum is the amplitude spectrum (length of the vector) squared, which is the same as the sum of the squared real and imaginary values of the spectrum. The spectrum of a sound (or any other waveform) can be obtained by a mathematical operation known as the Fourier transformation. Inverse Fourier transformation of a spectrum described by its amplitude and its phase angle can reconstruct the waveform. The waveform of a sound cannot be reconstructed from the power spectrum because it is an incomplete description of a sound.

All practical spectral analyses provide measures of the energy in certain (finite) frequency bands and integrated over a certain (finite) time. An approximation of the spectrum of sounds can be obtained by applying the sound to a bank of filters, the center frequencies of which are distributed over the range of frequencies of interest. The energy of the output of such filters displayed as a function of the filter's center frequencies is an approximation of the power spectrum. There is a limitation regarding the relationship between the width of the frequency bands within which the energy is obtained and the time over which the energy is integrated. Thus, obtaining measures of the energy within a narrow frequency band requires a longer observation time than obtaining measures of the energy within a broader band. This means that the product of time and bandwidth is a constant.

Frequency tuning of single neurons is prominent at all levels of the classical ascending auditory nervous system, including the auditory nerve and cells of the nuclei of the ascending auditory nervous system and those of the auditory cerebral cortex. Frequency tuning can be demonstrated in animal experiments using several different methods, but it has been studied most extensively using pure tones as stimuli. Frequency threshold curves that map the response areas of neurons with respect to frequency are the most commonly used descriptions of frequency selectivity in the auditory nervous system. The shape of such tuning curves obtained from cells in the different nuclei are different from those obtained from fibers of the auditory nerve. This is one sign of the transformation of frequency tuning that occurs in the classical ascending auditory pathways. The response to complex sounds undergoes more extensive transformations than the frequency threshold tuning curves obtained using pure tones as stimuli suggest. However, much more is known about responses to tones than to complex sounds. This chapter is devoted to the neural representation of simple sounds such as tones and clicks. Neural coding of complex sounds such as tones and broadband sounds, the frequencies or amplitudes of which vary at different rates, are the topic of Chapter 8.

The fact that the place and the temporal representation of frequency can be demonstrated throughout the auditory nervous system, however, does not determine which one (or both) of these two principles is the basis for frequency discrimination of sounds by the auditory nervous system. This is because the presence of a certain type of information in the nervous system does not mean that it is utilized in frequency discrimination. This question is discussed in more detail in Chapter 7.

FREQUENCY TUNING IN THE AUDITORY NERVOUS SYSTEM

The frequency tuning of auditory nerve fibers is a result of the frequency selectivity of the basilar membrane, while the coding of the temporal pattern of a sound is a result of the ability of hair cells to modulate the discharge pattern of single auditory nerve fibers with the waveform of the vibration of the basilar membrane (Fig. 6.1). Each point on the basilar membrane can be regarded as a band-pass filter that filters the sound that reaches the ear. The discharge pattern of single auditory nerve fibers therefore becomes modulated with a *filtered* version of the sound rather than the sound itself. This means that the temporal code of sounds also includes information about the spectral filtering of the basilar membrane. We discuss this later in this chapter.

FREQUENCY TUNING IN THE AUDITORY NERVE

Each auditory nerve fiber (Type I, see Chapters 1 and 5) innervates only one inner hair cell, and the discharges of a single auditory nerve fiber are thus controlled by the vibration of a small segment of the basilar membrane. This is the basis for the frequency selectivity of single auditory nerve fibers. Auditory nerve fibers discharge spontaneously in the absence of external sounds and increase their discharge rates when the vibration of the basilar membrane exceeds the threshold of the hair cell to which the nerve fiber in question connects. The lowest level of sound that produces a noticeable change in a fiber's discharge rate is regarded as the fiber's threshold. The threshold of a nerve fiber is lowest at a specific frequency and that is the fiber's *characteristic frequency* (*CF*). The frequency range of tones to which a single auditory nerve fiber responds widens with increasing sound intensity (Fig. 6.2).

A contour of the frequency–intensity range within which an auditory nerve fiber responds with a noticeable increase in its discharge rate is known as the nerve fiber's *frequency threshold curve* or *frequency tuning curve*. Frequency threshold curves have provided the most commonly used method to study the frequency selectivity of single auditory nerve fibers. When such frequency threshold curves are obtained for a sufficiently large number of nerve fibers, the result is a family of tuning curves that covers the entire range of hearing of the particular animal studied (Fig. 6.3). The range of hearing of different species differs; therefore, the set of tuning curves obtained in different animals will also differ. The shape of the tuning curves of auditory nerve fibers tuned to low frequencies is different from those tuned to high frequencies. Nerve fibers that are tuned to high frequencies have asymmetric tuning curves, with

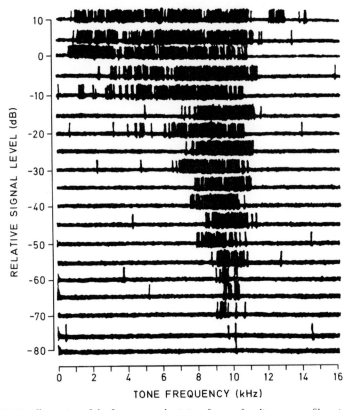

FIGURE 6.2 Illustration of the frequency selectivity of a set of auditory nerve fibers in a guinea pig. The nerve impulses elicited by a tone, the frequency of which is changed from low frequencies to 16 kHz (horizontal scale), are shown. The different rows represent responses to tones of different intensities (given in arbitrary decibel values) (from Evans, E. F. (1972). The frequency response and other properties of single fibers in the guinea-pig cochlear nerve. *J. Physiol* **226**:263–287).

the high-frequency skirt being very steep and the low-frequency skirt much less steep. Nerve fibers that are tuned to low frequencies have tuning curves that are more symmetrical.

Two-Tone Inhibition

When two tones are presented at the same time, specific interactions between the two tones may occur. For example, the response elicited by a tone at a fiber's CF can be inhibited (suppressed) by another tone when that tone is within a certain range of frequency and intensity (Fig. 6.4). The response areas of each auditory nerve fiber are thus surrounded by inhibitory frequency

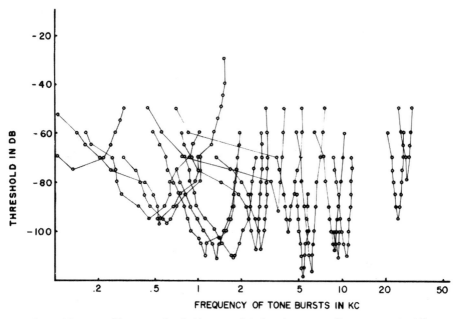

FIGURE 6.3   Typical frequency threshold curves of single auditory nerve fibers in a cat. The different curves show the thresholds of individual nerve fibers. The left-hand scale gives the thresholds in arbitrary decibel values (from Kiang, N. Y.-S., Watanabe, T., Thomas, E. C., and Clark, L. F. (1965). "Discharge Patterns of Single Fibers in the Cat's Auditory Nerve." MIT Press, Cambridge, MA).

response areas. The discharge rate of the response elicited by a tone within the fiber's response area can thus be decreased by applying a second tone with frequency and intensity within one of these inhibitory areas. Such inhibitory areas are usually located on each side of a fiber's (excitatory) response area.

Auditory System Is a Nonlinear Spectrum Analyzer

We have seen in Chapter 3 that the cochlea is a nonlinear frequency analyzer. The frequency selectivity of the basilar membrane of the cochlea is reflected in the frequency selectivity of single auditory nerve fibers and this makes it possible to study such aspects on the function of the cochlea as its nonlinearity. Recording from single auditory nerve fibers is technically easier than measuring basilar membrane vibrations and it may more accurately reflect the function of the cochlea as a frequency analyzer. In the following, we discuss the frequency selectivity of the cochlea and how it depends on the sound stimuli.

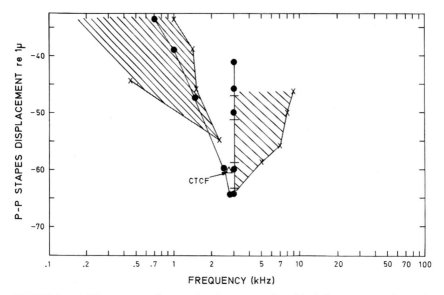

FIGURE 6.4 Inhibitory areas of a typical auditory nerve fiber (shaded) in a cat together with the frequency threshold curve (solid circles). The inhibitory areas were determined by presenting a constant tone at the characteristic frequency of the nerve fiber (CTCF) together with a tone, the frequency and intensity of which were varied to determine the threshold of a small decrease in the neural activity evoked by the constant tone (CTCF) (from Sachs, M. B., and Kiang, N. Y.-S. (1968). Two-tone inhibition in auditory nerve fibers. *J. Acoust. Soc. Am.* **43**:1120–1128).

Cochlear Frequency Selectivity Depends on the Intensity of the Sound

When the frequency threshold tuning curves of auditory nerve fibers were first obtained, at the end of the 1950s [84], it was a surprise that the tuning of auditory nerve fibers was very sharp, much sharper than the tuning of the basilar membrane as it was known at that time. Although it was believed that the tuning of the basilar membrane was the source of the frequency tuning of single auditory nerve fibers, the studies of the responses from single auditory nerve fibers suggested that some kind of sharpening of the basilar membrane tuning occurred before its vibrations were converted into a neural code. Several mechanisms for sharpening of neural tuning were suggested but none were ever supported by results of experimental studies. Eventually, much later, it was shown that the discrepancy between the sharpness of tuning between the basilar membrane and auditory nerve fibers was a result of the nonlinearity of the basilar membrane and the fact that early measurements of the vibration of the basilar membrane were done at very high sound levels and while the tuning curves of the auditory nerve fibers were obtained at very low sound levels.

When studies of the basilar membrane vibrations were done using a larger range of sound intensities it was shown that the tuning of the basilar membrane is much more acute at low sound intensities than at high sound intensities [82] (Fig. 6.5). When it became possible to measure the vibration of the basilar membrane at sound levels near the hearing threshold its tuning was found to be as sharp as the tuning curves of auditory nerve fibers. The sharpness of frequency threshold curves can be explained by the nonlinear behavior of the basilar membrane, where outer hair cells are active elements that sharpen basilar membrane tuning. That tuning of auditory nerve fibers broadens at high sound intensities was confirmed in several studies in the 1970s, when it

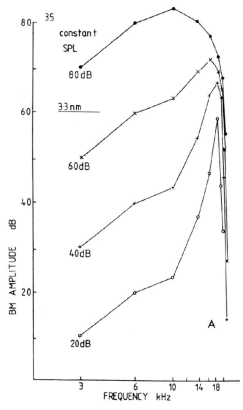

FIGURE 6.5   Vibration amplitude at a single point of the basilar membrane of a guinea pig obtained using pure tones as test sounds at four different intensities. The amplitude scale is normalized, and the individual curves would have coincided if the basilar membrane motion had been linear (from [82] with permission).

became possible to measure the frequency selectivity of single auditory nerve fibers at high sound levels. The frequency selectivity of auditory nerve fibers was derived by analyzing the discharge pattern in response to broadband noise using statistical signal analysis methods [9, 25, 27, 129, 132]. Comparison of the tuning of single auditory nerve fibers and the tuning of the basilar membrane (using noise stimuli) [132] (Fig. 6.6) shows remarkable similarities when the comparison is made at the same sound intensities. The two measures of tuning obtained at the same sound intensity are remarkably similar and they change in similarly over a large range of sound intensities.

> Some of these studies used noise as the stimulus and made use of the fact that the temporal pattern of discharges of single auditory nerve fibers is modulated by the waveform of low-frequency sounds. The use of broadbrand noise as the stimulus made it possible to determine the filter function of the basilar membrane in a large range of sound intensities. Analyzing the discharge pattern of single auditory nerve fibers [33, 129] thus yields measures of the spectral filtering that precedes impulse initiation in auditory nerve fibers. The results of such studies show that the tuning of auditory nerve fibers becomes broader with increasing sound intensity (Fig. 6.6) and the center frequency (characteristic frequency) of a nerve fiber shifts toward lower frequencies. The width of the tuning is expressed in "Q_{10dB}", which is the center frequency divided by the width measured 10 dB above the threshold, thus an inverse measure of the broadness of the tuning. The shift toward lower frequencies is gradual between sound intensities near the threshold and above the physiologic sound levels (approximately 75 dB above the threshold). These properties of auditory tuning are discussed in more detail in Chapter 7.

The tuning of the basilar membrane and that of auditory nerve fibers can be displayed in different ways. In Fig. 6.6 the tuning of an auditory nerve fiber and that of the basilar membrane (Fig. 6.5) are both shown in a comparable way. The ratio of the response of the nerve fiber and the vibration amplitude is shown (in decibels) for different sound intensities. If the frequency selectivity was a linear function, the individual curves for the different sound intensities would coincide. They obviously do not and this is an indication of cochlear nonlinearity. The fact that the curves of the response to high sound intensities appear below the curves of the response to sounds of lower intensities is a result of the gain of the cochlear amplifier being lower for high sound intensities. The shift of the individual curves can also be interpreted as a sign of the amplitude compression that occurs in the cochlea.

It is not only the width of the tuning of the basilar membrane and auditory nerve fibers that change with sound intensity, but also the frequency to which the basilar membrane and auditory nerve fibers are tuned shifts when the stimulus intensity is changed (Figs. 6.5 and 6.6). Already Rhode's measurements of the basilar membrane tuning [Chapter 3] indicated such a shift and, before that, electrophysiological studies of cochlear tuning had demonstrated

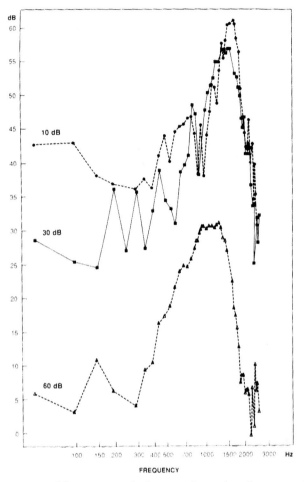

FIGURE 6.6   Estimates of frequency transfer function of a single auditory nerve fiber in a rat at different stimulus intensities (given in decibels SPL), obtained by Fourier transforming cross-correlograms of the responses to low-pass-filtered pseudorandom noise (3400-Hz cutoff). The amplitude is normalized to show the ratio (in decibels) between the Fourier transformed cross-correlograms and the sound pressure. The individual curves would have coincided if the cochlear filtering and neural conduction had been linear (from Møller, A. R. (1999). Review of the roles of temporal and place coding of frequency in speech discrimination. *Acta Otolaryngol.* **119**:424–430, with permission from Cambridge University Press).

such a shift in the location of maximal vibration amplitude. Thus, it was probably Honrubia and Ward [74] who were the first to show results that indicated that the tuning of the basilar membrane is affected by sound intensity.

These investigators showed that the location on the basilar membrane where the cochlear microphonic (CM) response recorded from the scala media was largest shifted more than 4 mm toward the base of the cochlea when the sound intensity was increased from 60 to 100 dB, thus demonstrating that the location of the maximal vibration amplitude shifts toward the base of the cochlea when the sound intensity is increased. This corresponds to a shift in the frequencies (of several octaves) to which a certain point is tuned when the sound level is varied. Later studies showed less shift in frequency tuning of the basilar membrane as a function of sound intensity was less and the shift was different in the low-frequency range compared with the high-frequency range.

The advances in our understanding of the function of the frequency selectivity of the ear and the auditory nerve fibers described above led to elimination of the need to involve any neural sharpening mechanism to explain the sharpness of tuning curves of auditory nerve fibers.

Frequency Selectivity of Single Auditory Nerve Fibers is Physiologically Vulnerable

Recordings from single auditory nerve fibers provided some of the earliest indications that the frequency selectivity of the basilar membrane is physiologically vulnerable. Thus, Evans showed in 1975 [32] that the frequency tuning curves lost their tip in anoxia-exposed animals (see Fig. 3.6). Similar changes were seen after poisoning of the cochlea with, for instance, furosemide (a diuretic that is ototoxic). These results, however, were then interpreted as a sign of the presence of a "second filter" that would normally sharpen the tuning of the basilar membrane. As has been described in Chapter 3 these changes in frequency tuning are caused by loss of the function of the outer hair cells that normally act as "motors".

Functional Importance of Cochlear Nonlinearity

The widening of the cochlear tuning with increasing sound intensity may serve to adapt the auditory frequency analyzer to perform optimally over a large range of sound intensities. Sharp spectral filters, such as the cochlear filters at low sound intensities, improve the signal-to-noise ratio and may thus improve the detection of weak sounds in noise. Broader filters, such as cochlear filters at higher sound intensities, are better suited for processing fast changes in the amplitude of a sound and may be important in discriminating complex sounds. Thus, changes in the width of the cochlear filters with changes in sound intensity may make the cochlea function optimally both for weak sounds, where detection is important, as well as for louder sounds, where temporal resolution is important. The fact that the frequency to which auditory nerve fibers are tuned is not only dependant on the location on the basilar membrane of the hair cells they innervate but also on the intensity of sounds implies the place principle may not be sufficiently robust to explain frequency discrimination. (This is discussed in more detail in Chapter 7).

FREQUENCY TUNING IN NUCLEI OF THE ASCENDING AUDITORY PATHWAY

When studied using conventional methods (frequency threshold tuning curves), practically all cells in all of the nuclei of the ascending auditory pathway, including the auditory cerebral cortex, show clear frequency selectivity. Frequency selectivity thus seems to be a prominent feature of the responses of single nerve cells in all the nuclei of the classical ascending auditory pathway. Most of the cells in the cochlear nuclei have tuning curves the shapes of which are similar to those of auditory nerve fibers (Fig. 6.7A), but some cells have tuning curves of different shapes (Fig. 6.7B). The shapes of tuning curves of cells of more centrally located auditory nuclei vary more. The difference is greatest in neurons in the auditory cortex but large variations in the shape of frequency tuning curves is also seen in neurons of the superior olivary complex (Fig. 6.8), the inferior colliculus (Fig. 6.9), and the medial geniculate body. The diversity of the shapes of the frequency tuning curves from different nuclei is most likely a result of convergence of many nerve fibers on a single nerve cell and the interplay between inhibitory and excitatory input to a neuron. The convergence of excitatory input may result in broadening of the tuning of a nerve cell that receives its input from many excitatory nerve fibers that are tuned to different frequencies. Tuning of many neurons in the inferior colliculus is much sharper than tuning of auditory nerve fibers (Fig. 6.9). That may be the result of interplay between inhibitory input and excitatory input that sharpen the tuning by mechanisms known as lateral inhibition.[3] Sharpening of frequency tuning has been demonstrated in neurons of the medial geniculate body (MGB) by gamma-amino-butyric acid (GABA) mediated inhibition.[4] [219]

---

[3] Lateral inhibition is a term borrowed from vision studies to explain enhancement of contrast and it is used in connection with the somatosensory system to explain sharpening of sensory response areas on the skin.

[4] Gamma amino buturic acid (GABA) is a common inhibitory neurotransmitter in the central nervous system.

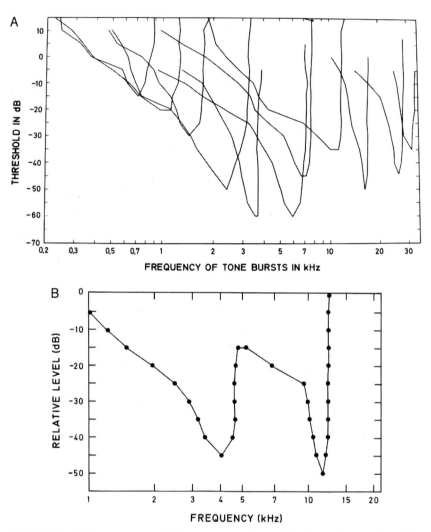

FIGURE 6.7   (A) Frequency threshold tuning curves from cells in the cochlear nucleus of the rat. Response curves that are similar to those of auditory nerve fibers (from Møller, A. R. (1969a). Unit responses in the cochlear nucleus of the rat to pure tones. *Acta Physiol. Scand.* 75:530–541). (B) Frequency threshold tuning curves with a different shape from cells in the cochlear nucleus of the rat (from 131).

PRIMARY AUDITORY CORTEX

The responses of neurons of the auditory cortex are affected by anesthesia, which hampers studies of such neurons. Results of recordings from anesthe-

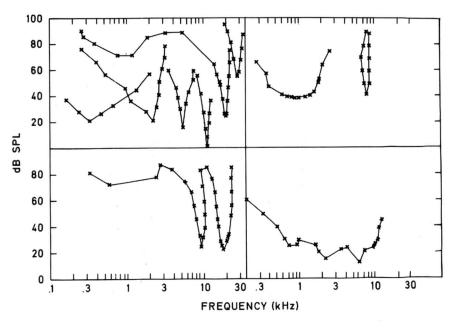

FIGURE 6.8 Examples of frequency tuning curves with different shapes obtained from neurons of the superior olivary complex of the cat (from Guinan, J. J., Guinan, S. S., and Norris, B. E. (1972). Single auditory units in the superior olivary complex. I. Responses of sounds and classifications based on physiological properties. *Int. J. Neurosci.* 4:101–120).

tized animals may not truly reflect the normal condition, and the tuning and the variability in tuning in the auditory nervous system may be different in awake animals. The effect of anesthesia on responses from nerve cells and fiber tracts is larger the more central the structure is located from which recordings are made. This means that the validity of experimental results obtained in anesthetized animals from central structures is more questionable than that of results from more peripheral structures.

FREQUENCY TUNING DEPENDS ON MEASURES OF NEURAL DISCHARGES

The most common ways to study frequency tuning of auditory nerve fibers has been to obtain frequency threshold curves using pure tones. When different measures of neural activity are used, the frequency tuning of auditory nerve fibers appears different from threshold tuning curves. Thus, the shapes of the curves showing a nerve fiber's firing rate as a function of the frequency of a

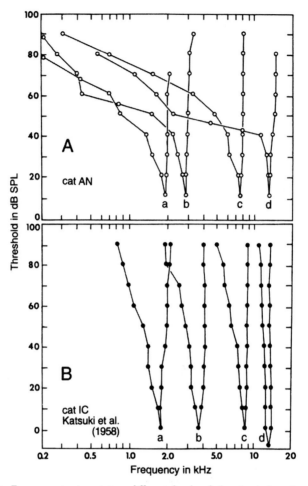

FIGURE 6.9 Frequency tuning at two different levels of the classical auditory pathways. (A) The auditory nerve. (B) Sample of narrowly tuned neurons in the inferior colliculus (Reprinted from Suga, N. (1995). Sharpening of frequency tuning by inhibition in the central auditory system—Tribute to Yasuji Katsuki. *Neurosci. Res.* **21**:287–299, Copyright 1995 with permission from Elsevier Science).

tone stimulus is different from those of frequency tuning curves of auditory nerve fibers (Fig. 6.10A). Yet another method to determine the frequency selectivity of an auditory nerve fiber determines the sound level required to evoke a certain increase in the firing rate of a single auditory nerve fiber (iso-rate curves) (Fig. 6.10B). These differences in the frequency tuning of single auditory nerve fibers that result from the use of different methods to describe

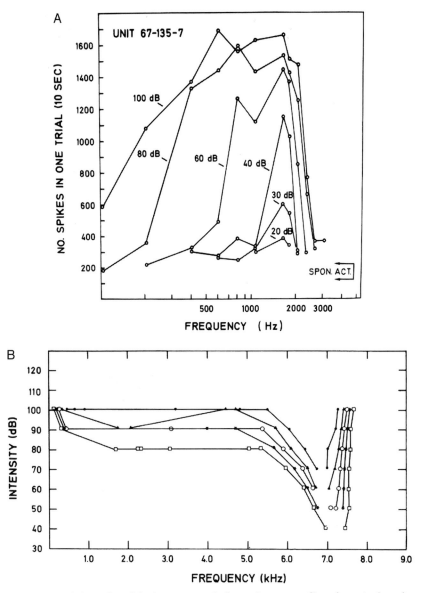

FIGURE 6.10 (A) Number of discharges per trial of an auditory nerve fiber of a squirrel monkey stimulated by tones of 10 s duration, shown as a function of the frequency of the tones. The different curves represent sounds of different intensities (in arbitrary decibels) (from Rose, J. E., Hind, J. E., Anderson, D. J., and Brugge, J. F. (1971). Some effects of stimulus intensity on response of auditory fibers in the squirrel monkey. *J. Neurophysiol.* 34:685–699). (B) Iso-rate curves of the responses from an auditory nerve fiber of a squirrel monkey (from Geisler, C. D., Rhode, W. S., and Kennedy, D. T. (1974). The responses to tonal stimuli of single auditory nerve fibers and their relationship to basilar membrane motion in the squirrel monkey. *J. Neurophysiol.* 37:1156–1172).

the discharges of single auditory nerve fibers are the consequence of the nonlinear vibration of the basilar membrane (Chapter 3). Also, the nonlinear properties of the neural transduction in hair cells make the conversion of the mechanical stimulation of hair cells into the discharge rate of single auditory nerve fibers nonlinear. Insufficient understanding of how the firing rate of single auditory nerve fibers are related to the displacement of the basilar membrane makes it difficult to interpret the results of such studies. In a few studies phase locking of neural discharges has been used to determine the frequency selectivity of auditory nerve fibers (page 160).

FREQUENCY TUNING CAN BE MODIFIED BY SOUND

It has recently been shown that the frequency selectivity of cortical neurons is modified by the corticofugal system, which consists of descending connections from the auditory cortex to the thalamus. Thus, Suga and his co-workers [242] found that inactivation of cortical neurons affected the tuning of subcortical neurons. Such changes in tuning occurred in neurons in the inferior colliculus and the medial geniculate body, which were tuned to different frequencies than those cortical neurons from which the descending fibers originated (Fig. 6.11). Since the tuning of cortical neurons depends on the tuning of neurons in the medial geniculate body, these descending connections complete a closed feedback loop that may adjust the frequency selectivity based on the sounds that reach the ear. This means that neural tuning along the neural axis of the classical ascending auditory pathways not only is different in the different nuclei and the cerebral cortex, but also that tuning can be modified in various ways through neural plasticity.

TONOTOPIC ORGANIZATION IN THE NUCLEI OF THE ASCENDING AUDITORY PATHWAY

The different nerve cells of the ascending auditory nervous system are organized anatomically in an orderly fashion according to the frequency to which they are tuned and nerve cells tuned to similar frequencies are located anatomically close to each other. This is known as tonotopic organization. Maps showing the frequency to which neurons are tuned can be drawn on the surface of nuclei as well as in sections of the nuclei of the classical ascending auditory pathways (Fig. 6.12). Also, the auditory cortex is anatomically organized according to the frequency to which neurons are tuned (tonotopic organization). The functional importance of the tonotopic organization is unknown but

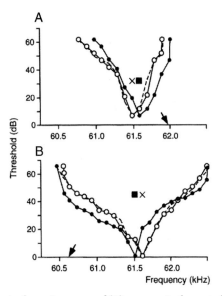

FIGURE 6.11 Changes in the tuning curves of (A) a neuron in the ventral division of the medial geniculate body, (B) a neuron in the dorsoposterior division of the inferior colliculus of moustached bats after focal inactivation of cortical neurons, the CF's of which are indicated by arrows. The tuning curves obtained before inactivation (open circles) are shown together with tuning curves obtained during inactivation (solid circles) and after recovery from inactivation (dashed lines). The inactivation was done by injecting a local anesthetic (lidocaine) in a location of the auditory cortex (from Zhang, Y., Suga, N., and Yan, J. (1997). Corticofugal modulation of frequency processing in bat auditory system. *Nature* **387**:900–903).

its prominence and consistency have supported the hypothesis that frequency tuning plays an important role for auditory discrimination.

TIME INTERVAL CODING

Temporal frequency coding has been studied to a lesser degree than frequency tuning in the auditory nervous system and many questions regarding the importance of temporal coding of sounds remain unanswered. The experimental methods used to study coding of the temporal pattern of sounds reveal that phase locking is most prominent in the auditory nerve and it is less prominent in structures more centrally located in the ascending auditory pathways. It is generally assumed that temporal coding of frequency deteriorates as it travels up the neural axis because of limitations in synaptic transmission of temporal information (synaptic jitter). It is therefore assumed that the temporal code must be converted into a code that can be preserved in the

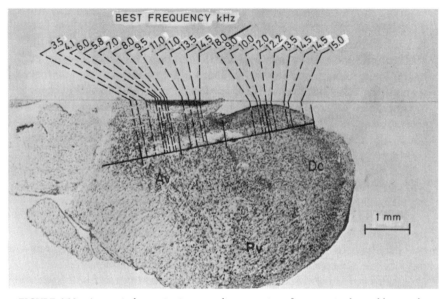

FIGURE 6.12   Anatomical organization according to tuning of neurons in the cochlear nucleus
in the cat. Abbreviations: Dc, dorsal cochlear nucleus; Pv, posterior ventral cochlear nucleus;
Av, anterior ventral cochlear nucleus (from Rose, J. E., Galambos, R., and J. R. Hughes. Microelectrode Studies of the Cochlear Nuclei in the Cat. Johns Hopkins Medical Journal 104 (1959),
211–251, © Johns Hopkins University Press).

ascending auditory pathway, but little is known about such decoding of temporal information. However, synaptic transmission can also enhance temporal
precision. This has been demonstrated to occur in neurons with many synapses
but its functional role is unknown [119].

These factors have made the temporal coding of the frequency or spectrum
of sounds less attractive as an explanation of frequency discrimination in the
auditory system. However, such reasoning is based on experimental results of
recordings from single nerve cells processed in a way that is fundamentally
different from the way the nervous system processes sensory information.
More recent studies, however, have shown indications that temporal coding
may play a much greater role in discrimination of complex sounds that earlier
believed. These matters will be discussed in detail in Chapter 7. In this chapter
we discuss experimental results on temporal coding of sounds in the auditory
nerve and nuclei of the auditory nervous system.

NEURAL CODING OF PERIODIC SOUNDS

The time locking of neural discharges to the waveform of a sound is known
as *phase locking*. It means that more nerve impulses are delivered at a certain

phase of the sound. Averaging the recorded neural activity to many cycles of a tone is necessary to demonstrate phase locking to a pure tone. Practically, that is done by compiling a *period histogram* of the responses. For that, the duration of one period of the sound is divided into a series of bins and the number of nerve impulses that fall into each bin is counted.

PHASE LOCKING IN THE AUDITORY NERVE

The discharges of single nerve fibers are *time locked* to the waveform of a sound that is within the fiber's response area, at least for frequencies below 5000 Hz, but probably even for higher frequencies. Period histograms of the responses to low-frequency tones have the shape of half-wave-rectified sine waves (Fig. 6.13).

Phase locking of the discharges of single auditory nerve fibers to complex periodic sounds can also be demonstrated. Thus, period histograms of the response to sounds that are the sums of several pure sine waves (tones) of different frequencies have forms similar to the wave shapes of the half-wave-

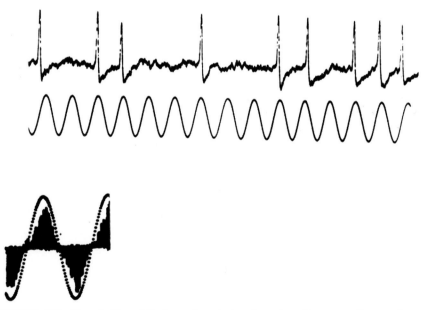

FIGURE 6.13   Phase locking of discharges in a single guinea pig auditory nerve fiber to a low-frequency tone (300 Hz) near threshold (from Arthur, R. M., Pfeiffer, R. R., and Suga, N. (1971). Properties of "two tone inhibition" in primary auditory neurons. *J. Physiol. (Lond.)* **212**:593–609).

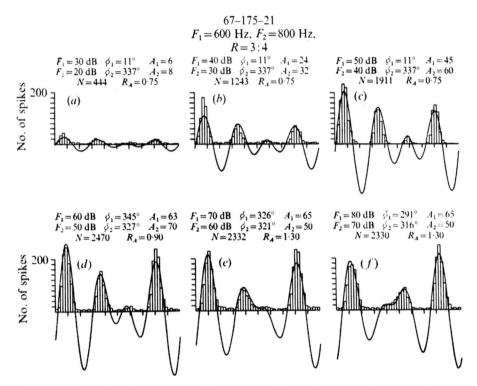

FIGURE 6.14 Period histograms of discharges in a single auditory nerve fiber of a squirrel monkey to stimulation with two tones of different frequencies that were locked together with a frequency ratio of 3 : 4 and an amplitude ratio of 10 dB. The different histograms represent the responses to this sound when the intensity was varied over a 50-dB range (modified from Rose, J. E., Hind, J. E., Anderson, D. J., and Brugge, J. F. (1971). Some effects of stimulus intensity on response of auditory fibers in the squirrel monkey. *J. Neurophysiol.* 34:685–699).

rectified sound waves (Fig. 6.14).[5] The different sine waves must be multiples of each other to get a waveform that repeats itself accurately, and the period histograms are compiled over the period of such a waveform. It is thus not the sound itself that activates auditory nerve fibers, but the sound filtered by the basilar membrane—this means that auditory nerve fibers phase-lock to a

[5] The histogram is a half-wave-rectified version of the basilar membrane vibration because cochlear hair cells are only excited when the basilar membrane is deflected in one direction, namely toward the scala vestibuli (as discussed earlier in Chapter 3). This, however, is an oversimplification. Some hair cells are in fact excited when the basilar membrane is deflected in the opposite direction and some are excited when the velocity of the basilar membrane is highest. The reason for that complexity in excitation of the inner hair cells is the active role of outer hair cells, the motion of which contribute to excitation of the inner hair cells, and a viscoelastic coupling between the basilar membrane and the inner hair cells ([158, 244] and Chapter 3).

band-pass-filtered sound. Band-pass filtering alters the waveform of broad-band sounds.

PHASE LOCKING IN NUCLEI OF THE ASCENDING AUDITORY PATHWAY

Phase locking to pure tones is prominent in many cells of the cochlear nucleus, more so in the ventral cochlear nucleus than the dorsal cochlear nucleus. Time locking to pure tones, and particularly to repetitive clicks, can also be observed in many cells of the inferior colliculus and the medial geniculate body. The upper frequency of phase locking is lower than it is in the auditory nerve. In the medial geniculate it is rarely observed at rates higher than 800 clicks per second.

PHASE LOCKING OF COMPLEX SOUNDS

Phase locking of the discharges of auditory nerve fibers can be demonstrated in response not only to pure tones, but also in response to complex sounds. Complex sounds, such a speech sounds, contain several periodic or quasiperiodic components. The fundamental (vocal cord) frequency is one component and other are damped oscillations the frequency of which are that of the vowel formants.[6] In order to determine the formant frequencies on the basis of the temporal pattern it is necessary that each of these damped oscillations are coded independently in a different population of auditory nerve fibers. The spectral selectivity of the basilar membrane makes that possible because it divides the audible spectrum into suitable slices before the waveform is coded in the discharge pattern of auditory nerve fibers. This means that the periodicity of each of the vowel formants are coded in different populations of auditory nerve fibers. This is known as "synchrony capture" and it enables different populations of auditory nerve fibers to carry the periodicity of different spectral components of a sound. This separation of spectral components may be the most important feature of the frequency selectivity of the basilar membrane and that is discussed in more detail in Chapter 7.

The increase in the width of the cochlear filter with increasing stimulus intensity may impair the separation of vowel formants before coding the wave-

---

[6] The spectrum of a vowel has several peaks, known as formants. Formants are the results of the acoustic properties of the vocal tract and the frequencies of the formants uniquely characterize a vowel. In the time domain, each formant contributes a damped oscillation to the total waveform of a vowel. The frequencies of these damped oscillations are the formant frequencies, and these damped oscillations are repeated with the frequency of the vocal cords, i.e., the fundamental frequency of the vowel in question.

form and thus impair the preservation of phase locking to individual formant frequencies. Such deterioration of frequency acuity of the cochlear filtering may be one reason why speech discrimination is impaired when the sound intensity is raised above a certain value. Absence of the acoustic middle ear reflex, which results in the input to the cochlea being greater than normal, has been shown to cause impairment of speech discrimination at high sound intensities (see Chapter 12) and that may be a result of widening of the cochlear filters impairing "synchrony capture".

Phase locking of discharges of auditory nerve fibers also occurs in response to broadband noise [9, 25, 129, 132] and has been used to determine the properties of the cochlear filters over a large range of sound intensities (see p. 160, and Fig. 6.6). These studies have demonstrated phase locking of auditory nerve impulses over a much larger range of sound intensities than the range over which the average discharge rate increases with increasing sound intensity [33, 129]. The discharge rates of most auditory nerve fibers show a saturation at sound levels as low as 20–30 dB above their threshold. Thus, the average discharge rate is essentially constant in response to sounds in the entire physiological range of sound levels, but phase locking can be demonstrated over the entire physiological range of sound intensities (Fig. 6.15).

PRESERVATION OF THE TEMPORAL CODE OF FREQUENCY

While frequency tuning is prominent throughout the ascending auditory pathway, it has been believed that phase locking deteriorates as the information

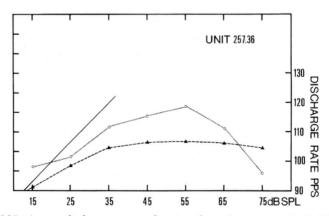

FIGURE 6.15 Average discharge rate as a function of stimulus intensity (dashed line) of an auditory nerve fiber in a rat together with a measure of the fiber's ability to phase lock to the stimulus sound (low-pass-filtered noise), shown as a function sound intensity.

travels along the ascending auditory pathway. The reason given is that synaptic transmission implies a certain amount of jitter, which "blurs" the time pattern of the neural code.[7] It has therefore been assumed that the phase-locked train of neural discharges must be converted into a spike-rate code or a spatial code in order to be preserved in synaptic transmission until it can be interpreted by higher centers of the nervous system. However, that assumes that only few nerve fibers terminate on a nerve cell. Nerve cells on which many fibers terminate act as spatial integrators and may in fact increase the temporal precision of phase locking.

That such enhancement of temporal precision in firing can occur has been demonstrated in some nerve cells in the cochlear nucleus which fire with great temporal precision in response to transient stimulation such as clicks and tone bursts [119] (Fig. 6.16). These neurons respond with greater temporal precision than auditory nerve fibers which means that the temporal precision has increased rather than decreased as a result of synaptic transmission. Many nerve fibers are assumed to terminate on the neurons that respond with such great precision response. They thus act as signal averagers and not only compensate for synaptic jitter but even increase the accuracy of temporal coding of the waveform of the sound stimuli. As we discuss in Chapter 8, it is not the rate of tone bursts as such that these neurons are tuned to but, rather, these neurons are tuned to the silent interval between sounds. However, the fact that these neurons respond with a greater temporal precision than auditory nerve fibers demonstrate that spatial integration in the nervous system can improve temporal precision. This is opposite to synaptic jitter—these neurons respond to transient stimulation with a greater precision than that of their input (from auditory nerve fibers).

How Does the Nervous System Extract Information from the Temporal Code of Neural Discharges?

The consistency of the organization of neurons according to the frequency to which they respond best (tonotopic organization) suggests that frequency (spectral) tuning is important for extraction of spectral information about a sound. It is much more difficult to understand how the nervous system can decode temporal information so that it may be used for discrimination of

[7] Jitter means that the time at which nerve impulses occur vary randomly, which produces a lack of temporal precision. Thus, phase locking to the waveform of low-frequency sounds does not occur precisely to a certain phase of the sound. Only the average number of nerve impulses is higher at a certain phase of a sound than at other phases. The random variation around that mean value is a result of the probabilistic nature of synaptic transmission, which occurs in hair cells and cells in the various nuclei of the ascending auditory pathway.

UNIT 63.6   UNIT 65.1

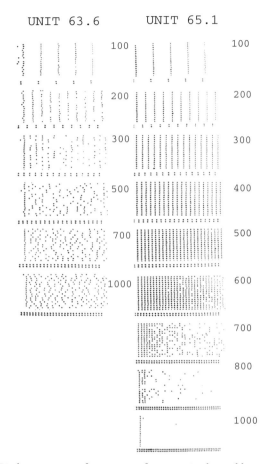

FIGURE 6.16 Discharge pattern of two types of neurons in the cochlear nucleus of a rat in response to clicks of different repetition rates. Each nerve impulse is indicated by a dot and the stimulus clicks are indicated by pairs of dots below the responses. The stimulus clicks were presented in 50-ms-long bursts (from Møller, A. R. (1969b). Unit responses in the rat cochlear nucleus to repetitive transient sounds. *Acta Physiol. Scand.* 75:542–551).

frequency. We know that the temporal pattern (waveform) is coded in the discharge pattern of single auditory nerve fibers and that it can be recovered experimentally. That, however, does not prove that information about the frequency of sound is actually extracted from the phase-locked neural responses. (We will discuss the importance of the place and temporal coding of frequency in detail in Chapter 7).

It is assumed that the auditory nerve supplies the nervous system with a neural code that is phase locked to the time pattern of the vibration of the basilar membrane, thus band-pass-filtered versions of the sound that reaches the ear. In order to use that code to determine the frequency of a sound, the nervous system must determine the interval between nerve impulses that are time locked to individual waves of a sound. Models have been proposed to explain the ability of the nervous system to determine the time between the arrival of a sound at the two ears that is the basis for directional hearing. The most common of these models is the one proposed by Jeffress [78] for determination of intraaural differences in sound localization (described in Chapter 9).

Determining the time interval between two waves of a sound wave is more complex than determining the time difference between neural activity that originates in the two ears. However, similar principles and similar neural circuitry could be used for decoding temporal information in sounds. Licklider [107] proposed a variation of Jeffress's model especially for detecting the time intervals between individual waves of sound. This model is based on autocorrelation analysis to determine the intervals between sound waves.

> Autocorrelation analysis is similar to cross-correlation analysis described in Chapter 9 for analysis of the delay between the sounds that reach the two ears. When autocorrelation analysis of a signal is done it is common to sample the signal, multiply the amplitude of the signal and a replica of the signal, sample by sample, and then add the values. That process is repeated after the signal and its replica are shifted one sample relative to the other. This is then continued for as many delays as is required. The resulting autocorrelation function appears as a series of values that are the function of the delay. This is the way autocorrelation analysis is done using digital computers. However, a neural system that performs autocorrelation analysis does not obtain the correlation point by point as is done by a computer. Instead, the nervous system is assumed to use an array of multipliers, one for each delay. The output of each multiplier neuron provides a continuous signal that is the correlation at one delay as a function of time.

The autocorrelation model for decoding time intervals requires a set of delays and multipliers (or coincidence detectors), thus similar to Jeffress's model. The range of time intervals that must be determined is much longer, however, than the delays between the sounds that reach the two ears at different azimuths. The delays in the variable delay lines required for frequency determination of sounds in the audible range is much longer than those required for directional hearing. For decoding temporal information in sounds the frequency of which is higher than 1000 Hz, axons of different lengths may serve as the required variable delay lines. Each axon is assumed to be connected to a nerve cell that also receives input from another axon of a different length, the nerve cell acting as a coincidence detector. Such an array of axons of different length may be located in the cochlear nucleus, while the neurons that act as coincidence detectors may be located in the superior olivary nuclei [57]. Langner and Schreiner [101] found evidence from studies in cats that some neurons in the inferior colliculus have different delays thus providing the delay lines needed for determining the frequency of sounds of relatively high frequency. Delays longer than 1 ms can probably not be accomplished using axons of different lengths to produce delays because it would require very long axons that are not available.

The human ear can discriminate changes of approximately 2 Hz in a 1000-Hz tone. This means 1000 Hz and 1002 Hz can be differentiated. The difference in the length of one period of a 1000-Hz and a 1002-Hz tone is 2/1000 the period length of a 1000-Hz tone, therefore 2 μs. This is thus about the same time difference as can be discriminated in the arrival time of sounds at the two ears, but longer delays are also required for periodicity detection than for determining intraaural time differences.

To determine the pitch of vowel sounds by using variable delay lines and coincidence detectors, delays of at least 10 ms would be required. Such long times cannot be detected by a coincidence detector because that long of a delay cannot be accomplished using axons of different lengths. The delays necessary to explain echolocation in bats are also much longer than the intraaural delay that can be generated by axons. The delays associated with echolocalization in bats are between 0.4 and 18 ms. Delays of that length have been assumed to be accomplished by an "inhibitory gate" that has a variable time [221]. Similar mechanisms may be used for determining of the frequency of low-frequency sound such as the fundamental frequency of vowel sounds.

ANATOMICAL LOCATION OF NEURONS THAT DETECT TIME INTERVALS

It has been suggested that neurons in the superior olivary complex may have the ability to detect time intervals. Neurons in the superior olivary complex acting as coincidence detectors can discriminate time differences between the arrival of sounds at the two ears. For that, the neurons receive input from both side's cochlear nuclei, which may contain the axons of variable lengths that act as variable delay lines. Some of these neurons can detect very small time differences, on the order of 1 μs. If neurons in the MSO should act as detectors of frequency they must receive input from neurons in the cochlear nucleus on the same side. The two inputs to MSO neurons must arrive at different delays in order to be used in autocorrelation analysis. This could be used as the basis for determining the intervals between sound waves similarly to intraaural time differences but if the variable delay lines are the same axons in the cochlear nuclei that also are used for determining time differences between sounds that arrive at the two ears only sound with frequencies higher than approximately 1000 Hz could be analyzed.

Sounds of lower frequencies may be analyzed in other nuclei of the ascending auditory pathways. Thus the ventral nucleus of the lateral lemniscus (VNLL) in the mammalian auditory system seems to specialize in coding temporal information, as shown in the echolocating bat and dolphin [19] and Langner and Schreiner [101] presented evidence that variable delays needed for determining the frequency of a sound from the temporal pattern of nerve impulses may exist in the inferior colliculus (IC). They also found that many neurons in the IC have an intrinsic periodicity of firing (like an oscillator) which they suggested might be used as a time base and a variable delay. The common intervals in the firing of neurons that have such intrinsic periodicity in their firing pattern were approximately 0.4 ms and were little affected by the physical characteristics of the stimulus such as intensity and frequency of a pure tone.

Other evidence of the ability of the nervous system to detect time intervals between sounds in a single channel comes from studies of the flying bat. Suga and his co-workers showed that echolocating flying bats measure the time interval between the emission of a sound and the arrival of the echo to determine the distance to an object. They found that the auditory cortex of the flying bat makes a map from the measured difference between time of the emission of a sound and the arrival of the echo at the bat's ear [220]. A similar neural mechanism could perhaps convert the frequency of a sound into a place code that could be used for frequency discrimination. The delays of an echo that are of interest are between 0.4 and 18 ms. Thus much longer than the intraaural delays. The delays necessary for echolocation are assumed to be created by an "inhibitory gate" that has a variable time [221].

The neurons in the cochlear nucleus that respond to transient sounds with a single discharge and that require a certain duration of silence before they can fire again [119] may be an example of neurons that use inhibition to determine the time interval between sounds. This may be similar to the inhibition-based variable delays suggested by O'Neill and Suga [166, 167] for explaining the bat's discrimination of delays between transmittal and receipt of an echo (see Chapter 8).

CHAPTER 7

Is Temporal Code or Place Code the Basis for Frequency Discrimination?

ABSTRACT

1. The cochlea delivers a code to the auditory nervous system that, if properly deciphered, yields information about both the (power) spectrum and the waveform (periodicity) of a sound, but it is not known which of these two representations is used for frequency discrimination.
2. The frequency selectivity of the basilar membrane is the basis for the place principle of frequency discrimination. Coding of the temporal pattern of sounds in the discharge pattern of auditory nerve fibers is the basis for the temporal principle of frequency discrimination.
3. Because place coding is affected by the sound intensity, it may not be sufficiently robust to explain auditory frequency discrimination.
4. Studies of the neural coding of vowels in the cat's auditory nerve show a higher degree of robustness of the temporal code compared with the place code.
5. Studies of patients with injuries to the auditory nerve also support that temporal coding is important for discrimination of complex sounds.
6. Temporal coding is (practically) independent of sound intensity but this hypothesis has been disputed because it may lack robustness concerning

transmission along the auditory neural axis and there is uncertainty about temporal information decoding.

7. The exact mechanisms of decoding the temporal frequency code are unknown, but similar neural circuits, as those decoding directional information, may decode temporal frequency information.

8. The most important function of cochlear frequency selectivity may be that it prepares sounds for temporal coding by dividing the spectra of complex sounds into (narrow) bands before their conversion into a temporal code.

INTRODUCTION

It is the auditory system's ability to discriminate frequency that makes it possible to hear the difference between two tones of different frequency. More importantly, it is also frequency discrimination that makes it possible to hear the difference between complex sounds with different spectra. It is assumed that auditory frequency discrimination is essential for discrimination of vowels because vowels are uniquely described by the frequency location of the peaks in their spectrum. Discrimination of tones on the basis of their frequency and discrimination of complex sound such as vowels from their spectrum may use different neural mechanisms but it is generally assumed that frequency discrimination of both tones and complex sounds rely either on the frequency analysis of the basilar membrane (place principle) and/or on neural analysis of the code of the sounds' temporal pattern (temporal principle).[1] While there is ample evidence that both these two representations of frequency are coded in the auditory nervous system, it is not known which one of these two principles is used by the auditory system in the discrimination of natural sounds or for the discrimination of unnatural sounds, such as pure tones in experiments done in the laboratory under more or less normal conditions. It may be that the place and the temporal principle of frequency discrimination may be used in parallel by the auditory system for discrimination of sounds of different kinds. The fact that the frequency is so clearly represented in the nervous auditory system according to the place principle does not mean that it is important for discrimination of the frequency of sounds. Its primary function could be to divide the frequency spectra of complex sounds into discrete bands that are then coded temporally. There are several reasons why the importance of the temporal representation of sounds may have escaped investigators.

Designing experiments to determine if it is the place principle or the temporal principle that is the basis for frequency discrimination is difficult because

---

[1] When the spectra of sounds are discussed in this chapter it refers to the power spectrum.

the spectral and temporal properties of sounds are closely linked together and the temporal pattern of a sound cannot be manipulated experimentally without also altering its spectrum. In order to be a candidate for frequency discrimination, the coding must be robust, which means that it must not be affected by sound intensity and the frequency coding must be preserved as the information travels up the neural axis to a location where it is decoded. One way of assessing the possible role of either one of these two principles for frequency discrimination is therefore to consider how robust the coding of place and time is with regard to being independent of sound intensity. Another way is to study how place and temporal information are preserved as they travel up the neural axis. It is possible to design experiments to answer these two questions, but it is more difficult to study how these two kinds of information are decoded in the nervous system.

STUDIES OF CODING OF SYNTHETIC VOWELS

We first discuss how coding of formant frequencies is preserved over a range of sound intensities and then discuss in more general terms how cochlea frequency analysis depends on sound intensities.

PRESERVATION OF THE PLACE CODE

The fact that neurons in all parts of the classical ascending auditory pathways respond best to a certain sound frequency and are anatomically organized according to the frequency to which they respond best (tonotopic organization) indicates that the place representation of sounds in the cochlea is maintained throughout the ascending classical auditory nervous system. The tonotopic organization of nuclei of the auditory system including the cerebral cortex can be regarded as a projection of the basilar membrane of the cochlea onto these neural structures. This has been taken as an indication that the nervous system uses place information as a basis for frequency discrimination.

Studies that show tonotopic frequency representations throughout the auditory nervous system were done by determining the threshold of neurons to tones of different frequencies. This means that the tonotopic maps reflect situations near the threshold of hearing, thus different from normal listening conditions. If the central nervous system uses the place code as a basis for frequency discrimination of complex sounds, the frequency representation must also exist for sounds above threshold. Thus, the discharge rate of auditory nerve fibers in an ensemble of nerve fibers must increase with increasing stimulus intensity over a large range of stimulus intensities in order to code the spectrum envelope of broadband sounds such as that of vowels at physiological sound levels. This is not supported by studies of the neural representation of vowels in the responses from single auditory nerve fibers. Studies by Sachs and co-workers [196] of the effect of sound intensity on coding of vowel sounds in the auditory nerve have shown that the discharge rates of large populations of auditory nerve fibers reproduces the formant frequencies of vowels only at low sound intensities. Thus the discharge rates of many auditory nerve fibers collected in the same animal and plotted as a function of the CF have distinct peaks that correspond to the formants of the vowels only when the vowels were presented at low intensity (Fig. 7.1). These peaks became less distinct as the sound intensity was increased and at physiologic sound levels these peaks were poorly defined (Fig. 7.1). Spectral separation of vowel formants based on the (average) discharge rates of single auditory nerve fibers thus becomes poor at sound intensities in the range of conversational speech. While the spectral envelope of vowels was represented in the discharge of ensembles of nerve fibers at low sound intensities the place representation of vowel formants deteriorated when the sound intensity was raised from near-threshold values to physiologic sound levels. This means that the place representation of vowel formants cannot be regarded as sufficient to discriminate formant frequencies, which is necessary for discrimination of vowels.

PRESERVATION OF THE TEMPORAL CODE

In studies of the temporal code of synthetic vowels by Sachs and co-workers [239] it was found that the temporal structure of such sounds was coded in the discharge pattern of auditory nerve fibers over a large range of sound intensities. These investigators computed period histograms of auditory nerve fiber responses to synthetic vowels (similar to those used in the study illustrated in Fig. 7.1).

> Histograms of the response to synthetic vowels (Fig. 7.2) show the distribution of discharges over one period of the fundamental frequency of the vowel and they reveal the periodicity of the oscillation within the frequency band that is represented by the response area of the fiber from which recordings are made. This information can be extracted through Fourier-transformation of the histograms (Fig. 7.2). The discrete Fourier transforms were computed of histograms with duration of one pitch period. The computed spectra therefore show harmonics of the fundamental frequency of the vowel and the formants appear as peaks in the envelope of these spectra. The histograms from neurons with the CF near the frequency of a formant will show a periodic pattern with the frequency of the formant. The fundamental frequency of the vowel, the response of which is illustrated in Fig. 7.2, was 128 Hz. The frequency of the first formant was 768 Hz (6th harmonics of the fundamental frequency). The frequency of the second formant was 1152 Hz (9th harmonic) and that of the third formant was 2432 Hz (19th harmonic). The CF of the nerve fiber,

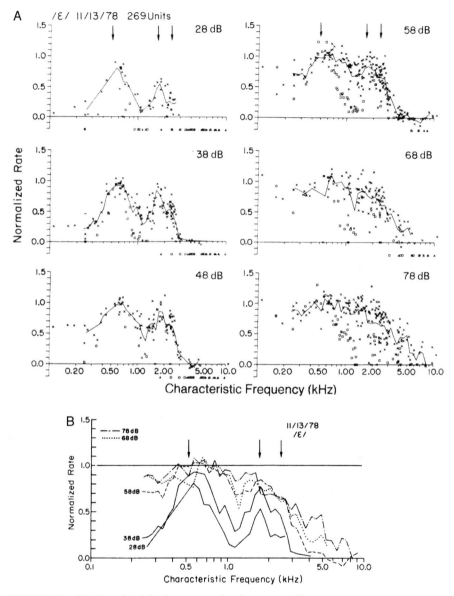

FIGURE 7.1   (A) Normalized discharge rates of auditory nerve fibers in a cat in response to a synthetic vowel /ε/ presented at different sound intensities (in decibels SPL). The arrows mark the frequency of the three vowel formants (from [196]). (B) Averaged normalized discharge rates from A (from Sachs, M. B., and Young, E. D. (1979). Encoding of steady-state vowels in the auditory nerve: Representation in terms of discharge rate. *J. Acoust. Soc. Am.* **66:**470–479).

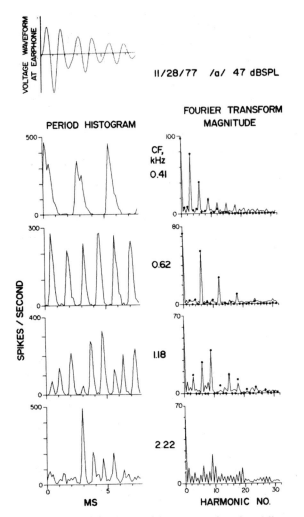

FIGURE 7.2 Period histograms (left column) of the responses from four different auditory nerve fibers of a cat in response to stimulation with a synthetic vowel /a/. Right column shows Fourier transforms of these histograms. The electrical signal applied to the earphone is shown on top (from Young, E. D., and Sachs, M. B. (1979). Representation of steady-state vowels in the temporal aspects of the discharge patterns of populations of auditory nerve fibers. *J. Acoust. Soc. Am.* **66**:1381–1403).

the response of which is shown in the second histogram from top, was 620 Hz, thus close to that of the first formant. Consequently, the periodicity seen in the histogram is close that frequency and the 6th harmonic is dominant in the Fourier-transformed histogram. The CF of the fiber, whose response is shown in the third histogram from top, was 1180 Hz, thus close to that of the second formant, and the periodicity in the histograms is consequently close to that of the frequency of the second formant, thus the 9th harmonic dominate in the Fourier transform of the histogram. That the second formant also is represented in the Fourier transform of the histogram can be explained by the asymmetry of the response area of auditory nerve fibers (Chapter 6). The bottom histogram showing the response from a nerve fiber tuned to a frequency close to the third formant shows a spectral peak at the frequency of the third formant (the 19th harmonic) but lower formants are also represented in the response. A vowel can be regarded as the result of spectral filtering of the air pulses generated by the vibration of the vocal cords. The filter is the vocal tract and it is the properties of that which result in the spectral peaks known as formants. The discharge pattern of an auditory nerve fiber is controlled by the energy within the response area of the nerve fiber. That process is equivalent to spectral filtering. It is thus a filtered version of the vowel spectrum that excites auditory nerve fibers.

When information such as that displayed in Fig. 7.2 was compiled for different stimulus intensities, it was found that the formants were coded in the time pattern of the discharges of single auditory nerve fibers over a large range of stimulus intensities (Fig. 7.3) for three different vowels. That means that the time pattern is more robust with regard to the intensity of the sound than the coding of the spectral envelope of a sound (place principle), which supports the hypothesis that temporal analysis is more important for discrimination of vowels than their spectrum. This has been supported in more recent studies which showed that the discharge rates of auditory nerve fibers in cats with cochlear hearing loss from acoustic trauma is insufficient for discrimination of synthetic vowels on the basis of their spectrum (the frequency of the second formant), yet the temporal code contained information that would allow such discrimination [117].

Implications for Cochlear Implants

That speech discrimination is more dependent on the time pattern of sounds than on their spectrum is encouraging for the success of cochlear implants. Cochlear spectral analysis cannot easily be mimicked by cochlear implants because they must rely on a small number of stimulating electrodes, but exciting auditory nerve fibers in accordance with the time pattern of sounds is possible. Experience from cochlear implants showing that satisfactory speech discrimination can be achieved by devices that have only a few channels supports the hypothesis that temporal coding of frequency is important for speech discrimination and perhaps sufficient. Such devices could not possibly provide place representation of sounds that would be sufficient for discrimination of speech sounds such as vowels. The fact that increasing the number of channels of cochlear implants increases speech discrimination supports the hypothesis that spectral separation serves to improve temporal coding. The implication would be that little improvement would be achieved by increasing the spectral separation beyond a certain number.

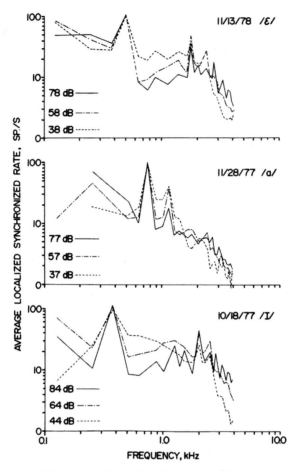

FIGURE 7.3 Averaged Fourier transforms of the histograms of the response to three different synthetic vowels presented at three different sound intensities (given in decibels SPL) (from Sachs, M. B., and Young, E. D. (1979). Encoding of steady-state vowels in the auditory nerve: Representation in terms of discharge rate. *J. Acoust. Soc. Am.* **66**:470–479).

BASES FOR PRESERVATION OF THE TEMPORAL CODE OF FREQUENCY

That temporal information is available in the discharge pattern of auditory nerve fibers is not sufficient to show that it is available for frequency discrimination, but it must also be preserved in the ascending auditory nervous system until it can be decoded. There is an ongoing debate on whether temporal information can be preserved as it ascends through the auditory nervous system from the cochlea to the anatomical location of the ascending auditory pathway where the temporal code is converted into a different neural code.

Synaptic transmission is assumed to impair the temporal code and it has therefore been assumed that the temporal code must be converted into a different code at a peripheral level of the auditory nervous system. This assumption seems to be supported by experiments that show that the discharge pattern of neurons in more centrally located nuclei and the cerebral cortex do not follow the temporal pattern of the stimulus sound (phase locking, see Chapter 6) as well as more peripherally located neurons do, indicating that the temporal coding of frequency deteriorates as the information travels up the neural axis. It has been assumed that synaptic transmission is the cause of this deterioration by adding "jitter" to the temporal code (see Chapter 6). That the timing information deteriorates as early as in the synaptic transmission of hair cells has been demonstrated in experiments by Weiss and co-workers [93], but investigators differ in their interpretation of results obtained from single nerve fibers, which show that phase locking to tones decreases with increasing frequency above approximately 2–3 kHz and it has been difficult to demonstrate phase locking to tones above 5 kHz. Experimental methods of studies of phase locking in single neurons, however, have technical limitations regarding the ability to demonstrate phase locking at high frequencies. It may be that the temporal information is effectively coded in the auditory nerve at much higher frequencies than 5000 Hz and it is just not detectable by the experimental methods presently used.

Another factor that may have been overlooked when the results of recordings from single nerve fibers are interpreted is related to the fact that the activity in only one nerve cell is observed at a time. When many nerve fibers converge on one nerve cell such as occurs in the cochlear nucleus, the result is spatial integration of neural activity, which can enhance temporal precision and thus counteract its deterioration by synaptic jitter. Only a few experiments have addressed that question. In one study [119], improvement of temporal precision of coding of the waveform of sounds has been demonstrated in recordings from certain cells of the cochlear nucleus. These nerve cells fire more precisely than auditory nerve fibers in response to transient sounds (Fig. 6.16, Chapter 6). This shows that precision of timing is not only preserved, but even improved through synaptic transmission. Thus, while phase locking of neural discharges in auditory nerve fibers decreases gradually above a certain frequency, spatial integration in neurons of the nervous system may enhance the temporal coding to an extent that compensates for the decrease of phase locking in the auditory nerve.

Perhaps the best proof of preservation of accurate timing information in the auditory nervous system comes from coding of the temporal information, which is the basis for our ability to discriminate the direction of a sound source (discussed in Chapter 8). This ability is at least partly based on the difference in the arrival time of a sound at the two ears. This obviously requires timing information to be preserved with great precision until it can be decoded.

The results of studies of binaural hearing thus show convincingly that the auditory nervous system can preserve and detect very small time intervals (or rather very small differences in time intervals). Time differences in the order of 5–10 μs between the arrival of sounds at the two ears can be detected.

The discrimination of the difference between the arrival times of sounds at the two ears probably occurs in the neurons of the medial superior olivary complex. These very small time differences must thus have been preserved in the coding of sounds in the auditory nerve and the cochlear nucleus in order for such small time differences to be preserved in the auditory nervous system for binaural hearing. The same auditory neural circuitry is used for transmission of information about frequency and it may therefore be assumed that the timing of such information also can be preserved with great precision. The importance of synaptic jitter as an obstacle in preserving the temporal coding is thus probably overstated.

> The discharges of single auditory nerve fibers and neurons of the ascending auditory pathway are probabilistic events, which means that it cannot be predicted exactly if or when a nerve fiber will fire. Only the probability can be determined. Thus, when a nerve fiber is activated by sound, its *average* discharge rate will increase, and the probability that a discharge will occur at a certain time relative to presentation of a sound will increase. In a similar way, the probability of firing may vary along the waveform of a sound. This means that the likelihood that a nerve fiber or a nerve cell will discharge is higher at a certain phase of a low-frequency tone. This is the basis for the neural coding of the periodicity of sounds.
>
> The probabilistic nature of the discharges of auditory nerve fibers "blurs" the temporal pattern of the neural code of sounds and thus limits the phase locking of high-frequency sounds. This variability makes it necessary to average the discharges in the response from a single nerve fiber to many repeated presentations of the same stimuli. The nervous system, however, does not use this method to reduce statistical variability but instead integrates the discharge patterns of many nerve fibers to obtain a stable response. Thus this (natural) spatial integration can reduce the statistical variability of neural discharges in response to a single stimulus presentation while recordings from only one nerve fiber require the responses of many stimuli to be added to obtain a similar reduction in the statistical variability of the discharges. This difference in the way experimental data are processed and the way the central nervous system extracts information must be considered when experimental data from single nerve cells are evaluated.

ROBUSTNESS OF THE PLACE CODE OF FREQUENCY

It is known that frequency discrimination is robust and largely independent of the intensity of a sound [213]. If the place code of frequency is the basis for frequency discrimination, it must be robust and not change noticeably over a large range of sound intensities. The relation between the location of the maximal deflection of the basilar membrane and the frequency (or spectrum) of a sound must thus not be affected by the intensity of a sound if frequency discrimination is based on the place principle. However, there is ample evidence from experiments in animals that the center frequency of tuning of the basilar membrane is affected by the intensity of the stimulus sound. We have already discussed that the representation of formants in the discharge pattern of single auditory nerve fibers deteriorate with increasing sound intensity (page 184). Other experiments have convincingly shown that the location of the maximal vibration amplitude of the basilar membrane not only depends on the frequency (spectrum) of a sound, but also on the sound's intensity, as has been discussed earlier in this book (Chapters 3 and 6). Similar dependence on sound intensity is evidenced from recordings from single auditory nerve fibers [66, 129] (Fig. 7.4). Since the tuning of neurons throughout the auditory nervous system is based on the tuning of the basilar membrane, the tonotopic representation of sounds that is prominent throughout the ascending auditory nervous system most likely shows similar intensity dependence, but that has not been studied specifically.

The fact that frequency tuning changes with the intensity of a sound means that the place principle lacks the robustness that is assumed to be necessary to explain psychoacoustic findings regarding discrimination of frequency. That cochlear frequency analysis is intensity dependent makes the place principle an unlikely candidate for the basis of auditory frequency discrimination. As

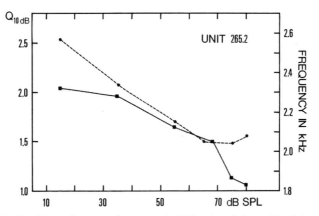

FIGURE 7.4   The shift in the center frequency (solid lines) and the width of the tuning of a single auditory nerve fiber (dashed line) in the auditory nerve of a rat as a function of the stimulus intensity. The width is given a "Q_{10dB}," which is the center frequency divided by the width at 10 dB above the peak (from Møller, A. R. (1977). Frequency selectivity of single auditory-nerve fibers in response to broadband noise stimuli. *J. Acoust. Soc. Am.* **62**:135–142).

Zwislocki has stated, "Therefore, if the intensity-dependent shift in the cochlear excitation maximum found in gerbils has a counterpart in human cochleae, as appears likely, *the excitation maximum cannot constitute an adequate physiological code for pitch.*" [245].

It has been suggested that it might not be the peak of the envelope of vibration of the basilar membrane that is important for frequency discrimination, but instead the entire envelope or the edges (slopes) of the envelope. It has also been suggested that frequency discrimination may rely on the steep high-frequency slope of tuning curves of single auditory nerve fibers. That would correspond to the slopes (skirts) of the frequency tuning of a single point on the basilar membrane. It has been claimed that the location of the skirts of the frequency tuning curves of the basilar membrane might vary less than the location of the peak of the basilar membrane motion when the sound intensity is changed. The high-frequency skirts of the frequency threshold tuning curves of cells in the cochlear nucleus are extremely steep [118]. The shape of these tuning curves might therefore be used by the auditory system to detect differences in the spectrum of sounds and thus be the basis for frequency discrimination according to the place principle. However, the frequency threshold tuning curves of single auditory nerve fibers and cochlear nucleus cells may not reflect the function of the auditory system under normal conditions because frequency threshold tuning curves are obtained by determining the threshold to pure tones presented in a quiet background.

Tuning curves of auditory nerve fibers obtained by using noise stimuli at intensities within the physiologic range of hearing (Fig. 6.6, Chapter 6) are more representative for the function of the auditory system under normal conditions than frequency threshold curves. The slope of the high-frequency skirts of such functions changes with sound intensity similarly to the shift in the frequency of the peak of frequency tuning curves. The same is the case for the mechanical tuning curves of the cochlea (Fig. 6.5, Chapter 6).

IMPORTANCE OF FREQUENCY ANALYSIS IN THE COCHLEA

The studies in animals on the coding of synthetic vowels in the auditory nerve described above indicated that the cochlear spectral analysis is not important for speech discrimination (p. 183). That hypothesis is supported by the results of studies of individuals who show signs of change in the wave motion on the basilar membrane—from a traveling wave motion to a partial standing wave motion (recall from Chapter 3 that the normal traveling wave motion of the basilar membrane, the basis for cochlear frequency selectivity, is the basis for the place principle of frequency discrimination). These abnormalities became evident in recordings of the click-evoked responses from the exposed intracranial portion of the auditory nerve in patients undergoing neurosurgical operations [141]. Some of these patients had abnormal responses consisting of a prolonged series of waves that followed the compound action potentials (CAP) normally seen in such responses (see Chapter 10). These oscillations in the CAP were interpreted to indicate the normal traveling wave pattern on the basilar membrane was (partly) replaced by a standing wave motion [141]. A standing wave motion on the basilar membrane cannot provide spectrum analysis of the cochlea, but it would not affect the temporal coding of sounds in the auditory nerve. The fact that these individuals did not have any noticeable impairment in their speech discrimination supports the hypothesis that cochlear spectrum analysis is not important for speech discrimination.

IMPORTANCE OF TEMPORAL COHERENCY OF AUDITORY NERVE FIRING FOR SPEECH DISCRIMINATION

It is well known that speech discrimination is more impaired in patients with hearing loss from injury of the auditory nerve than in individuals with the same threshold elevation from cochlea injury. This is evident from studies of patients with acoustic tumors and in patients in whom the auditory nerve has been injured by surgical manipulations (see Chapter 15). Such patients have a varying degree of hearing loss but the speech discrimination is always decreased more than expected from the pure tone audiogram.

Injury to the auditory nerve is associated with a decrease in neural conduction velocity in the nerve. The decrease in neural conduction velocity is usually not uniform in the fibers of the auditory nerve, resulting in an increased temporal dispersion of the neural activity, which impairs the temporal coherence of the discharges in auditory nerve fibers. If frequency discrimination is based on temporal coding of sounds in the auditory nerve, information from different locations along the basilar membrane must appear temporally coherent when it enters the central nervous system so that the interval between individual sound waves can be determined accurately. It is therefore likely that such impairment of temporal coherence of neural activity in the auditory nerve is the cause of the impairment of speech discrimination in patients with injuries to the auditory nerve [137]. That temporal coherence of the nerve impulses that reach the cochlear nucleus is important is also supported by morphological studies that show that the diameters of different auditory nerve fibers normally are very similar [210], thus indicating that the conduction velocity among different auditory nerve fibers varies very little. The poor speech discrimination associated with auditory nerve injuries may thus be a result of impaired coherence of auditory nerve impulses indicating that temporal coding is important for speech discrimination.

Acoustic tumors and surgical manipulation of the auditory nerve could also injure the efferent fibers, which would cause a change in the function of the outer hair cells. However, the effect on hearing function from severance of the efferent bundle (done in connection with vestibular neurectomy) has been found to be minimal [197].

DECODING TEMPORAL INFORMATION

The fact that interaural time intervals of approximately 1 μs can be detected by the auditory nervous system means that the nervous system has the ability to detect very small time intervals. If this capability was applied to monaural sounds it would mean that phase locking would occur up to very high frequencies. The task of performing temporal analysis of sound is, however, different from that of directional hearing. Temporal analysis of sound requires that the interval between sound waves is determined while directional hearing is based on determination of the difference in arrival time at the two ears. Decoding the temporal information of frequency thus requires slightly different neural circuitry but evidence has been presented that such neural circuitry exists in the same area of the brain that is known to decode the time difference between the arrival of sounds at the two ears (the medial superior olivary complex, MSO).

COCHLEAR SPECTRAL FILTERING MAY BE IMPORTANT FOR TEMPORAL CODING

The spectral filtering in the cochlea divides the audible spectrum into narrow portions before the sound is coded into a pattern of nerve impulses in the auditory nerve. This means that the temporal pattern within limited parts of the spectrum becomes coded in different populations of nerve fibers. The frequency of the sound within a narrow frequency band can be determined by measuring the time interval between individual waves of the filtered waveform that is coded in the pattern of neural discharges of individual auditory nerve fibers. While cochlear frequency analysis may play a minor role in frequency discrimination, its main importance may thus be in dividing the spectrum of sounds into frequency bands of suitable width before the sound is coded into the discharge pattern of individual auditory nerve fibers. This division reduces the requirements regarding coding of details of the waveform of complex sounds. Without such separation of the spectrum of natural sounds into narrow frequency bands, coding of the temporal pattern would require coding of fine details of the time pattern of sounds in the discharge pattern of auditory nerve fibers, which likely would exceed the limits for neural coding. The separation also reduces the demand on the neural circuitry that decodes the temporal information. Coding of the "raw" sound wave would likely overwhelm any neural discriminator of temporal information.

As an example, analysis of the temporal pattern of vowel sounds is more likely to provide accurate information about formant frequencies if the temporal analysis is performed separately in narrow bands, each of which contain no more than one formant. This phenomenon is known as "synchrony capture." When only one formant is contained in such frequency bands, the output of such a filter is a damped oscillation, the frequency of which is the formant frequency. The formant frequency can thus be determined accurately by measuring the interval between two waves of the output of such filters, which simplifies decoding of temporal information about frequency.

Spectral selectivity in the cochlea deteriorates at high sound intensities and in individuals with injured cochleae. Such decreased spectral separation may impair "synchrony capture" and the typical decrease in speech discrimination in such situations may be a result of impairment of temporal coding of frequency because of the widening of cochlear tuning. Absence of the acoustic middle ear reflex, which results in less amplitude compression before sounds reach the cochlea, also reduces speech discrimination at high sound intensities, probably because of widening of the cochlear filters, because of the larger sound input to the cochlea (Chapters 3 and 12).

That division of the spectrum of complex sounds into bands prior to analysis by the auditory nervous system is important for speech discrimination is supported by the observation that different bands of the speech spectrum contribute independently to speech discrimination. That observation was made early in speech research by Harvey Fletcher at the Bell Telephone Laboratories ([38], see also [4]).

A DUPLEX HYPOTHESIS OF FREQUENCY DISCRIMINATION

The studies described above agree that the temporal principle for frequency discrimination is more important for frequency discrimination in connection with complex sounds than the place principle and that the spectral selectivity in the cochlea probably has its greatest importance as a preprocessor of sounds before they are coded in the auditory nerve, which facilitates temporal processing of sounds. That place coding can be important for frequency discrimination was recognized in 1949 by Wever, who presented the volley theory (described in his book [228]), which suggested that both place and temporal coding were used for frequency discrimination. Wever suggested that temporal coding was most important at low frequencies and that place coding of frequency was most important at high frequencies. In the mid-frequency range both principles would work side by side in frequency discrimination.

Spectral analysis in the cochlea may be essential for frequency discrimination of high-frequency sounds because temporal analysis seems to be restricted to low frequencies, although the frequency limit for temporal analysis is not known. However, several psychoacoustic studies indicate that temporal analysis is important for frequency discrimination in the entire audible frequency range.

Coding of Complex Sounds

ABSTRACT

1. Auditory nerve fibers and cells in the nuclei of the classical ascending auditory pathway respond poorly to steady-state sounds. The discharge rate of most neurons reaches a plateau far below the physiologic range of sound intensities.

2. Sounds that change rapidly in frequency or amplitude are coded in the discharge pattern over a larger range of stimulus intensities compared to constant sounds or sounds with slowly varying frequency or intensity.

3. The response to complex sounds (the frequency or intensity of which changes) cannot be predicted from knowledge about the response to steady sounds or tone bursts.

4. Cells in the cochlear nucleus (CN) and the inferior colliculus (IC) respond vigorously to changes in the intensity of a stimulus sound. A 1-dB change in amplitude of a stimulus sound can result in a nearly 100% change in discharge rate of some cells of the CN and the IC.

197

5. The coding of the envelope of amplitude-modulated (AM) sounds is more prominent in the discharge pattern of cells in the cochlear nucleus compared with auditory nerve fibers.

6. The responses of auditory nerve fibers and cells in the nuclei of the classical ascending auditory pathways are phase locked to the modulation waveform of AM sounds over a large range of sound intensities.

7. Auditory nerve fibers have low-pass-shaped modulation transfer functions.

8. The modulation transfer functions of some CN cells have a low-pass shape for low stimulus intensities, which changes to a band-pass shape in the physiologic range of sound intensities. The modulation transfer function for some neurons in the CN is narrow and such cells may be regarded to be tuned to a certain modulation frequency.

9. The response of nerve cells in the nuclei of the classical ascending auditory pathway to changes in the frequency of tones depends on how fast the frequency of the tones changes.

10. The frequency range within which neurons in the CN respond to tones (response area) becomes narrower when the frequency of the tones is changed rapidly. Above a certain (high) rate of change, the response area again widens.

11. Thus coding of changes in both frequency and amplitude of sounds is greatly enhanced in the nuclei of the classical ascending auditory pathway as the information travels up the neural axis of the auditory system. Additionally, the dynamic range of neural coding of changes in frequency (spectrum) or amplitude of sounds is much larger than the dynamic range for coding of steady sounds.

INTRODUCTION

Most studies of coding of sound in the classical ascending auditory pathways have employed simple sounds such as pure tones and clicks. The recorded responses have been analyzed by determining the threshold of firing (frequency threshold curves or tuning curves) and the distribution of nerve impulses during and after the presentation of tone bursts [poststimulus time (PST) histograms].

Chapter 6 described the frequency representation in the classical auditory nervous system and discussed the place principle and the temporal principle of coding of frequency (spectrum). The most used stimuli for studies of these properties of the response from single auditory nerve fibers and cells in the nuclei of the classical ascending auditory pathways have been simple sounds such as tone bursts and steady broadband noise. Natural sounds such as speech sounds, however, are much more complex than the tones, clicks, and noise sounds used in the studies described in Chapter 6.

Natural sounds have broad spectra that change more or less rapidly. Changes in frequency and amplitude are prominent features of natural sounds. These changes in the amplitude and spectrum carry the information in such sounds as speech sounds. Many studies have demonstrated that changes in the frequency and amplitude of sounds are enhanced in the discharge pattern of the nuclei of the classical ascending auditory pathway. That is an indication that the nervous system transforms sounds and enhances aspects of sounds that are rich in information and suppresses the representation of sounds that carry little or no information. A steady sound such as a pure tone does not provide any information after it has been switched on, except perhaps what information might be provided in its duration. Many nerve cells in nuclei of the auditory system only respond to a tone when it is turned on and when it is turned off, which indicates that the auditory nervous system "filters" sound with regard to its information contents.

The choice of stimuli for studying of the function of the auditory nervous system has been affected more by technical issues concerning the generation and description of the sounds than by how well they represent natural sounds. Tone bursts have been the most commonly used stimuli in studying the auditory system but these stimuli are specialized sounds that differ from natural communication sounds. Pure tones are easy to generate and to describe, whereas it was more difficult to generate complex sounds before the development of computer systems. Currently it is possible to synthesize nearly any kind of sound on inexpensive laboratory computers.

Complex sounds are also more difficult to describe than pure tones, another factor that has detracted investigators from using complex sounds. Amplitude-modulated (AM) sounds and tones, the frequencies of which change at different rates, resemble natural sounds such as speech but are easier to generate and describe. Although they are less complex than natural sounds, they are more appropriate for studying the transformation of information in the auditory system than pure tones. Natural sounds normally appear together with a background of other sounds, but stimuli used in studies of the auditory system are usually presented in a background of silence; this is another example of how unnatural the stimuli used to study the auditory nervous system are.

Most of the published studies have been done in anesthetized animals. Anesthesia affects the responses to a degree that depends on the type of anesthesia and the anatomical location of the structure from which recordings are made. Generally, the effect of anesthesia is greater in the central parts of the auditory pathway than it is in more peripheral structures.

We begin this chapter by describing the response of auditory nerve fibers and cells of the nuclei of the classical ascending auditory pathway to steady tones and tone bursts because that has been the traditional way to study the function of the auditory system. We then proceed to describe how various parts of the classical auditory nervous system respond to complex sounds such as AM sounds and tones with rapidly varying frequencies.

RESPONSE TO TONE BURSTS

Neurons of the auditory system are commonly characterized and classified by their response to tone bursts. The classification based on the use of tone bursts as stimuli provide little insight into how the auditory system responds to natural sounds, but it is reviewed here because of its extensive use in earlier studies.

AUDITORY NERVE

Histograms of the response of single auditory nerve fibers to short bursts of tones or broadband noise reveal an initial high rate of firing followed by an exponential decrease in firing rate (Fig. 8.1). When the sound is switched off, the discharge rate falls below the fiber's firing rate in silence (the spontaneous rate). The firing gradually returns to the rate it had before the tone was switched on. Poststimulus time (PST) histograms such as those seen in Fig. 8.1 are compiled by adding the number of discharges from a single auditory nerve fiber to many presentations of the same sound and PST histograms therefore represent the average firing pattern of a nerve fiber. When no sound is presented, only the spontaneous activity is present. The discharge rate immediately after the stimulus sound has been switched off is lower than the spontaneous discharge rate. Note that the histograms in Fig. 8.1 only cover the intensity range up to approximately 40 dB above threshold. This is well below what is regarded as the physiological range of sound intensities (approximately 50–75 dB above threshold of hearing).

Practically all auditory nerve fibers have spontaneous activity similar to the example shown in Fig. 8.1. The spontaneous discharge rate may be a result of mechanical stimulation of hair cells by vibrations of the cochlear fluid that is not induced by outside sounds or as a result of random release of quanta of neurochemicals at the hair cell–nerve fiber synapse.

The discharge rate of auditory nerve fibers in response to continous tones increases as the sound intensity is raised above the fiber's threshold, but the discharge rates for most nerve fibers reach a plateau at sound intensities well below physiologic levels (Fig. 8.2A). Thus, most auditory nerve fibers have a small dynamic range in response to continous tones. Similar results are obtained when the discharge rate is determined from PST histograms. A small population

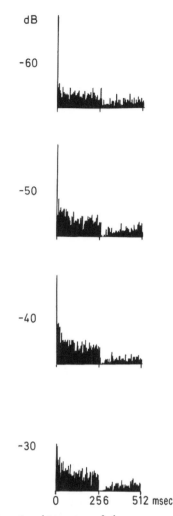

FIGURE 8.1 Poststimulus time histograms of the responses to 250-ms-long bursts of a 5.8 kHz tone from a typical auditory nerve fiber in a cat. The stimulus level is given in arbitrary decibel values. Obviously, the threshold of this fiber is slightly lower than −60 dB (from Kiang, N. Y.-S., Watanabe, T., Thomas, E. C., and Clark, L. F. (1965). "Discharge Patterns of Single Fibers in the Cat's Auditory Nerve." MIT Press, Cambridge, MA).

of auditory nerve fibers respond with increased discharge rates over a large range of sound intensities (Fig. 8.2B) [106, 162]. The dynamic range of auditory nerve fibers is partly related to their spontaneous activity. Fibers with low spontaneous rates tend to have a larger dynamic range than fibers with high

A

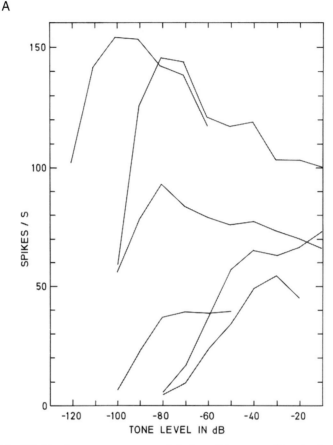

FIGURE 8.2 (A) Stimulus–response curves of single auditory nerve fibers in the cat. The discharge rate is shown as a function of the stimulus intensity for continous tones at the CF of the fiber from which the recordings were made. The sound level is given in arbitrary decibel values. The threshold is slightly below −100 dB for the fibers studied, except the top left curve, where it is approximately −120 dB (modified from Kiang, N. Y.-S., Watanabe, T., Thomas, E. C., and Clark, L. F. (1965). "Discharge Patterns of Single Fibers in the Cat's Auditory Nerve." MIT Press, Cambridge, MA). (B) Stimulus–response curves for three different auditory nerve fibers in a guinea pig to tones at the fiber's CF. Squares show the response from a nerve fiber with a low threshold (below the threshold of compound action potentials, CAP) and a high spontaneous activity (84.4 spikes/s). Solid circles show the response from a nerve fiber with a threshold near the CAP threshold and low spontaneous activity (0.2 spikes/s). The open circles represent the response of a fiber with threshold near that of the CAP and no spontaneous firings. Reprinted from *Hear. Res.,* Volume 55, Müller, M., Robertson, D., and Yates, G. K., Rate-versus-level functions of primary auditory nerve fibres: Evidence of square law behavior of all fibre categories in the guinea pig, pp. 50–56, copyright 1991, with permission from Elsevier Science.

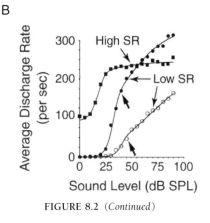

FIGURE 8.2 *(Continued)*

spontaneous activity (Fig. 8.2B). These neurons may thus communicate information about the intensity of a sound over most of the audible intensity range (from threshold to 80–90 dB HL). Fibers with high spontaneous activity and low threshold have a small dynamic range and saturate 20 to 30 dB above threshold.

COCHLEAR NUCLEUS

While the shapes of the PST histograms of the discharges of different auditory nerve fibers in response to tone bursts are similar, the shapes of the PST histograms of nerve cells in the cochlear nucleus (CN) to tone bursts vary between cells. The shapes of such PST histograms have been used to classify the response pattern of CN neurons. The best known and widely used classification of the response pattern of cells of the CN is the one published by Pfeiffer, [172] who divided the neurons in the CN into four different groups according to their response to tone bursts (Fig. 8.3). While this classification separates nerve cells according to their response to tone bursts, it is doubtful whether neurons with similar PST histograms also show similar responses to complex sounds.

OTHER NUCLEI OF THE CLASSICAL ASCENDING AUDITORY PATHWAY

The response to tone bursts from neurons in nuclei that are located more centrally vary within wide limits. Many cells respond best to the onset of a

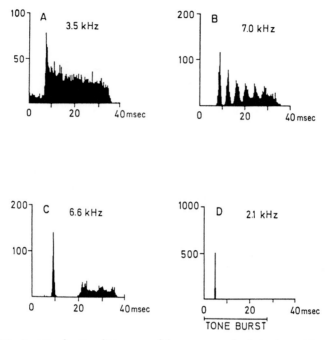

FIGURE 8.3 Poststimulus time histograms of the responses of cells in the cochlear nucleus of cats to tone bursts. Each histogram represents one class of units. (A) Primarylike; (B) chopper; (C) pause; (D) onset (from Pfeiffer, R. R. (1966). Classification of response patterns of spike discharges for units in the cochlear nucleus: Tone-burst stimulation. *Exp. Brain Res.* 1:220–235).

tone burst, indicating a preference for transient sounds. Since many cells respond poorly to continuous sounds, obtaining records of their discharge rate as a function of sound intensity is difficult.

CODING OF SMALL CHANGES IN AMPLITUDE

While the response to tone bursts may be regarded as an illustration of the response to changes in sound intensity, the changes that tone bursts represent are far greater than those of natural sounds. The responses to tone bursts do not provide information about how small changes in sound intensity are coded in the auditory nerve and nuclei of the auditory nervous system. The response to a small rapid increase or decrease in the intensity of a tone can be illustrated by observing the discharges of a neuron in the CN in response to a continuous sound, the intensity of which is increased and decreased stepwise (Fig. 8.4). It is seen that a small increase in the intensity of the sound results in a large

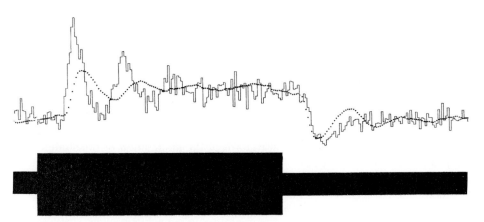

FIGURE 8.4 Period histogram of the response from a cell in the cochlear nucleus of a rat to tones the intensity of which was changed up and down in a stepwise fashion. The dots show the calculated response obtained from the response to a tone that was amplitude modulated by pseudorandom noise (from Møller, A. R. (1979). Coding of increments and decrements in stimuli intensity in single units in the cochlear nucleus of the rat. *J. Neurosci. Res.* 4:1–8. Copyright 1979, Wiley-Liss, Inc., a subsidiary of John Wiley & Sons, Inc.).

but brief increase in the discharge rate. When the sound intensity is again decreased, the discharge rate decreases briefly below its steady-state discharge rate.

The response to changes in a sound's intensity (Fig. 8.4) illustrate how neural discharges respond to small stepwise changes in a sound's intensity but it does not provide information about the effect of the rate with which the intensity of a sound is varied. That can be studied in experiments where tones or noise that are amplitude modulated with a sinusoidal waveform are used as stimuli. Amplitude-modulated sounds are similar to many natural sounds such as speech sounds and communication and warning sounds made by various animals. Recordings of the responses from single auditory nerve fibers and cells of the nuclei of the classical ascending auditory pathways to AM sounds thus yield results that reflect processing of natural sounds more closely than the responses to tone bursts or sounds the amplitudes of which are changed stepwise. (The sounds used to obtain the results in Fig. 8.4 may be regarded as a tone that is modulated by a rectangular waveform.)

AMPLITUDE-MODULATED SOUNDS

Systematic studies have been published of the responses of single auditory nerve fibers to amplitude-modulated tones and noise [18, 40, 83, 85, 128,

170, 204] and from cells in the CN [41, 42, 43, 44, 122, 123, 127, 128, 180, 243] and the IC [178, 179]. A few studies of the responses from neurons of the auditory cortex have also been published [26, 71, 72, 199]. We begin by describing results of studies in the CN because they are the most extensive.

Cochlear Nucleus

The discharge pattern of single nerve cells in the CN in response to sinusoidal AM tones is modulated by the waveform of the modulation of the stimulus sound, as is seen from modulation period histograms) of the discharges in response to AM tones (Fig. 8.5).[1] Neurons in the CN reproduce the (sinusoidal) modulation waveform of AM tones and noises faithfully in their discharge pattern over a large range of modulation frequencies (Fig. 8.6) [122].

> Several measures of the modulation of the neural discharges have been used to describe the coding of AM sounds in the discharge pattern of auditory nerve fibers and cells of auditory nuclei. One such measure is the relative modulation of modulation period histograms of the discharges, i.e., the ratio of the amplitude of the modulation of the histograms and the modulation of the stimulus sounds. An example of that is shown in Fig. 8.6, where the solid line in the graph shows the relative modulation, expressed in decibels (the scale to the left of the graph). The dashed line is the phase angle between the modulation of the sound and that of the histograms. Such graphs are known as Bode plots of modulation transfer functions. Zero dB on that scale corresponds to a modulation of 100% of the histograms. Zero degrees means that a sine wave that fits the histogram falls exactly over the sine wave of the modulation. A negative phase angle means that the modulation of the histogram lags the modulation of the sound.
>
> Another frequently used measure of modulation of neural discharges, introduced by Goldberg and Brown [57], is called the synchronization index. It is defined as the fraction of the nerve impulses that occur phase locked to the sound (or the envelope of an AM sound).

The modulation of the discharges of the cell depicted in Fig. 8.6 is largest at a modulation frequency of approximately 300 Hz. A modulation of a histogram of -10 dB means that the histogram has been modulated 30%. If the modulation depth of the stimulus sound is 20%, as it was in the experiment depicted in Fig. 8.6, a 30% modulation of the histograms indicates that the histograms were modulated more than the stimulus sound was. This can be expressed as a modulation gain (of approximately 4 dB). When modulation transfer functions are divided by the depth of the modulation of the stimulus tone (carrier), they describe the gain with regard to the reproduction of the modulation waveform in the discharge pattern of a nerve cell. Bode plots of

---

[1] Modulation period histograms show the distribution of nerve impulses over one period of the modulation similarly to the period histograms used to study coding of the waveform of a sound in the discharge pattern of single auditory nerve fibers (phase locking; see Fig. 6.13 in Chapter 6).

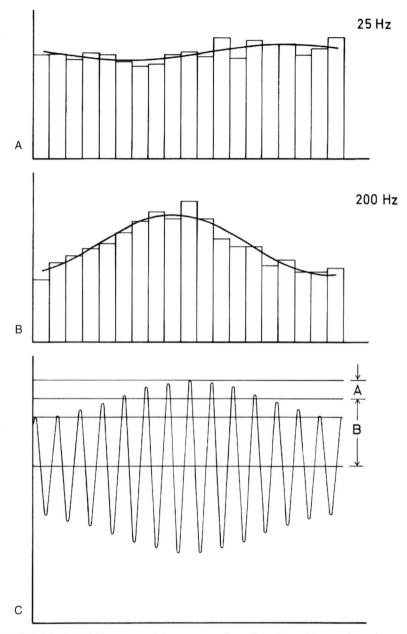

FIGURE 8.5 Period histograms of the response of a cell in the cochlear nucleus of a rat to amplitude-modulated tones. The frequency of the tone was 15 kHz, equal to the cell's CF. Histograms of the response to two different modulation frequencies. (A) Modulation frequency 25 Hz. (B) Modulation frequency 200 Hz. (C) One period of the modulated sound. A is the modulation; B is the mean amplitude of the stimulus. A/B is the modulation depth.

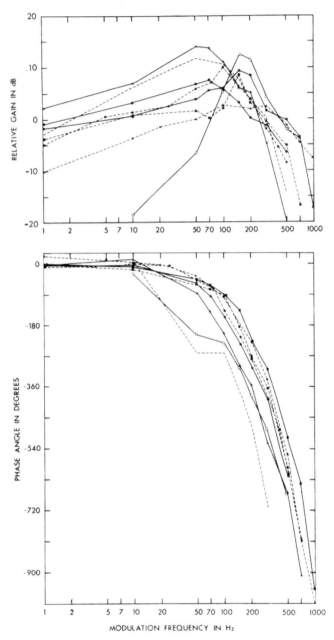

FIGURE 8.7   Modulation transfer functions (gain functions) of eight typical cells in the cochlear nucleus of a rat. The frequency of the carrier tones was equal to the CF of the units (ranging from 950 Hz to 30 kHz), and the sound intensity was 20 dB above the threshold of the cells (from Møller, A. R. (1972). Coding of amplitude and frequency modulated sounds in the cochlear nucleus of the rat. *Acta Physiol. Scand.* **86**:223–238).

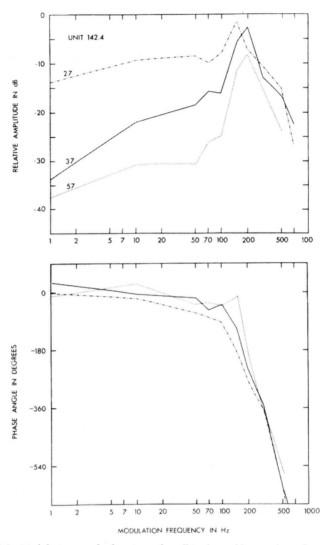

FIGURE 8.8 Modulation transfer functions of a cell in the cochlear nucleus of a rat where the different curves represent different sound intensities, given in decibels SPL (approximately the same as decibels above threshold). The modulation depth was 20% and the frequency of the carrier tone was equal to the cell's CF (15.2 kHz) (from Møller, A. R. (1972). Coding of amplitude and frequency modulated sounds in the cochlear nucleus of the rat. *Acta Physiol. Scand.* **86**:223–238).

intensity of an excitatory tone causes an increase in the discharge rate (Fig. 8.9).

The results shown in the previous graphs were obtained when the stimuli were continuous tones that were amplitude modulated with either a sinusoidal waveform or pseudorandom noise.

Responses to Sounds That Are Modulated by Noise

Using nonperiodic sounds for modulation of tones or noise has several advantages in studying the coding of the amplitude modulation of sounds. Natural sounds are often modulated with nonsinusoidal waveforms. The response to sinusoidal modulated tones or noise only represents the response to one modulation frequency at a time and neural discharges may phase-lock to the modulation waveform. To avoid these problems and get a more representative stimulus, some investigators have used broadband signals to modulate tones or noise [123]. When low-pass-filtered noise is used to modulate a sound such as a tone, the entire modulation transfer function can be obtained from a single recording. Pseudorandom noise repeats itself periodically and analysis of the response from single cells to sounds that are amplitude modulated with pseudorandom noise can be done by compiling a period histogram of the responses over one period of the pseudorandom noise. These histograms are then cross-correlated with the waveform of the modulation (the pseudorandom noise). Such cross-correlograms are estimates of the impulse response of the system under test and the Fourier transforms of such correlograms thus yield an estimate of the frequency transfer function. When pseudorandom noise is used to modulate a sound (tone or noise) the resulting cross spectra are estimates of the modulation transfer function and thus similar to the modulation transfer functions that are obtained by compiling the results from using sinusoidal modulations with different frequencies of the modulation. Studies in the CN have shown that the modulation transfer functions that are obtained using these two different methods are similar[123].

Some investigators have used bursts of tones where part of the bursts were sinusoidal amplitude modulated [42, 43]. When tone bursts were used, a poststimulus time histogram (PSTH) was made of the response covering the entire presentation of the sound (150 ms, [42]) (Fig. 8.10). These investigators confirmed that the modulation transfer function changed from a low-pass shape to a band-pass shape when the sound level was increased [43].

Frisina and co-workers [42] compared the modulation gain of CN cells that responded differently to tone bursts. Using the classification described by Pfeiffer [172] (Fig. 8.3) these investigators found that the modulation gain at 150 Hz was highest for "on"-type units, where it was approximately 2.7 dB at 50 dB [42], and less for "chopper"-type (-7.5 dB) units. Primarylike cells had modulation gains of approximately -9.4 dB. These results may be interpreted to show that the further CN cells are from being primarylike the better they encode amplitude modulation. Primarylike neurons receive fewer auditory nerve fibers than other types of CN nerve cells. Therefore, the ability to encode the modulation waveform of an AM sound may thus be related to the number of auditory nerve fibers that terminate on a cell.

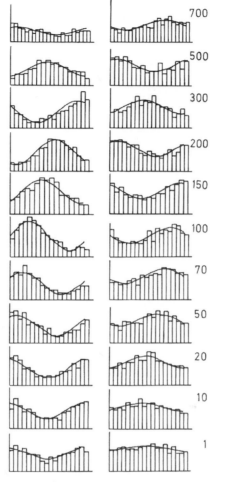

FIGURE 8.9   Period histograms of the response to amplitude-modulated tones of a cell in the cochlear nucleus of a rat. Two tones were presented simultaneously, one excitatory (at CF = 4.5 kHz) and the other (5.5 kHz) located in the cell's inhibitory response area. The histograms in the left-hand column were obtained when the excitatory tone was modulated and the inhibitor was unmodulated; right-hand column histograms were obtained when the inhibitory tone was modulated and the excitatory tone was unmodulated. The modulation frequency is given on the histograms (from Møller, A. R. (1975b). Dynamic properties of excitation and inhibition in the cochlear nucleus. *Acta Physiol. Scand.* **93**:442–454).

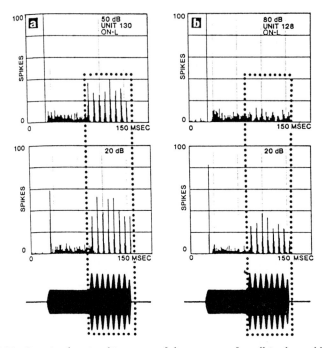

FIGURE 8.10   Poststimulus time histograms of the response of a cell in the cochlear nucleus of a cat to tone bursts the last half of which were amplitude modulated. The response to two different sound intensities are shown. Reprinted from *Hear. Res.* **44**, Frisina, R. D., Smith, R. L., and Chamberlain, S. C., Encoding of amplitude modulation in the gerbil cochlear nucleus. I. A hierarchy of enhancement, pp. 99–122, copyright (1990), with permission from Elsevier Science.

Auditory Nerve Fibers

The discharge pattern of single auditory nerve fibers may be modulated by the envelope of AM tones, but the modulation wave form is reproduced in the discharge pattern to a lesser extent than what occurs in CN units for the same sounds [83, 128, 170]. The modulation transfer functions of auditory nerve fibers are low-pass functions with cut-off frequencies in the range of 1000 Hz (Fig. 8.11). The cut-off frequency for fibers with a high CF is higher than for fibers with a low CF. Auditory nerve fibers reproduce the modulation waveform in their discharge pattern to a greater extent for sounds of low intensity (Fig. 8.12) [18]. Fibers that have a low spontaneous rate have the strongest coding of the modulation waveform of AM tones at CF (Fig. 8.12) [18, 40]. Fibers with a medium-high spontaneous rate and fibers with high spontaneous rate reproduced the envelope of AM tones to a lesser degree and the influence of the sound intensity on the reproduction of the modulation varies considerably among different nerve fibers.

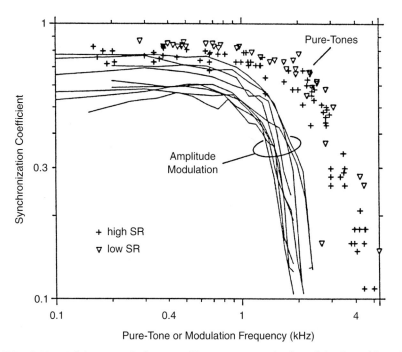

FIGURE 8.11 Modulation transfer functions of the response to amplitude-modulated sound for auditory nerve fibers together with the degree of phase locking (synchronization index) for pure tones as a function of frequency for auditory nerve fibers (from Joris, P. X., and Yin, T. C. T. (1992). Responses to amplitude-modulated tones in the auditory nerve of the cat. *J. Acoust. Soc. Am.* 91:215–232).

Comparison between Auditory Nerve Fibers and Cells in the Cochlear Nucleus

The modulation of the discharge rate of cells in the CN by the modulation waveform is larger than that in the auditory nerve fiber for the same AM sound (Fig. 8.13). Thus cells in the CN, besides extending the encoding of the envelope of AM sounds to a larger range of sound intensities, also enhance coding of changes in the intensity of a sound.

That cells in the CN amplify the response to small changes in amplitude of a sound may seem paradoxical, but it is probably a result of the fact that cells of the CN receive input from many auditory nerve fibers. Convergence of primary nerve fibers onto CN cells causes averaging of inputs from many auditory nerve fibers by CN cells which enhances small changes in the discharge rate of auditory nerve fibers similarly to the signal averaging techniques used to recover evoked potentials from a background of other signals (noise). The modulation gain seems to be related to the number of auditory nerve fibers

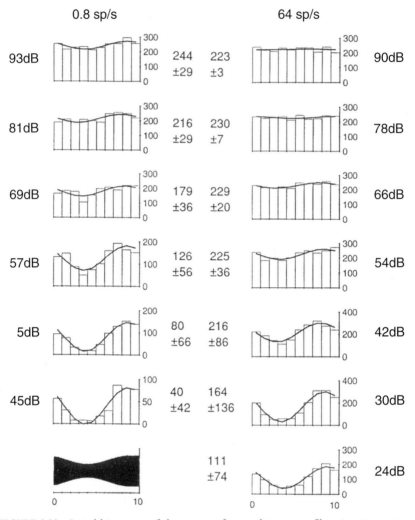

FIGURE 8.12 Period histograms of the response from auditory nerve fibers in guinea pigs to amplitude-modulated sounds at different sound intensities. The two rows of histograms are from two nerve fibers with different spontaneous activity (left column: 0.8 spikes/s and right column: 64 spikes/s (from Cooper, N. P., Robertson, D., and Yates, G. K. (1993). Cochlear nerve fiber responses to amplitude-modulated stimuli: Variations with spontaneous rate and other response characteristics. *J. Neurophysiol.* 70:370–386).

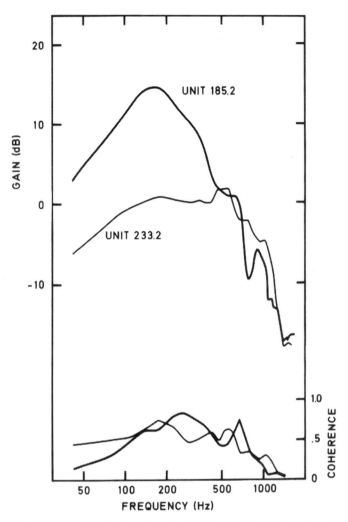

FIGURE 8.13 Modulation transfer functions of a cell in the cochlear nucleus of a rat (CN, upper thick curves) and of an auditory nerve fiber (AN, lower thin curve). The vertical scale shows gain (in decibels). The modulation transfer functions were obtained from analysis of the response to tones that were amplitude-modulated with pseudorandom noise. The bottom curves are coherence functions which show the degree of significance of the gain functions (from Møller, A. R. (1976). Dynamic properties of primary auditory fibers compared with cells in the cochlear nucleus. *Acta Physiol. Scand.* **98**:157–167).

that converge onto a CN cell [42, 43]. The larger dynamic range of the response to amplitude modulation of CN cells compared with auditory nerve fibers may also, partly, be a result of the fact that CN cells receive both excitatory and inhibitory input from the auditory nerve.

Effect of Masking Noise

Most studies of the neural coding of amplitude modulation have been done in the quiet. Few studies have explored the effect of a background sound on the activity of auditory nerve fibers and cells in the CN. Rhode and Greenberg [181, 182] showed that the phase locking of both auditory nerve fibers and CN cells is relatively resistant to background noise. Frisina *et al.* [40] showed that background noise could increase or decrease the reproduction of the modulation waveforms in the discharge pattern of auditory nerve fibers depending on the spontaneous activity of the fibers. In the CN, background noise can enhance the reproduction of amplitude modulation in cells in the dorsal CN (DCN) while the reproduction of the modulation of amplitude modulated sounds is little affected by background noise in the ventral CN (VCN).

Superior Olivary Complex

The reproduction of amplitude modulation in neurons in the superior olivary complex has not been systematically studied.

Inferior Colliculus

In an early study Erulkar *et al.* [30] showed that some cells in the inferior colliculus (IC) preferentially responded to AM tones. The cell depicted in Fig. 8.14 only responded (with a single discharge) at a certain phase of the modulation. Later systematic studies confirmed that coding of AM sounds in neurons in the IC is prominent [179]. The modulation transfer functions of cells in the IC change from low-pass functions to band-pass functions when the stimulus intensity is increased (Fig. 8.15), thus, similar to cells in the CN, but the peak in the modulation transfer function occurs at a lower modulation frequency than in the CN. The variations in the response pattern between different cells in the IC are larger than in the CN. The responses to the modulation waveform are robust and adding masking noise affects the shape of the modulation transfer functions only to a small extent even when the noise level exceeds the sound level of the AM tones from which the response is derived (Fig. 8.16).

The average discharge rate of cells in the IC is affected by the amplitude modulation. The discharge rate increases with the modulation frequency and

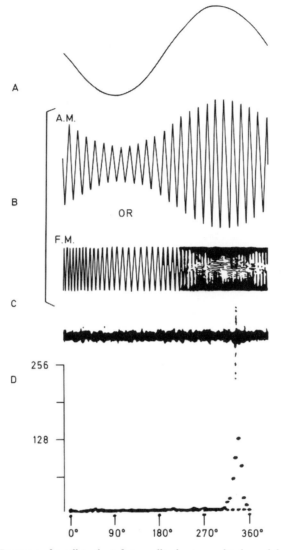

FIGURE 8.14 Response of a cell in the inferior colliculus to amplitude-modulated tones. The modulation was sinusoidal and the tone was modulated 50% (from [30]).

reaches a maximum around 500 Hz for sinusoidal AM tones [101]. This modulation frequency far exceeds the frequency at which the modulation waveform is reproduced to the greatest extent. The range of modulation fre-

822/3   Cf 18.8 kHz

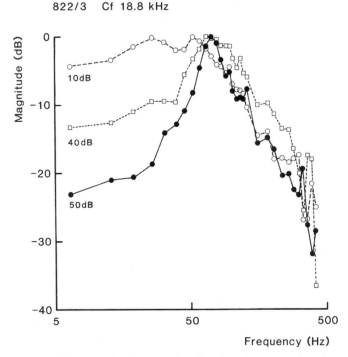

FIGURE 8.15   Modulation transfer functions of a cell in the inferior colliculus of a rat obtained at three different sound intensities (10, 40 and 50 dB above the cell's threshold). The transfer function changes from low-pass to band-pass with increasing sound intensity. Reprinted from *Hear. Res.*, Volume 27, Rees, A., and Møller, A. R., Stimulus properties influencing the responses of inferior colliculus neurons to amplitude-modulated sounds, pp. 129–144, copyright 1987, with permission from Elsevier Science.

quencies where the modulation waveform is best reproduced in the discharge pattern is 10 and 30 Hz.

Auditory Cortex

Few systematic studies have been done on coding of AM sounds in the neurons of the auditory cerebral cortex. In one study, Schreiner and Urbas [199] showed that the modulation transfer functions of cortical neurons were band-pass type (Fig. 8.17). These investigators used both sinusoidal modulation and modulation with a rectangular waveform. It is interesting that the response is approximately the same for ipsilateral and contralateral stimulation. The question about the representation of the uncrossed classical ascending auditory pathway has

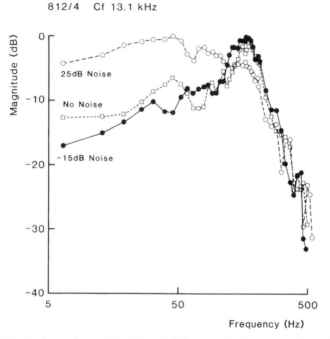

FIGURE 8.16   Similar graphs as in Fig. 8.23 with different levels of background noise. Reprinted from *Hear. Res.*, Volume 27, Rees, A., and Møller, A. R., Stimulus properties influencing the responses of inferior colliculus neurons to amplitude-modulated sounds, pp. 129–144, copyright 1987, with permission from Elsevier Science.

been debated in many fields of auditory research and the conclusions differ. We return to this debate several times in the following chapters.

Efferent Nerve Fibers

Efferent fibers of the auditory nerve also code the modulation waveform of AM sounds in a way that depends on the frequency of the modulation. Efferent fibers respond best to the modulation between modulation frequencies of 100–140 Hz. A tone that is amplitude modulated 30% produced approximately 100% modulation of the discharges in efferent fibers, thus a modulation gain of approximately 3. This is more than the double of the greatest modulation gain of afferent auditory nerve fibers [63]. The reason that the modulation is reproduced to a greater extent in efferent fibers than in afferent auditory nerve fibers is probably a result of the greater degree of spatial averaging of the information that occurs in the efferent fibers.

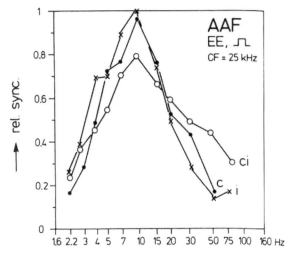

FIGURE 8.17 Modulation transfer functions obtained in the anterior auditory field (AAF) of the auditory cerebral cortex of a cat to contralateral (c), ipsilateral (i), and bilateral (ci) stimulation with amplitude-modulated tones. Reprinted from *Hear. Res.,* Volume 21, Schreiner, C. E., and Urbas, J. V., Representation of amplitude modulation in the auditory cortex of the cat. I. The anterior auditory field (AAF), pp. 227–242, copyright 1986, with permission from Elsevier Science.

Nerve Cells in the Auditory Periphery Code Both the Modulation and the Periodicity of the Carrier of AM Sounds

When low-frequency tones are amplitude modulated, both the modulation (envelope) and the periodicity of the tone are coded in the time pattern of the discharges of auditory nerve fibers and CN cells. This can be demonstrated by autocorrelation analysis of the discharge pattern in response to AM tones [120]. If these findings apply to speech sounds, it would mean that both the fundamental frequency of a vowel and the formant frequencies are coded in the temporal pattern of single auditory neurons in the CN. Again, we do not know if the information is used by the central nervous system in discrimination of sounds; we can only determine if the information is present or not.

RESPONSE TO TONES WITH CHANGING FREQUENCY

We have seen in a previous chapter that fibers of the auditory nerve only respond to continuous tones within a certain range of frequencies (Fig. 6.2,

Chapter 6) and that the frequency selectivity of auditory nerve fibers depends on the intensity of the stimulus sound (Fig. 6.6, Chapter 6). When the frequency and intensity of a stimulus tone is within the response area of an auditory nerve fiber, or a cell in a nucleus of the classical ascending auditory pathways, its discharge rate becomes a function of the frequency of the tone. That can be visualized by using a tone the frequency of which is varied and compiling a histogram of the distribution of discharges as a function of the frequency of the tone (Fig. 8.18). Such histograms describe the response areas of an auditory fiber or of a nerve cell in an auditory nucleus. When the frequency of the stimulus tone is varied slowly, such histograms show a similar frequency range of response, to those obtained using steady tones (as shown in Chapter 6, Fig. 6.10). They do not have the same shape as those of inverted-frequency tuning curves, but such histograms are broader and their tips are less sharp. The higher the stimulus intensity, the wider the frequency range over which the neurons respond.

EFFECT OF RATE OF CHANGE IN THE FREQUENCY OF TONE STIMULI

The effect of the rate of change in the frequency of a tone stimulus on the response of single auditory nerve is different from that of cells in the nuclei of the classical ascending auditory pathway. The response area of auditory nerve fibers is little affected by the rate of change of the frequency of a stimulus tone but the response area of single nerve cells in the CN and other nuclei of the classical ascending auditory system can change radically when the rate with which the frequency of a sound is changed.

Auditory Nerve

The responses of single auditory nerve fibers of the cat to tones the frequency of which was changed at different rates show a slight increase in the height of the histograms with little change in the shape of the histograms as the rate of change was increased [203] (Fig. 8.19). This means that the tuning of single auditory nerve fibers is little affected by the rate of change in the frequency of a continous tone. The fact that the shape of the tuning curves is not affected by the rate of change in the frequency of the stimulus tone can be demonstrated when the histograms are displayed on a normalized scale (Fig. 8.20).

Cochlear Nucleus

The response of neurons in the CN to tones the frequency of which changes rapidly is different from that of auditory nerve fibers. Thus, a histogram of the

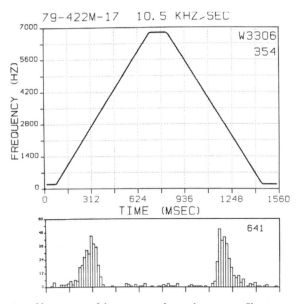

FIGURE 8.18 Period histograms of the response of an auditory nerve fiber in a cat (lower graph) to tones the frequency of which was varied up and down (upper graph) in a range that extended over the range in which the fiber responded. Reprinted from *Hear. Res.*, Volume 4, Sinex, D. G., and Geisler, C. D., Auditory-nerve fiber responses to frequency-modulated tones, pp. 127–148, copyright 1981, with permission from Elsevier Science.

responses become narrower and higher when the rate of change in frequency of the tone is increased from a low rate [125] (Fig. 8.21). This means that the shape of the response area becomes narrower when the frequency of a stimulus tone changes rapidly and more nerve impulses were delivered when the frequency of the stimulus tone is closer to the cell's characteristic frequency. The total number of nerve impulses, however, is only slightly dependent on the rate of change in frequency of the stimulus tone, which means that it is mostly a redistribution of nerve impulses that occurs when the rate of change in the frequency of the stimulus tone is increased.

The height of the histograms does not continue to grow as the rate of change in frequency is increased but reaches a maximum height when the frequency of the stimulus tone is changed at a certain rate (Fig. 8.22). The height of the peaks in the histograms of the response to tones increases more for tones between 45 and 65 dB above threshold than for tones of lower intensity (25 dB above threshold). The rate of change at which the histograms reach their maximal height depends on the direction of change in frequency of the stimulus tone (rising compared with falling frequency). In many nerve

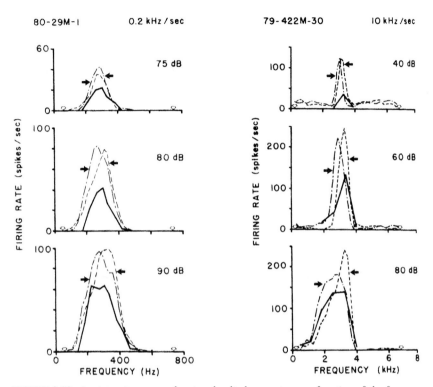

FIGURE 8.19 Iso-intensity curves showing the discharge rates as a function of the frequency of the tone stimulus at different intensities (given by legend numbers). Solid lines: steady tones; dashed lines: response to tones the frequency of which was changed at a rate of 0.2 kHz/s (left column); and 10 kHz/s (Right column). Reprinted from *Hear. Res.*, Volume 4, Sinex, D. G., and Geisler, C. D., Auditory-nerve fiber responses to frequency-modulated tones, pp. 127–148, copyright 1981, with permission from Elsevier Science.

cells the maximum height occur at a lower rate of change in frequency for falling frequency compared with rising frequency. Tones with falling frequency also lead to taller histograms than tones with rising frequency. The effect of the direction of the change in frequency is most pronounced when the frequency of a stimulus tone is changed at a high rate.

Different nerve cells in the CN respond differently to tones the frequency of which is varied rapidly, but all cells that have been studied showed enhancement of their frequency selectivity when the rate of change in frequency is increased up to a certain value [125]. Some nerve cells in the CN have a high rate of spontaneous activity and such cells may not change their firing rate noticeably in response to tones with constant or slowly varying frequency.

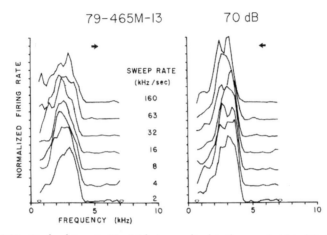

FIGURE 8.20 Similar data as in Fig. 8.5 but normalized to the same height of the histograms. (Left column) The change in the frequency of the tones was from low to high. (Right column) Frequency change from high to low. Reprinted from *J. Neurophysiol.*, Volume 64, Yin, T. C. T., and Chan, J. C. K., Interaural time sensitivity in medial superior olive of cat, pp. 465–488, copyright 1990, with permission from Elsevier Science.

Such neurons may, however, respond vigorously to tones with rapidly varying frequency and show pronounced frequency selectivity (Fig. 8.23). Neurons that display little frequency selectivity when stimulated with tones with steady or slowly varying frequency often respond markedly different to tones of increasing and decreasing frequency.

When a tone at CF is superimposed on the tone the frequency of which is changed, some nerve cells show little frequency selectivity when the frequency of the variable tone is changed slowly, but they show clear and pronounced frequency selectivity when the frequency is changed at a high rate (Fig. 8.24). Some nerve cells in the CN show mainly inhibition of their spontaneous firing when stimulated with pure tones of constant frequency. When the frequency

---

FIGURE 8.21 Period histograms of a cell in the cochlear nucleus of a rat in response to tones the frequency of which was varied between 5 and 25 kHz at different rates. (A and C) Slow rate. (B and D) Fast rate. The top histograms (A and C, slow rate) show the responses obtained when the duration of a full cycle was 10 s and the lower histograms (B and D, fast rate) show the responses obtained when the duration of a complete cycle was 156 ms. The change in the frequency of the stimulus tone was accomplished by having a trapezoidal waveform control the frequency of the sound generator (E). The two left-hand graphs (A and B) are histograms of a full cycle of the modulation and the right-hand graphs (C and D) show the details between the vertical lines in the left-hand graphs (from Møller, A. R. (1974b). Coding of sounds with rapidly varying spectrum in the cochlear nucleus. *J. Acoust. Soc. Am.* **55**:631–640).

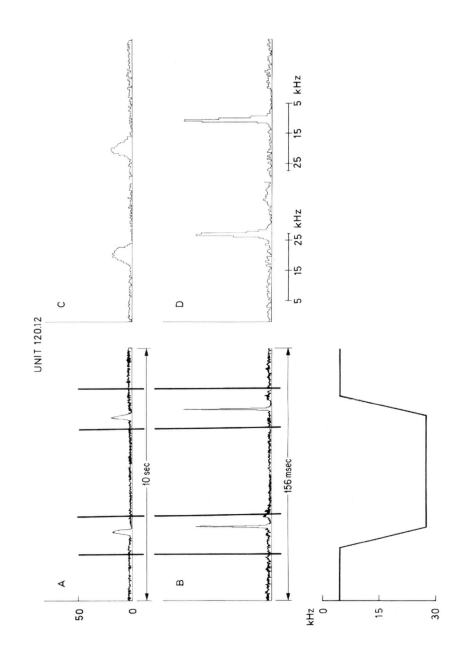

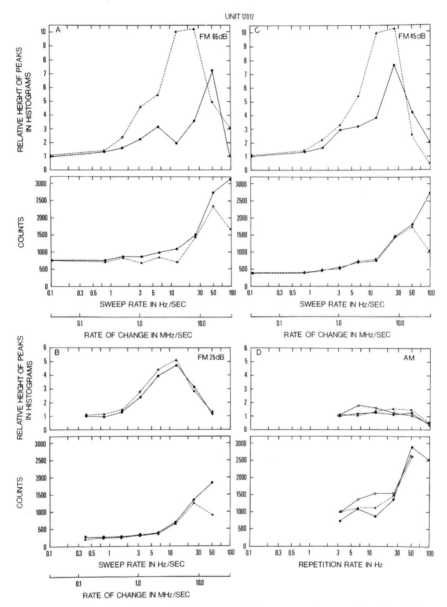

FIGURE 8.22 The relative height of the histograms of the response of a nerve cell in the cochlear nucleus of a rat (CF of 22 kHz) in relation to the rate of change of the frequency of the tones used as stimuli. The height of the histogram peaks of the response obtained when the frequency is changed at a slow rate was set to the value of 1.0. The height of the histograms of the response to tones of increasing frequency (solid lines) is different from those to tones of decreasing frequency (dashed lines). The results depicted in the three graphs marked FM (A, B, and C) were obtained at three different stimulus intensities (65, 45, and 25 dB above the threshold for that cell). The number of spikes in the peaks of the histograms is shown in the graphs below. The graphs labeled D show the response to amplitude-modulated sounds (modified from Møller, A. R. (1971). Unit responses in the rat cochlear nucleus to tones of rapidly varying frequency and amplitude. *Acta Physiol. Scand.* **81**:540–556).

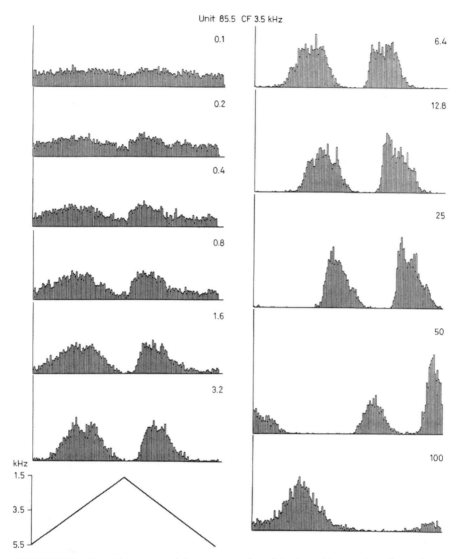

FIGURE 8.23 Period histograms of the responses of a cell in the cochlear nucleus of a rat that show little frequency selectivity to tones of slowly varying frequency but a pronounced frequency selectivity when the frequency of the tones are varied rapidly. The rate of frequency change was varied by changing the rate by which the cycle of change in frequency of the stimulus tones are repeated (given by legend numbers). The cell's CF was approximately 3.5 kHz and the stimulus intensity was 47 dB SPL (from Møller, A. R. (1983a). "Auditory Physiology." Academic Press, New York).

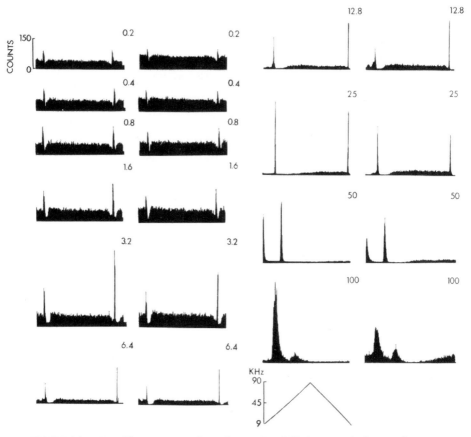

FIGURE 8.24   Period histograms similar to those in Fig. 8.23. A tone with changing frequency, 25 dB above threshold, was presented in a quiet background (first and third column from left), and when presented together with a constant tone (at CF and 32 dB above threshold) (second and fourth columns). The frequency was changed between 9 and 90 kHz (see insert below). The rate of frequency change was varied by changing the rate by which the cycle of change in frequency of the stimulus tones are repeated (given by legend numbers) (from Møller, A. R. (1971). Unit responses in the rat cochlear nucleus to tones of rapidly varying frequency and amplitude. *Acta Physiol. Scand.* **81:**540–556).

of the tone is changed at a high rate, such units change their response pattern to become excitatory with similar enhancement of the responses as seen in other cells (Fig. 8.25).

Comparison between Cochlear Nucleus and Auditory Nerve Fibers

The responses from cells in the CN to tones with rapidly changing frequencies are different from the responses to similar sounds from auditory nerve fibers.

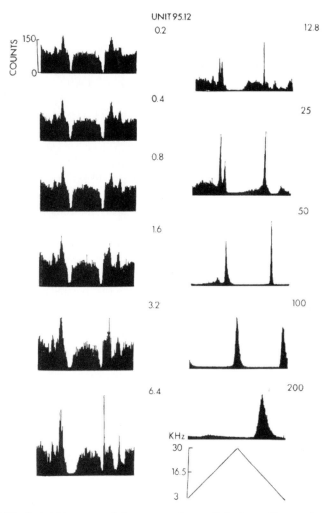

FIGURE 8.25   Period histograms of the response from cells in the cochlear nucleus of rats to tones the frequency of which was varied at different rates (similar to Figs. 8.23 and 8.24). This cell had a high spontaneous rate and responded to tones with slowly varying frequency mainly with inhibition of its spontaneous firing. The rate of frequency change was varied by changing the rate by which the cycle of change in frequency of the stimulus tones are repeated (given by legend numbers) (from Møller, A. R. (1971). Unit responses in the rat cochlear nucleus to tones of rapidly varying frequency and amplitude. *Acta Physiol. Scand.* **81**:540–556).

In the cochlear nucleus the response changes often dramatically when the rate of change in frequency is increased [125] whereas the response from auditory nerve fibers is little affected by the rate of change of the frequency of a tone

[203]. However, the study of auditory nerve fibers used tones where frequency of which was changed at lower rates than the tones used in the study of the responses from cells in the CN. The peaks of the histograms of the responses from nerve cells in the CN reached their maximal height for rates of change larger than 1 MHz/s [121, 125] but the study of auditory nerve fibers was limited to rates of frequency changes less than 0.2 MHz/s. If the study of auditory nerve fibers had covered a similar range of rates of frequency change as used in the studies of the responses from cells in the CN, a similar sharpening of the response area for auditory nerve fibers may have been seen. Nevertheless, it seems likely that at least some of the sharpening of the tuning in the CN does not just reflect properties of auditory nerve fibers. It is more likely that part of the enhancement in the response from CN cells is a result of interactions between inhibition and excitation in the auditory neurons that converge on individual CN cells. This assumption is supported by findings by Britt and Starr [10], who reported on intracellular recordings from cells in the CN to tones of changing frequency. The sharpening of the response areas of CN cells is yet another demonstration of the complexity of the auditory nervous system and the extension of the information processing that occurs as the information travels up the neural axis of the classical auditory nervous system.

Other Nuclei of the Classical Ascending Auditory Pathways

The pronounced enhancement of the response to tones of rapidly changing frequency thus occurs at the earliest station of auditory neural processing (i.e., the CN) and represents the beginning of processing of sound. The response from neurons in other nuclei of the classical ascending auditory pathways to sounds of rapidly changing frequency has not been studied to the same extent as it has in the CN. It is therefore not known if this sharpening of the response to tones with rapidly changing frequency is preserved or further enhanced as the information travels along the neural axis toward the auditory cortex.

Auditory Cortex

One of the first studies of the responsiveness of nerve cells in the auditory system to tones with rapidly changing frequency [229] showed that such neurons often did not respond to steady tones but only responded to rapid changes in frequency. Neurons in the auditory cortex also responded preferentially to one phase of the waveform of frequency modulation of pure tones that were sinusoidally frequency modulated (Fig. 8.26A), but the response adapted rapidly to steady tones (Fig. 8.26B). These results were obtained from cells in the primary auditory cortex in unanesthetized cats [229].

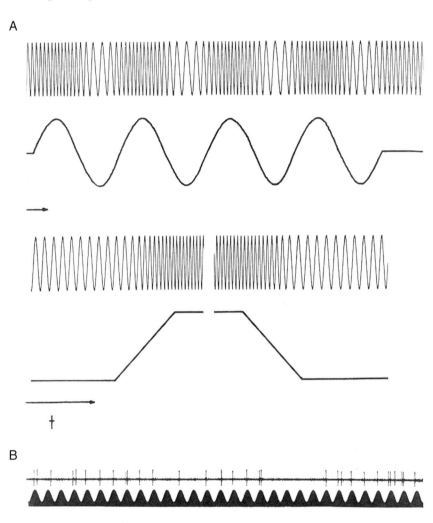

FIGURE 8.26 (A) Illustration of a sinusoidal frequency-modulated tone and a ramp-modulated tone. (B) Response from a cell in the auditory cerebral cortex in response to a tone at 11.6 kHz and 75 dB SPL that was within the cell's response area. The tone was frequency modulated as indicated by the modulation shown below the discharges (the frequency changed ±4.75%). The time bar is 1 s, the amplitude calibration is 2 mV. (A and B are from Whitfield, I. C., and Evans, E. F. (1965). Responses of auditory cortical neurons to stimuli of changing frequency. *J. Neurophysiol.* **28**:655–672).

More recently, more extensive studies of the responses from nerve cells in the primary auditory cortex (AI) of the barbiturate-anesthetized cat [72, 73] showed a preference to sounds of rapidly changing frequency. Most neurons responded preferentially to tones the frequency of which changed at rates of 1 MHz/s or higher. These neurons thus seem to prefer rates of change in frequency that are in the same range as the most preferred range of neurons in the CN (cf. Fig. 8.22). It is known that barbiturate anesthesia suppresses neural activity in the auditory cortex and, therefore, these results must be interpreted carefully. However, the results of this study [71, 72] indicate that preference for rapid change in frequency may occur throughout the classical ascending auditory pathway.

NEURAL SELECTIVITY TO OTHER TEMPORAL PATTERNS OF SOUNDS

Most studies of coding of sounds in the auditory nervous system have focused on the frequency or spectrum of sounds. However, features that are not related to the spectrum of sounds are also coded in the neuronal response in the auditory nervous system.

TUNING TO DURATION OF SOUNDS

Neurons that respond in accordance with the duration of a sound were described in the IC of flying bats [13]. These neurons fired more nerve impulses when sounds of a specific duration were presented (Fig. 8.27). Most neurons in the IC responded only to the beginning of long tone bursts, but these neurons did not respond at all to tones of long duration. The duration to which the neurons responded with the largest number of discharges (best duration) varied from 1 to 30 ms. Application of bicuculline, a $GABA_A$ antagonist, eliminated the duration tuning, thus indicating that $GABA_A$ ergic inhibition is involved in the duration selectivity of these neurons.

TUNING TO TIME INTERVALS UNRELATED TO SPECTRUM

Some neurons in the CN respond to transient sounds with a single discharge, and the discharges follow repetitive stimuli up to a certain rate, above which they cease to fire [119]. These neurons that only fire once for each sound and

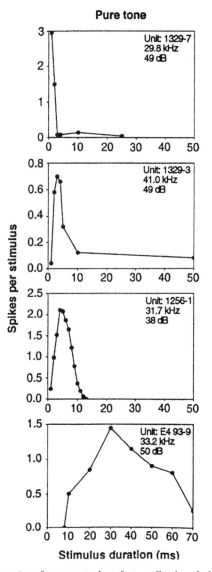

FIGURE 8.27 Duration tuning of neurons in the inferior colliculus of a bat (the big brown bat). The graphs show the responses of four different nerve cells to tone bursts of different duration (horizontal axis). Reprinted with permission from Casseday, J. H., Ehrlich, D., and Covey, E. Neural tuning for sound duration: Role of inhibitory mechanisms in the inferior colliculus. *Science* 264:847–850. Copyright 1994 American Association for the Advancement of Science.

cease firing above a certain rate function like they were "tuned" to the time pattern of a sound independent on the spectrum of the stimulus sound (Fig. 6.16). This is supported by the fact that reversing the polarity of every other click (from condensation to rarefaction) changes the spectrum but does not change these neurons' response pattern. These neurons also respond to high-frequency tone bursts with one discharge for each tone burst, again an indication that their tuning is not related to the spectrum of the sound to which they respond. These neurons thus replicate a precise timing, and since they only fire once for each sound, and cease firing above a certain rate, they become "tuned" to the time pattern of a sound in a similar way as auditory nerve fibers are tuned to the frequency (spectrum) of a sound (Fig. 6.16). When tone bursts of different duration are used as stimuli it becomes evident that the upper rate at which these "transient" neurons fire is related to the duration of the silent period between sounds (Fig. 8.28). This indicates that these CN cells are thus not "tuned" to the repetition rate of sounds but rather to the silent interval between two sounds. That specificity to the silent period between two sounds may be a result of inhibition that is released a certain time after a cell is activated.

> The selectivity to duration of silence may be similar to the inhibition-based variable delays which Suga and coworkers suggested could explain the bat's ability to discriminate delays between transmittal of a sound and receiving of an echo. Some neurons in the IC of the flying bat respond preferentially to sound with certain intervals. Yan and Suga [232] showed that the response of such "delay-tuned" neurons is enhanced by electrical stimulation of neurons in the auditory cortex. This means that a descending tract (corticofugal system, see Chapter 5 and Fig. 5.11) can affect the tuning to time intervals of neurons in the IC. In other studies the same authors [233] showed that neurons in the medial geniculate body are more sharply tuned to delays than the delay-tuned neurons in the IC. The corticofugal descending system together with the classical ascending auditory system constitute a closed feedback loop that may produce self-adjustment of some forms of sound analysis. These results were obtained in the bat, but similar analysis may occur in the auditory system of other mammals including humans.

Such duration tuning may be important for echolocating bats because they change the duration of the echolocation sound under different circumstances. It may also be important for discriminating communication sounds including speech sounds, which have a variety of durations. It may also help to filter out nonspeech background sounds where the duration of silence may be random- or outside the duration-range of speech sounds.

SPATIAL ORGANIZATION TO FEATURES OTHER THAN FREQUENCY

Tonotopic organization of neurons is prominent from the auditory nerve to the primary auditory cortex. However, neurons in the nuclei of the classical

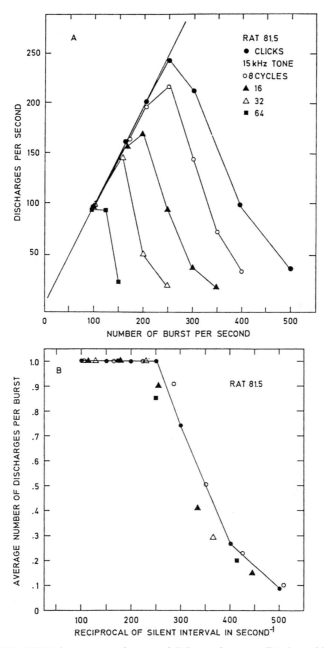

FIGURE 8.28 (A) Discharge rate as a function of click rate of a nerve cell in the cochlear nucleus that responds to each transient sound in the same way as that illustrated in Fig. 5.16, right column. Response to tone bursts of different duration (given in number of cycles of the 15-kHz stimulus tone) are shown. (B) Same data as in the upper graph replotted as a function of the inverse of the silent period (from Møller, A. R. (1969b). Unit responses in the rat cochlear nucleus to repetitive transient sounds. *Acta Physiol. Scand.* 75:542–551).

ascending auditory pathway are also spatially organized with regard to other features of sounds such as responding with a sustained burst of discharges to tone bursts or only responding to the onset, having mainly inhibitory or mainly excitatory response or having monotonic vs nonmonotonic intensity functions. Such organization, in addition to tonotopic organization has been demonstrated in the various divisions of the medial geniculate body [184]. Suga and his co-workers have demonstrated spatial organization of the cerebral cortex to a variety of features of complex sounds [219, 220]. He and his coworkers have studied the coding of complex sounds in flying bats extensively and these studies have revealed, among other findings, that the surface of the cortex is not only organized on in accordance with the frequency of sounds (tonotopic organization), but also other features of sounds have their own areas on the cortex.

> The flying bat emits short bursts of high-frequency tones. The flying bat uses these sounds for navigation as well as for localizing prey (echolocation). They determine the distance to an object that reflects their sound by determining the time it takes for the echo to arrive at their ear. This delay is represented on the surface of the bat's auditory cortex. The frequency of the reflected sound is slightly different from the emitted sound and that difference is directly related to the velocity of the bat relative to the reflecting object (Doppler shift).

Suga and his co-workers [214] have shown that changes in the frequency of sounds are specifically represented in the auditory cerebral cortex of the flying bat (Fig. 8.29). Neurons in certain anatomical areas of the cerebral cortex respond only to tones the frequency of which changes rapidly and neurons that respond to tones of constant frequency are separated spatially form those neurons. The navigational sounds of the bat are rich in harmonics and the different harmonics are represented in different areas of the auditory cortex. The bat's cortex thus "topographically" maps changes in frequency that correspond to the velocity of the bat in relation to the object, which reflects its navigational sound. The delay between the emission and receiving of the reflected sound corresponds to the distance to an object. The amplitude of the reflected sound (echo) is also represented on the surface of the auditory cortex. The amplitude of the echo is a measure of the subtended angle to the reflecting object, thus, together with the distance to the object, a measure of the relative size of the object. This means that different areas of the auditory cortex are devoted to processing different features of a sound, of which frequency (tonotopicity) is only one.

The features of the sounds described above are important for the bat but it is not difficult to recognize similarities between the bats' navigational sounds and communication sounds of many mammals. The importance of changes in frequency and amplitude and time interval between sounds for discrimination of sounds is similar for different species. The auditory system of other mam-

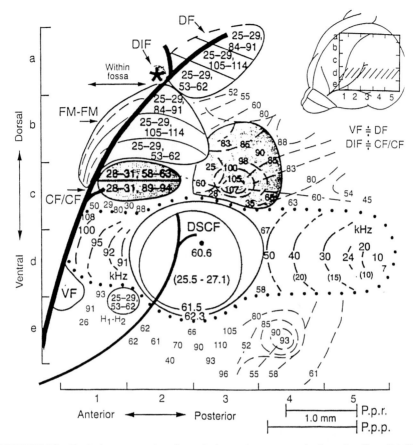

FIGURE 8.29 Cortical representation of sounds that are important to the flying bat (from [214]).

mals, including humans, may therefore perform similar kinds of analysis of sounds and it may be expected that the changes in frequency and amplitude of sounds and intervals between sounds are represented on the surface of the cortex of other mammals including humans.

CODING OF SOUND INTENSITY

Coding of sound intensity is much less obvious than coding of frequency and it has been much less studied despite the fact that it is a prominent feature of sound perception. Little is known about how sound intensity is coded in the auditory nervous system and it is not known how the threshold of hearing is established.

Coding of the intensity of a sound has been assumed to be related to the discharge rate in auditory nerve fibers. However, the discharge rate of most auditory nerve fibers does not increase as a function of stimulus intensity for continuous tones but reaches an asymptotic level (plateau) at stimulus levels that vary from fiber to fiber. In most fibers, such a plateau is reached only 20–40 dB above the threshold of the nerve fibers (Fig. 8.2A). The sound level at which the discharge rate does not increase with stimulus intensity is different for different frequencies relative to a fiber's CF. The discharge rate increases with increasing stimulus intensity at a steeper rate for tones the frequency of which are lower than the CF than for tones the frequency of which are higher than the CF. The maximal obtainable discharge rate is low in response to tones with frequencies well above the CF of a fiber [194]. Only a few auditory nerve fibers have discharge rates that increase monotonically with increasing stimulus intensity over a large range of stimulus intensities (Fig. 8.2B).

> The response patterns of neurons of the nuclei in the classical ascending auditory pathway are affected by anesthesia and yet only a few studies have been done in awake animals. While the effect of commonly used surgical anesthesia on the responses from auditory nerve fibers is small, anesthesia can have a noticeable effect on the response from cells in the CN [35] and the effect increases along the neural axis of the classical ascending auditory pathways. The results obtained in experimental animals under anesthesia may therefore not be a valid description of the response in awake animals.

CONCLUSIONS

Processing of sound in the auditory nervous system is complex and the results obtained using simple sounds such as pure tones do not describe the function of the system to other sounds. Studies done at threshold sound levels cannot be used as a basis for understanding the function of the auditory system at physiologic sound levels.

Enhancement of the response to sound with rapidly changing frequency or spectra, as demonstrated in the response of cells in the nuclei of the ascending auditory pathways, may be an example of a more general principle of signal processing in the auditory nervous system. The auditory system thus emphasizes important information and discards features that contain little or no information. This is important for achieving optimal processing of communication sounds and possibly in directional hearing, where psychoacoustic studies of the precedence effect indicate that the direction to a sound source is discriminated on the basis of the initial transient component of a sound [225]. The enhancement of the neural representation of changes in frequency (or spectrum) and amplitude that have been demonstrated to occur in the nuclei of the classical ascending auditory pathways is important because such sounds carry important survival information. The descending auditory systems optimize auditory analysis by adjusting the analysis in the nuclei of the classical ascending auditory pathway.

Hearing with Two Ears

ABSTRACT

1. Hearing with two ears (binaural hearing) is better that hearing with one ear (monaural hearing) in several ways.
   A. Hearing with two ears makes it possible to determine the direction to a sound source. Humans can discriminate a difference in azimuths (horizontal plane) of approximately 2° and of approximately 7° in the vertical plane.
   B. Hearing with two ears improves discrimination of sounds in noise.
   C. Hearing with two ears aids in selective listening to one speaker in an environment where several speakers are speaking at the same time.
2. Directional hearing is based on the difference between the neural activity evoked by sound that reaches the two ears.
3. The physical basis of directional hearing in the horizontal plane is the difference in the arrival time and the difference in the intensity of sounds at the two ears, both factors being a function of the azimuth.

243

4. The difference in the arrival time of sounds at the two ears is a result of the different travel time of sounds that exist at azimuths different from 0° and 180°.

5. The difference in the intensity of the sounds at the two ears is a result of the shadow and baffle effects of the head.

6. The delay between the arrival of sounds at the two ears can be detected by neurons that receive input from both ears with different delays and act as coincidence detectors.

7. The neural processing of interaural intensity difference is more complex and less studied than interaural time differences.

8. Neural processing of interaural time and intensity differences creates maps in the brain that represent auditory space.

9. The physical basis for directional hearing in the vertical plane is related to the effect of elevation on the spectrum of the sounds that reaches the ear canal. The outer ears and the shape of the head are important for directional hearing in the vertical plane.

INTRODUCTION

Hearing with two ears is associated with directional hearing, an ability to determine the direction to a sound source. This has had enormous importance for many species of vertebrate animals during evolution. Many examples of the importance of directional hearing are obvious and have undoubtedly resulted from evolutionary pressure. Many animals depend totally on the ability to determine the direction of sounds emitted by other animals, which may be foes or prey. The owl is one example of an animal that relies on sound in identifying its common food.

> The importance of directional hearing is not limited to birds and mammals. Thus, certain insects (night moths) that are prey for flying bats use directional hearing to avoid being eaten. These moths have only four cells that are sensitive to the echolocating sounds of bats. These cells are located on the thorax of the moth, two on each side. One pair has a low sound threshold and one pair has a high sound threshold. When only the most sensitive pair of these cells is activated, such as when a bat is far away, the moth turns away from the bat. This means that the moth must know the direction to the sound source, i.e., the bat. The moth obtains information about the direction to the sound source from the difference in the sound that reaches the two receptors. When the moth's high-threshold sound receptors are activated together with the most sensitive pair of receptors, indicating the bat is close to the moth, it closes its wings and falls to the ground. That is the fastest possible escape.

Binaural hearing is important not only because it makes it possible to determine the direction to a sound source. While many tasks can be done with only one ear, some cannot be performed at all. Many tasks are improved by listening with two ears and hearing with two ears is better overall than hearing with one ear. The discrimination of sounds in a noisy background is better with two ears than with one and hearing with two ears makes it possible to listen to one speaker in an environment where many people talk at the same time ("cocktail-party effect"). The advantage of hearing with two ears is greater if the masking noise in the two ears is different, for example, shifted by 180°. This "unmasking," caused by hearing with two ears, benefits from both a time (phase) difference between the masking sounds and from an intensity difference.

Hearing with two ears gives an impression of "space." Music, for example, is more enjoyable when perceived with two ears. This is perhaps most evident in listening to live music. Few people have the opportunity to do that and stereophonic reproduction of music is a good substitute because it produces the intensity and timing differences that are characteristic for binaural hearing.

Having two ears thus has advantages, but the advantages depend on both ears being equally sensitive. If hearing is impaired more in one ear than in the other ear, the advantages of hearing with two ears diminishes. People often become aware that they have an asymmetric hearing loss because they have difficulties in understanding speech where many other people are talking. Their ability to discriminate the target speaker is reduced due to the impaired directional hearing.

Electrophysiologic studies have been important in determining how the nervous system performs the task of determining the direction to a sound source. Making lesions in specific parts of the brain in animals has been used to find which parts of the brain are involved in specific tasks such as determining the direction to a sound source.

PHYSICAL BASIS FOR DIRECTIONAL HEARING

The ability to determine the direction of a sound source in the horizontal plane depends on the difference in the sound that reaches the two ears and to some extent on how the sound changes when the head is moved. The difference in intensity and the difference in time of arrival of a sound at the two ears are the physical bases for directional hearing, as described in Chapter 2. Differences in the spectrum of the sound at the two ears also play a role in binaural hearing and it is essential for directional hearing related to the vertical plane.

Both the differences in the arrival time and the difference in the intensity of the sound at the two ears are determined by the physical shape (acoustic properties) of the head and the outer ears, together with the direction of the sound source (Chapter 2, Figs. 2.3, 2.4, and 2.5). The sound arrives at the same time at the two ears when the head is facing the sound source (azimuth = 0°) and directly away from the sound source (azimuth = 180°). At any other azimuth, sounds reach the two ears at different times. Also, the sound intensity at the entrance of each of the ear canals is different except at 0° and 180° azimuth. The difference in the sound intensity has a more complex relationship to the azimuth than the interaural time difference (Chapter 2, Fig. 2.4). These differences in the sound that reaches the two ears are also the basis for improved speech discrimination in noise and the perception of "space," which is important when listening to sounds such as music.

In nature, a sound travels directly to the observer, whereas in rooms reflections from the walls, ceiling, and floor make sound reach the ears from many different directions. This complicates directional hearing. However, mainly the first sound that reaches the ears is used by the auditory system to determine the direction to a sound. This is known as the *precedence* effect and it reduces the effect of reflected waves that might cause confusion about the direction of a sound.

It is incompletely understood how we determine the direction to a sound in the vertical plane. Because the head is symmetric with regard to the medial plane, movements of the head in the vertical plane (elevation) does not create any time or intensity differences in the sound that reaches the two ears. The sounds that reach the two ears are therefore the same independent of the elevation of the sound source. However, the spectrum of a broadband sound that reaches the ear canal changes systematically with the elevation. This is in part due to the pinna (Chapter 2, Fig. 2.6) and is the physical basis for discrimination of the elevation of a sound source. The reason that we do so well in determining the elevation to a sound source overall is that most natural sounds are broadband sounds. Changes in the spectrum of a sound may also be responsible for our ability to determine the azimuth to a sound using only one ear.

BILATERAL INTERACTION IN THE AUDITORY SYSTEM

The neural substrate for detecting interaural time differences are neurons that receive input from both ears. This type of neuron can be found at several levels of the ascending auditory pathway. The most peripheral level at which this type of neuron has been found is in the superior olivary complex.

NEURAL MECHANISMS FOR DETECTION OF SMALL INTERAURAL TIME DIFFERENCES

Psychoacoustic experiments show that humans can detect interaural time differences of 5–10 μs under ideal conditions [223]. This corresponds to an angle of 1°–2° (the interaural time difference between arrival of sounds at the two ears for a 90° turn of the head is 650 μs) ([36] Chapter 2, Fig. 2.5). This means that neurophysiologic studies should be able to demonstrate a similar sensitivity to interaural time difference of neurons that are to be regarded as candidates for explaining the results of psychoacoustic studies.

Jeffress's Model of Binaural Hearing

Jeffress in 1948 [78] presented a hypothesis that describes how detection of interaural time differences may occur (Fig. 9.1). Jeffress's model consists of many neurons, each of which have two inputs, one from each ear. These neurons fire only when the inputs from the two ears arrive at the same time. Such neurons are known as "coincidence detectors." If the inputs from the two ears were delayed with the same amount (in the auditory nerve, cochlear nucleus, and in the fibertracts from the CN to the SOC), these neurons would fire only in response to sound that arrived at the two ears at the exact same time, thus coming from a source that is located directly in front or behind the observer. By delaying the neural activity from one ear more than that from the other ear sounds that arrive at the two ears with a certain time delay will evoke neural activity that occur at the same time at a coincidence neuron. If, e.g., the input from the left ear to such a coincidence neuron is delayed 50 μs more than that from the right ear, the two inputs will coincide when sounds arrive 50 μs earlier at the left ear than at the right year. This occurs when a sound source is located to the left of the midline, approximately 9° (90° corresponds to 650 μs) (see Chapter 2, Fig. 2.5). By making the delay from the two ears to different coincidence neurons different, sounds that arrive at the two ears at different intervals will activate different coincidence detectors. A specific neuron would then fire in response to sounds coming from a certain direction. The entire range of azimuths may be covered by delays that range from 0 to 650 μs. That range of delays could thus be realized by axons of different lengths and the entire range of azimuths can be covered.

This means that each coincidence neuron fires only when sound arrives at the two ears with a specific time delay corresponding to a certain azimuth. Which particular neuron fired would indicate to higher nervous centers what the delay was between the arrival of a sound at the left and the right ear and thus the azimuth of the sound source.

Correlation Models

The neural analysis of sounds with respect to the interaural time differences that Jeffress proposed is similar to cross-correlation analysis and investigators have found evidence that analysis similar to cross-correlation analysis occurs

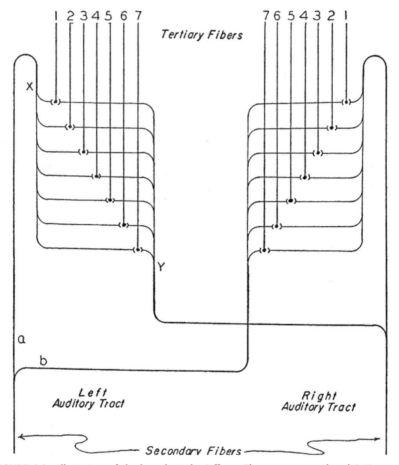

FIGURE 9.1 Illustration of the hypothesis by Jeffress. The neurons, numbered 1–7, receive separate ipsilateral and contralateral inputs, but the conduction time to reach these neurons varies according to the length of the axons that lead to these neurons. The two axons that lead to neuron 4 have the same length and so it would therefore respond when sounds arrive at the same time at the two ears. Neurons 1–3 would respond when sound to the left ear is delayed and neurons 5–7 will respond when the sound to the right ear is delayed (from Goldberg, J. M., and Brown, P. B. (1969). Response of binaural neurons of dog superior olivary complex to dichotic tonal stimuli: Some physiological mechanisms of sound localization. *J. Neurophysiol.* **32**:613–636).

in many different parts of the nervous system. The cross-correlation function is a measure of how similar two signals are when temporally shifted in relation to each other. If the two signals were identical, the correlation would have its highest value at zero delay. If the two signals were identical but temporally shifted, the correlation would have its highest value when they were shifted the same amount but in the reverse direction. If the two signals had components that were similar but shifted in time, in addition to a background of signals with no similarity, the cross-correlation would still have its maximal value at a delay that is equal to the time the two signals are shifted relative to each other. The time that the two signals are shifted in relation to each other can thus be determined by correlation analysis independent of the kind of signal.

> A common way of computing cross-correlation functions of two signals using digital computers consists of sampling the two signals, multiplying the amplitude of the two signals, sample by sample, and then adding the values. That process is repeated each time one of the two signals is shifted one sample relative to the other signal. The resulting cross-correlation function appears as series of values that are a function of the delay. However, a neural system that performs cross-correlation analysis does not obtain the correlation point by point, as is done when cross-correlation analysis is done by a computer. Instead, the nervous system is assumed to use an array of multipliers, one for each delay. The output of each multiplier neuron provides a continuous signal, which is the correlation at one delay as a function of time.

The correlation model of detection of interaural time differences is similar to the coincidence model of Jeffress. The only difference is that the coincidence detectors become multipliers in the correlation model. It seems unlikely that a neuron can perform accurate multiplication of its two inputs, but even approximations of multiplication will yield functions that have the basic properties of cross-correlation functions.

Coding of Directional Cues in the Superior Olivary Nucleus

Several investigators have found evidence that neurons in the superior olivary complex are selective with regard to interaural time intervals [235]. Some investigators of the responses of neurons in the medial superior olivary nucleus have found support for the hypothesis of Jeffress and identified neurons that function as "coincidence detectors" [57]. These investigators used low-frequency pure tones as stimuli and the neurons they studied responded with excitation to sounds presented to either ear (Excitatory–Excitatory, EE).

Histograms of the responses from a cell in the medial superior olivary nucleus that responds to stimulation of either ear illustrate the sensitivity of these neurons to interaural time differences. Histograms of the responses obtained when only the contralateral ear was stimulated (Fig. 9.2A) and only the ipsilateral ear (Fig. 9.2B) were compared with histograms obtained when both ears were stimulated with no delay between the sounds applied to the two ears (Fig. 9.2C) and when the sounds were presented at the two ears with different delays (Figs. 9.2D–9.2H). The results are summarized by curves showing the discharge rate and the vector strength as a function of the delay between the sounds at the two ears (Fig. 9.2I).[1]

---

[1] Vector strength is a quantitative measure that describes the phase locking of neural discharges to a specific waveform of stimulation.

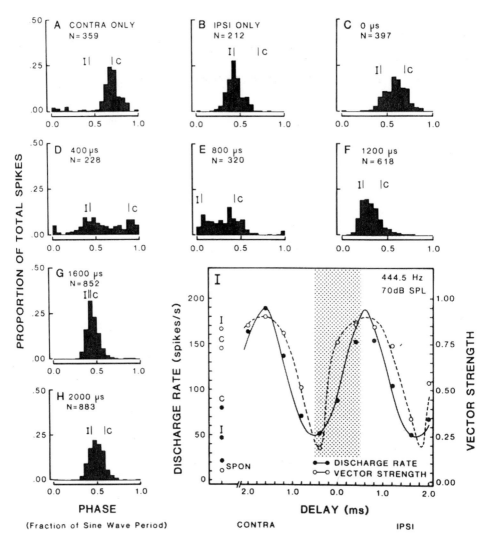

FIGURE 9.2 Period histograms of the response of a neuron in the medial superior olivary nucleus to a 444.5-Hz tone at 70 dB SL. The neuron responds to sound from both ears. (A) Histogram of the response to contralateral stimulation. (B) Response to ipsilateral stimulation. (C–H) Histograms of the response to bilateral stimulation where the delay between the sounds presented to the two ears was varied from 0 to 2000 μs. (I) Summary of the results from bilateral stimulation that shows the modulation of histograms of many nerve cells as a function of the delay between the sounds in the two ears. The modulation is expressed in vector strength where 1.0 is approximately equal to 100% modulation. (from Goldberg, J. M., and Brown, P. B. (1969). Response of binaural neurons of dog superior olivary complex to dichotic tonal stimuli: Some physiological mechanisms of sound localization. *J. Neurophysiol.* **32**:613–636).

The peak in the period histograms of the responses was delayed when the sound was applied to the contralateral ear compared with the histograms of the responses from the same cell to ipsilateral stimulation (Fig. 9.2). This observation was taken as an indication that the contralateral input to the neuron in question was delayed relative to the ipsilateral input, thus fulfilling one of the requirements of the hypothesis of coincidence detection for directional hearing, namely the existence of a delay in one of the inputs to neurons that receive input from both ears. Their results also indicated that the other requirement, namely the existence of a coincidence detector, is fulfilled in these neurons. Thus when sound was led to both ears, the neurons responded best when the sound to one ear was delayed by the amount that the period histograms of the monaural stimulation were shifted in relation to each other (Fig. 9.2).

Coding of Directional Cues in the Central Nucleus of the Inferior Colliculus

Also, neurons in the central nucleus of the inferior colliculus exhibit a similar response pattern as the neurons in the medial superior olivary nucleus, depicted in Fig. 9.2. Rose *et al.* [186] found neurons that responded best to specific interaural delays and different neurons had different preferred delays. Later, extensive studies by Yin and Kuwada [237] confirmed these studies and showed that many neurons in the central nucleus of the inferior colliculus responded best to binaural sounds when the sound in one ear was delayed a certain time relative to the sound in the other ear (Fig. 9.3).

Studies depicted in Figs. 9.2 and 9.3 used pure tones as stimuli, thus periodic signals. Recall that cross-correlation analysis of two periodic signals with the same frequency not only reveals the time the two signals are delayed in relation to each other but it also reveals the periodicity of the two signals. Thus the cross-correlation function of periodic signal has peaks between which the interval is equal to the period length of the signal. Using noise as stimuli instead of pure tones has the advantage that energy in a wide range of frequencies is presented at the same time. Yin and his colleagues [234] studied the response from neurons in the central nucleus of the inferior colliculus in the cat to broadband sounds presented at the two ears with different interaural delays (Fig. 9.4). Displays of the distributions of neurons in the central nucleus of the inferior colliculus as a function of delay to which they respond best shows that the delay of noise stimuli that gives the highest number of discharges falls within the physiologic range of interaural delays (Fig. 9.5), i.e., the range of delays produced when a sound source is located at different azimuths. This range is smaller for the cat than for humans because of the smaller head of the cat; these authors claim that it is approximately ± 400 μs in the cat.

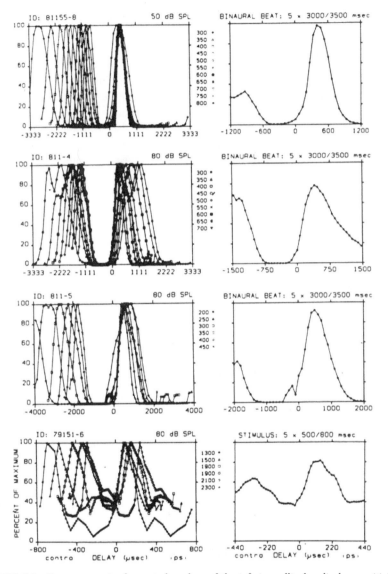

FIGURE 9.3   How neurons in the central nucleus of the inferior colliculus display sensitivity to interaural time delay. Each graph is from a single neuron. The individual curves in the left column show responses to tones of different frequency. The right-hand column shows the averaged delay curves for pure tones of different frequencies (from Yin, T. C. T., and Kuwada, S. (1984). Neuronal mechanisms of binaural interaction. *In* G. M. Edelman, W. E. Gall, and W. M. Crowan (Eds.), "Dynamic Aspects of Neocortical Function" (pp. 263–313). Wiley, New York).

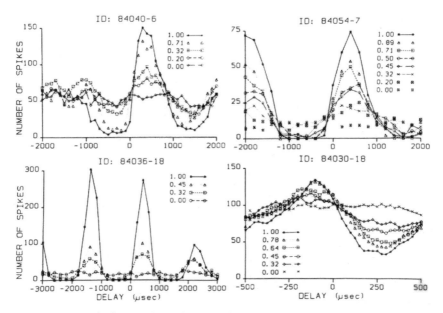

FIGURE 9.4 Results from similar studies as displayed in Figs. 9.2 and 9.3, but instead of using pure tones the results were obtained using broadband noise as stimuli. Responses from four different neurons are shown (from Yin, T. C. T., Chan, J. C. K., and Carney, L. H. (1987). Effects of interaural time delays of noise stimuli on low frequency cells in the cat's inferior colliculus. III. Evidence for cross-correlation. *J. Neurophysiol.* 58:562–583).

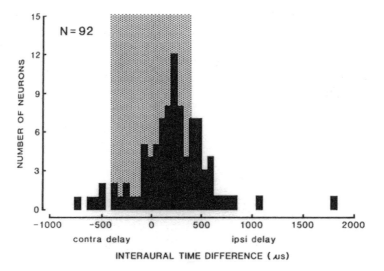

FIGURE 9.5 Distribution of interaural delays that gave the highest spike counts in 92 cells in the central nucleus of the inferior colliculus of the cat (from Yin, T. C. T., Chan, J. C. K., and Carney, L. H. (1987). Effects of interaural time delays of noise stimuli on low frequency cells in the cat's inferior colliculus. III. Evidence for cross-correlation. *J. Neurophysiol.* 58:562–583).

Some of the neurons studied displayed a periodicity in their response pattern to noise stimuli (Fig. 9.4), indicating that the cross-correlation of the neural activity from the two sides contained a periodic component, although the stimuli were broadband noise. That the discharge pattern of each auditory nerve fiber is excited by a band-pass-filtered version of the sound that reaches the ear explains the source of the periodicity in the discharge pattern of these neurons. Thus, band-pass filtered noise has a quasiperiodic waveform where the average period length is the inverse of the frequency to which the band-pass filter is tuned. Broadband sounds such as noise elicit a discharge pattern in single auditory nerve fibers that is phase locked to the band-pass-filtered version of the sound, the band pass filter being the cochlear filter. This explains the resemblance between the response to tones and noise in Figs. 9.3 and 9.4. The performance of these delay sensitive neurons to continuous sounds may thus be better explained by cross-correlation analysis than by coincidence detectors, but these two principles are in fact very similar.

It is not only continuous noise that causes such quasiperiodic excitation of nerve fibers, but also transient sounds lead to a damped oscillation (ringing) the frequency of which is equal to the frequency to which the basilar membrane is tuned. Any broadband sound may thus elicit a discharge pattern in auditory nerve fibers that is modulated by a quasiperiodic signal the frequency of which is equal to the frequency that the nerve fiber is tuned.

Auditory Cortex

The corpus callosum that connects the two sides of the auditory cortex is important for binaural hearing, for sound localization, and for fusing a sound image on the basis of sounds that reach the two ears [105]. Severing the corpus callosum impairs sound localization, which is even more impaired when one hemisphere (temporal lobe) is removed. There are some indications that elderly people have difficulties in fusing auditory images, which has been related to impairment of the connections between the auditory cortices.

Studies of the Barn Owl

Experiments in animals that are known to depend on their ability to discriminate direction of a sound source have supported the "variable delay and coincidence" hypothesis for determining the direction to a sound source. One such animal is the owl. Several investigators have studied one particular owl species, the barn owl. The nucleus, known as the mesencephalicus lateralis dorsalis (MLD), has been studied extensively in this animal. The MLD nucleus is believed to correspond to the inferior colliculus in mammals. Neurons in that nucleus have a high degree of interaural delay sensitivity (Fig. 9.6) with a similar periodicity in the response to noise, as shown in Figs. 9.4 and 9.5, from the central nucleus of the inferior colliculus of the cat. The fact that similar results can be obtained in a different species supports the generality of the results.

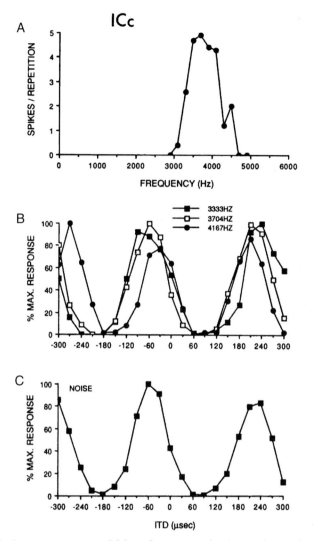

FIGURE 9.6   Sensitivity to interaural delays of a neuron in the MLD nucleus in the great horned owl. (A) Tuning of a neuron in the MLD. (B) Response to tonal stimulation with different interaural delays of that neuron. (C) Response to noise stimulation at different interaural delays (ITD) (from Volman, S. F., and Konishi, M. (1989). Spatial selectivity and binaural responses in the inferior colliculus of the great horned owl. *J. Neurosci.* 9:3083–3096).

Topographical Organization of Delay Sensitive Neurons

Neurons in the MLD of the barn owl are organized anatomically according to the interaural time difference to which they responded best [12, 222] (Fig. 9.7). Similar results were later found in neurons in the cat's medial superior olivary nucleus [235]. This organization is similar to that seen for frequency mapping, yet the significance of it is still being elucidated.

How Is the Temporal Code Preserved?

The use of the interaural time difference requires that the coding of the temporal pattern of the sound be preserved until the neural activity from the two ears can be compared. The demand on maintaining the temporal code is high because the differences in time of arrival of sounds at the two ears that can be detected is very small (5–10 μs), as shown in psychoacoustic experiments. It has been assumed that neurons, which compare the time of arrival of sounds at the two ears must be located peripherally in the ascending auditory nervous system because it is believed that temporal information is degraded in synaptic transmission. Some investigators, however, have found evidence that temporal information in fact may improve through synaptic transmission (see Chapter 7).

DETECTION OF INTERAURAL INTENSITY DIFFERENCES

While the interaural time difference varies in a regular way as a function of the azimuth, the interaural intensity difference is a much more complex function of azimuth (see Figs. 2.3 and 2.4 in Chapter 2). The difference in the intensity of the sounds that reach the two ears depends on the frequency of the sound, which complicates studies of the neural mechanisms that may be sensitive to interaural intensity differences.

Neurons that receive excitatory input from one ear and inhibitory input from the other ear are the most likely ones to be involved in detecting interaural intensity differences. Such neurons have been found in the superior olivary complex, mainly in the lateral superior olivary nucleus, in the dorsal nucleus of the lateral lemniscus, and the central nucleus of the inferior colliculus (Fig. 9.8).

Jeffress has suggested that intensity differences may be converted to time differences because the latency of the response of most neurons decreases with increasing sound intensity [78]. This would make it possible to use the same neural circuitry that detects interaural time differences for detecting interaural

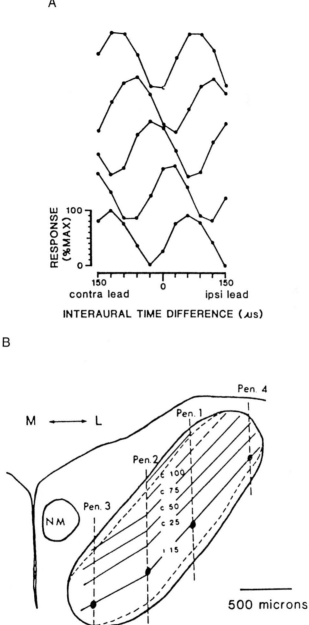

FIGURE 9.7 Topographical organization of interaural delay sensitivity in the nucleus laminaris in the barn owl (corresponding to the inferior colliculus in mammals). (A) Interaural delays for neurons at different depth. The curves show neurophonic responses, thus responses from many neurons but the results are similar to those in Figs. 9.5 and 9.6. (B) Anatomical localization of neurons with maximal response at different delays (from Sullivan, W. E. (1986). Processing of acoustic temporal patterns in barn owls and echolocating bats: Similar mechanisms for generation of neural place representations of auditory space. *Brain Behav. Evol.* **28**:109–121).

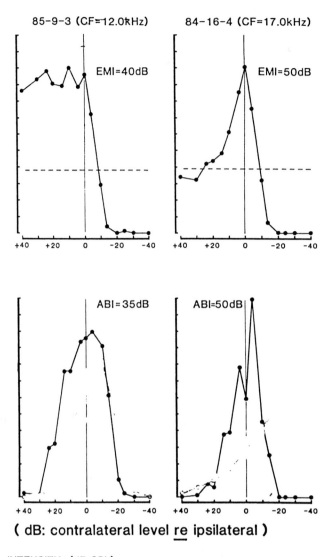

(dB: contralateral level <u>re</u> ipsilateral)

INTENSITY (dB SPL)

FIGURE 9.8 Illustration of the sensitivity of neurons in the central nucleus of the inferior colliculus in the cat to interaural intensity differences (from Irvine, D. R. F. (1987). Interaural intensity differences in the cat: Changes in sound pressure level at the two ears associated with azimuthal displacements in the frontal horizontal plane. *Hear. Res.* 26:267–286).

intensity differences. One would, however, think that intensity-dependent latency would add uncertainty to the detection of the interaural time differences. The assumption that the latency of the response decreases when the intensity of the stimulus is increased is, however, only valid for large changes in the intensity of sounds such as tone bursts, the intensities of which increase rapidly from below hearing threshold to a high intensity. The latency of small variations in the intensity of a sound such as the envelope of amplitude-modulated sounds does not change noticeably when the sound intensity is changed [126]. This means that the neural discharges evoked by the relatively small amplitude changes of most natural sounds will not change noticeably with the intensity of the sound. The intensity dependence on the latency of neural responses observed in experiments using tone bursts may thus be regarded mostly as an experimental artifact due to the use of sounds that do not resemble natural sounds.

DISCRIMINATION OF ELEVATION

So far, we have only discussed the directional hearing that is related to azimuth, thus only orientation in the horizontal plane. Little is known about the physical basis for discrimination of elevation. It is not possible to obtain information about the elevation of a sound source based on the difference between the sound that reaches the two ears because that is not affected by the elevation of a sound source. This is because the ears and the head are symmetrical around the midline in most animal species including humans. Despite that, psychophysical experiments show that we can discriminate approximately 7° in the vertical plane (elevation). It is assumed that it is based on changes in the spectrum of a sound as a function of the elevation of the sound source (Fig. 2.6). We discuss this in connection with perception of space.

REPRESENTATION OF AUDITORY SPACE

The ability of the auditory system to determine interaural time differences and interaural intensity differences is the basis for perception of space from sound. Mapping of a neuron's receptive field with regard to frequency and intensity is common (frequency tuning curves) (Chapter 6), but it is also possible to map a neuron's receptive field regarding the location of a sound source in space.

Neurons are anatomically organized with regard to the frequency they respond to best (tonotopic organization, see Chapter 6). Maps that relate auditory space also exist. Such maps are three-dimensional and therefore require information about the elevation besides the azimuth. The space maps are thus not a projection of the receptor epithelium such as tonotopic maps are but they are a result of processing of neural information regarding differences in the sounds that reach the two ears. Knudsen and his co-workers have called such maps "computational maps" to distinguish them from tonotopic maps, which are projections of an anatomical structure (the basilar membrane).

The regions in an auditory nucleus of the barn owl (the MLD nucleus), where maps of auditory space have been studied, differ from those that contain maps of the cochlea (tonotopic organization). Yet, the analysis of the spectrum of a sound is one factor that seems to be important for perception of space, particularly regarding elevation. How these two regions are connected remains to be elucidated.

Inferior Colliculus

Some investigators believe that the region of the MLD of the barn owl that contains a space map is homologous to the external nucleus of the inferior colliculus in mammals. The neurons that are located in the external nucleus of the inferior colliculus and which exhibit narrow receptive fields with regard to space most resemble the MLD space map. The tonotopically organized area of the MLD may then correspond to the central nucleus of the inferior colliculus in mammalians, which has been studied extensively. Recall that neurons in the central nucleus of the inferior colliculus are narrowly tuned and are tonotopically organized, while neurons in the external nucleus are broadly tuned and have no distinct tonotopical organization (Chapter 6). Other investigators [76] are skeptical about assuming such homology between the MLD and the inferior colliculus.

Superior Colliculus

Some neurons in the mammalian superior colliculus are sensitive to interaural intensity differences (Fig. 9.9A). The superior colliculus is normally not associated with auditory pathways but many of its neurons receive auditory input from the inferior colliculus. Neurons in the superior colliculus that respond to sound are organized topographically according to their sensitivity to interaural intensity difference (Fig. 9.9B). Visual and auditory perception of space are integrated in the superior colliculus, as the superior colliculus is involved in righting and movements of the eyes and of the head. The superior colliculus also coordinates other sensory inputs such as somatosensory input and input from the vestibular organ. The fact that many sensory systems converge onto the superior colliculus may explain why other senses can take over to create perception of space if one sense is impaired.

A
(82–42)

B
(82–31)

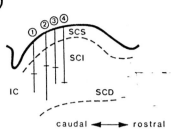

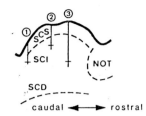

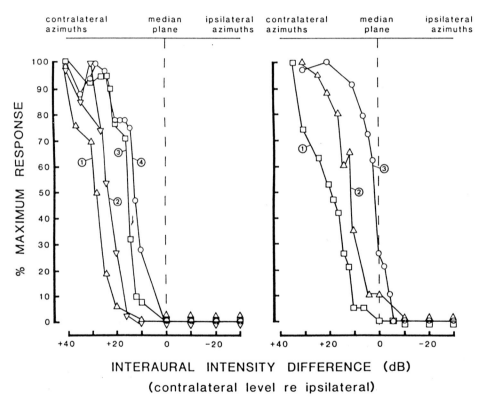

INTERAURAL INTENSITY DIFFERENCE (dB)

(contralateral level re ipsilateral)

FIGURE 9.9   Sensitivity to interaural intensity difference of neurons in the superior colliculus of the cat. (A) Normalized interaural intensity difference sensitivity of cells at different locations in the deep layer of the superior colliculus. The neurons were excited from one ear and inhibited from the other ear. The stimuli were noise bursts at 60 or 70 dB SPL. (B) Locations of the neurons depicted in A. Abbreviations: NOT, nucleus of the optic tract; SCD, deep layer of the superior colliculus (from Wise, L. Z., and Irvine, D. R. F. (1985). Topographic organization of interaural intensity difference sensitivity in deep layers of cat superior colliculus: Implications for auditory spatial representation. *J. Neurophysiol.* 54:185–211).

Some neurons in the superior colliculus respond to sound from a limited region of space. Such neurons have sharp peaked response curves regarding interaural time and interaural intensity (Fig. 9.10). These neurons are silent in response to binaural sounds outside these regions and to monaural sounds. That means that some neurons only respond when the sound comes from a certain place in space (three-dimensional direction). Such neurons may have been overlooked in conventional neurophysiologic experiments because it is common to search for neurons using monaural "search" sounds while a microelectrode is advanced through a nucleus. The neurons that showed the

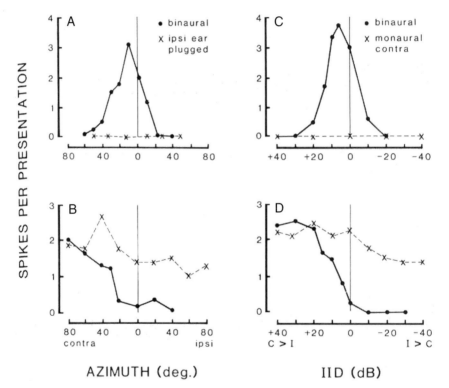

FIGURE 9.10 Responses from neurons in the deep layer of the superior colliculus of the cat. (A and B) Responses of two different cells to binaural sound and monaural (dashed lines, one ear plugged). (C and D) Response as a function of interaural intensity difference in two other neurons from the superior colliculus (from Irvine, D. R. F. (1992). Physiology of the auditory brainstem. *In* Popper, A. N., and Fay, R. R. (Eds.), "The Mammalian Auditory Pathway: Neurophysiology," (pp. 153–231), New York, Springer, and Middlebrooks, J. C. (1987). Binaural mechanisms of spatial tuning in the cat's superior colliculus distinguished using monaural occlusion. *J. Neurophysiol.* **57**:688–701).

greatest spatial selectivity (Figs. 9.10A and 9.10C) did not respond to monaural stimulation at all. The similarities between the responses of the neurons from the superior colliculus (Fig. 9.10) and those from the MLD in the barn owl are striking. The optic tectum in the barn owl (corresponding to the superior colliculus in mammalians) also contains maps of auditory space. As in mammals, most neurons in these nuclei receive input both from the visual and the auditory systems, as mentioned above.

CHAPTER 10

Electrical Potentials in the Auditory Nervous System

ABSTRACT

1. All neural structures of the ascending auditory pathway generate sound evoked electrical potentials that can be recorded by an electrode placed on the respective structure.

2. Compound action potentials (CAP) recorded directly from the intracranial portion of the auditory nerve in small animals are different from those recorded in humans because the eighth cranial nerve is longer in the humans than in small animals (2.5 cm in humans and approximately 0.8 cm in the cat).

3. In humans, the latency of the main negative peak of the CAP recorded with a monopolar electrode from the intracranial portion of the human auditory nerve is approximately one msec longer than that of the N_1 component of the action potential (AP) recorded from the ear.

4. Evoked potentials recorded with a bipolar electrode from a long nerve such as the human auditory nerve represent propagated neural activity.

5. The CAP recorded from the auditory nerve reflects synchronized activity in many nerve fibers and sustained activity does not result in any electrical potential.

6. If the propagated depolarization is brought to a halt, the potential recorded at that location reflects sustained as well as synchronized neural activity.

7. The responses recorded from the auditory nerve to continuos, low frequency sounds are the frequency following response (FFR).

8. The response recorded from the surface of a nucleus (such as the cochlear nucleus and the inferior colliculus) in responses to transient sounds has an initial positive-negative deflection that is followed by a slow deflection on which fast components are riding.

9. The initial positive-negative deflection of the response from a nucleus is generated by the termination of the nerve that serves as the input to the nucleus. The slow deflection that follows is generated by dendrites and the fast components riding on the slow wave are somaspikes generated by firings of nerve cells.

INTRODUCTION

Evoked potentials recorded directly from a nerve or a nucleus are known as near-field potentials, whereas the far-field potentials are the evoked potentials that can be recorded at a (large) distance from the active neural structures. The near-field potentials have large amplitudes and usually represent the neural activity in only one structure, whereas far-field potentials, such as the brainstem stem auditory evoked potentials (BAEP) that are discussed in Chapter 11, have small amplitudes and often contributions from many neural structures as well as muscles. Electrical potentials recorded directly from exposed structures of the ascending auditory pathways have helped to understand how far-field auditory evoked potentials are generated (see Chapter 11) and how the human auditory nervous system differs from that of commonly used experimental animals. Intracranial recordings have also contributed to insight into the normal and the pathologic functions of the auditory nerve and the cochlea. Recordings of evoked potentials generated by different parts of the auditory nervous system are also important in intraoperative neurophysiologic monitoring for reducing the risks of surgically induced injuries.

This chapter concerns the electrical potentials that can be recorded directly from structures of the classical ascending auditory pathway in response to sound stimulation. We first discuss evoked potentials recorded directly from the auditory nerve and then discuss responses recorded from nuclei of the ascending auditory pathway.

RECORDINGS FROM THE AUDITORY NERVE

STUDIES IN ANIMALS

The waveform of the compound action potentials (CAP) in response to click stimulation recorded from the intracranial portion of the eighth cranial nerve using a monopolar recording electrode typically has two negative peaks (N_1, N_2) (Fig. 10.1), thus similar to the AP recorded from the round window of the cochlea as described in Chapter 4.[1]

The length of the auditory nerve in animals in a small animal such as the cat is approximately 0.8 cm [45]. In the cat the latency of the N_1 in the response recorded from the auditory nerve in the internal auditory meatus is approximately 0.2 ms longer than that of the AP recorded from the round window (Fig. 10.2 [91]), which corresponds to the travel time in the auditory nerve from the ear to the recording site.

> Because the auditory nerve in small animals is very short, any recording site on the auditory nerve will be close to the cochlea and the cochlear nucleus. Potentials that originate in the cochlea and the cochlear nucleus are therefore conducted with little attenuation to the recording site by passive conduction in the VIIIth cranial nerve and the surrounding fluid. Intracranial recordings from the auditory nerve using a monopolar recording electrode will therefore not only yield potentials generated in the auditory nerve but also potentials that originate in the cochlea (mostly CM) and in the cochlear nucleus. These passively conducted potentials thus do not depend on the nerve being able to conduct propagated neural activity (depolarization). (Passive conduction is also the reason that recordings from the cochlea in small animals contain potentials that originate in the cochlear nucleus, as was discussed in Chapter 4.)
>
> The contributions from the ear and the cochlear nucleus to the responses recorded from the auditory nerve can be reduced by using bipolar recording techniques. Some investigators [171] have used a concentric electrode for recording from the intracranial portion of the auditory nerve to reduce the contamination of the neural response by the CM. However, a concentric electrode, consisting of a sleeve with an insulated wire inside, does not provide true bipolar recording because the two electrodes (the center core and the sleeve) do not have identical electrical properties. A concentric recording electrode is much more spatially selective than a monopolar electrode and the response recorded from the internal auditory meatus using a concentric electrode has no visible CM component (Fig. 10.2).

The most commonly used stimuli in connection with recordings of the CAP from the intracranial portion of the auditory nerve have been clicks or short bursts of tones or noise. Several studies have shown that the amplitude of the

---

[1] In this chapter, we use the notion compound action potentials (CAP) for the potentials recorded from the exposed auditory nerve, although they are similar to the potentials that are recorded from the round window of the cochlea, which are called action potentials (AP) (Chapter 4).

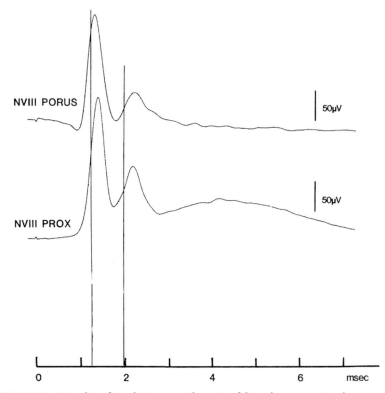

FIGURE 10.1 Recordings from the intracranial portion of the auditory nerve in a rhesus monkey, at two different positions, near the porus acousticus and near the brainstem. The stimuli were clicks presented at 107 dB PeSPL (peak equivalent sound pressure level) and at a rate of 10 pps. Reprinted from *Electroencephalogr. Clin. Neurophysiol.* **65**, Møller, A. R., and Burgess, J. E., Neural generators of the brain-stem auditory evoked potentials (BAEPs) in the rhesus monkey, pp. 361–372, copyright (1986), with permission from Elsevier Science.

CAP response increases with increasing stimulus level in a fashion similar to that of the AP recorded from the round window of the cochlea. The main reason for this is that the stimuli cause more nerve fibers to fire as the stimulus intensity is increased. The latency of the response decreases with increasing stimulus intensity, mainly because the generator potentials in the cochlear hair cells rise more rapidly at high stimulus intensities than at low stimulus intensities [126]. Cochlear nonlinearity also affects the latency differently at different stimulus intensities (see Chapter 3), which contributes to the dependence of the latency on the stimulus intensity [133]. The conduction velocity of nerve fibers and the synaptic delays are independent of the level of excitation and thus do not contribute to the intensity dependence of the latency of the CAP recorded from the auditory nerve.

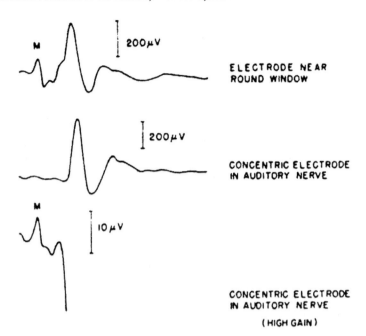

200μV

**ELECTRODE NEAR
ROUND WINDOW**

200μV

**CONCENTRIC ELECTRODE
IN AUDITORY NERVE**

10μV

**CONCENTRIC ELECTRODE
IN AUDITORY NERVE**

(HIGH GAIN)

FIGURE 10.2   Comparison between recording from the round window of the cochlea and from the intracranial portion of the auditory nerve in a cat using a concentric electrode. The stimulation was clicks. The M is the cochlear microphonic potential (from Peake, W. T., Goldstein, M. H., and Kiang, N. Y.-S. (1962). Responses of the auditory nerve to repetitive stimuli. *J. Acoust. Soc. Am.* **34**:562–570).

When the rate with which the stimuli (clicks or tone bursts) are presented is increased above a certain rate the amplitude of the CAP decreases (Fig. 10.3). Above a certain stimulus rate the responses elicited by the individual stimuli overlap, and the amplitude of the peaks may increase because the N_1 peak of one response coincides with the N_2 peak of the previous response (Fig. 10.3). When the rate of the stimulus presentation is increased beyond approximately 700 pulses per second (pps) the amplitude of the response decreases rapidly. The latency of the response increases slightly when the stimulus rate is increased.

Cut-End Potentials and the Response to Steady Sounds

When the VIIIth cranial nerve is severed, the response to transient stimulation recorded by a single electrode placed at the stump of the nerve is a single monophasic positive deflection rather than the double negative peaks (N_1 and N_2) normally seen when recording from the intact auditory nerve. Such "cut end" potential has been used to record potentials that reflect steady state neural activity in animal

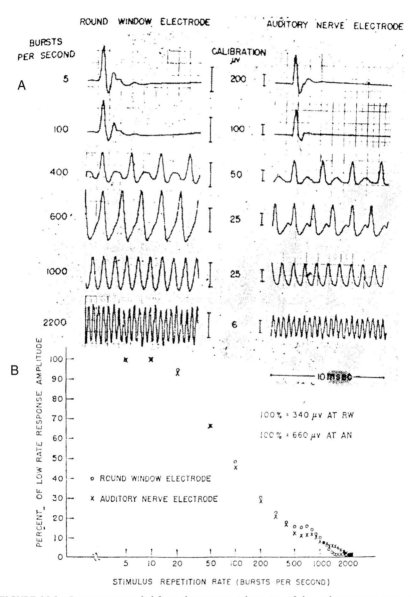

FIGURE 10.3 Responses recorded from the intracranial portion of the auditory nerve in a cat to clicks presented at different rates. (A) Waveform of the recorded response to 0.1-ms noise bursts. (B) Peak-to-peak amplitude of the responses to noise bursts (from Peake, W. T., Goldstein, M. H., and Kiang, N. Y.-S. (1962). Responses of the auditory nerve to repetitive stimuli. *J. Acoust. Soc. Am.* **34:**562–570).

experiments in studies of the function of the cochlea [23] (Fig. 10.4). To understand the advantages of the use the "cut-end" potentials in studies of the response from the auditory nerve it is useful to recall the relationship between the firing in single auditory nerve fibers and the response from a nerve recorded by a monopolar recording electrode. As shown by Goldstein [59] the response from a normal nerve is a convolution integral of the poststimulus histogram of the responses from single nerve fibers and the waveform of a single discharge of a nerve fiber, assuming that the waveforms of such discharges are all the same. Since the discharge of single nerve fibers are biphasic, the sum of randomly distributed nerve impulses will be zero. This is the reason that steady firing does not produce any evoked potentials when recording with a monopolar electrode from a long nerve. It was shown a long time ago that the nerve impulses are monophasic at the end of a nerve. That means that their mean value is different from zero, and consequently a steady potential can be recorded from the cut end of a nerve. The amplitude of such a potential is directly proportional to the firing rate of all fibers in a nerve.

Recordings from the cut end of the auditory nerve have been called "peristimulus compound action potentials" (PCAP) [23] (Fig. 10.5). These investigators created a nerve block in the intracranial portion of the VIIIth cranial nerve of anesthetized animals by applying pressure to the nerve. In such recordings a tone bursts elicit a steady potential throughout the duration of the tone, thus similar to poststimulus time histograms of neural activity in single nerve fibers [23] (Fig. 10.5).

RECORDING DIRECTLY FROM THE HUMAN AUDITORY NERVE

Several investigators [69, 145, 151, 209] reported at about the same time that the latency of the CAP recorded from the exposed intracranial portion of the

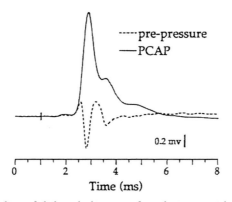

FIGURE 10.4 Recordings of click-evoked response from the intracranial potion of the auditory nerve in a chinchilla before (dashed lines) and after propagated neural activity was arrested by compression (solid line). The onset of the click is marked by a vertical line. Note that a positive potential is shown as an upward deflection in this graph (from Doucet, J. R., and Relkin, E. M. (1995). The perstimulus compound action potential: A new method for recording a compound potential from the chinchilla auditory nerve. *Aud. Neurosci.* 1:151–168).

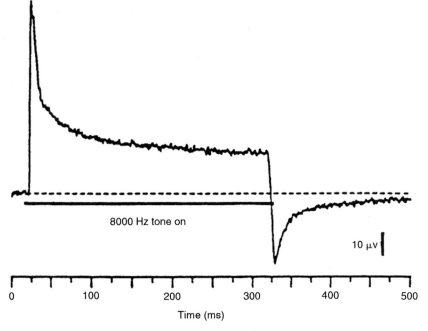

8000 Hz tone on

10 μv

0   100   200   300   400   500

Time (ms)

FIGURE 10.5   The PCAP recorded in response to a tone burst, 308 ms long at 20 dB. The broken level is the potential that is recorded between the stimulation (baseline). Note that positivity is an upward deflection (from Doucet, J. R., and Relkin, E. M. (1995). The perstimulus compound action potential: A new method for recording a compound potential from the chinchilla auditory nerve. *Aud. Neurosci.* **1**:151–168).

auditory nerve in human is longer than it is in animals when similarly recorded. The latency of the main negative peak of the CAP recorded from the intracranial portion of the auditory nerve in response to loud clicks is approximately 2.8 ms in response to loud clicks [152]. This is because the VIIIth cranial nerve in humans is 2.5 cm [97]. Thus much longer than in animals such as the cat (approximately 0.8 cm, [45]). The latency of the CAP recorded from the intracranial portion of the human auditory nerve is approximately 1 ms longer than the AP component of the ECoG recorded from the ear. [Compare that to a difference of approximately 0.2 ms in the cat (Fig. 10.2)].

Historical Background

It was probably Ruben and Walker [190] who first reported on recordings from the exposed intracranial portion of the VIIIth cranial nerve. These investigators recorded click-evoked CAPs from the auditory nerve during sectioning of the VIIIth nerve for Ménière's disease using a retromastoid approach to the cerebellopontine angle (Fig. 10.6). The waveform of the recorded potentials was complex with several peaks and valleys. Ruben and his coauthor suggested that the responses had contributions from cells of the cochlear nucleus. Examination of their recordings (Fig. 10.6) indicates that the intracranially recorded CAP had a longer latency in humans than in the cat, but the authors did not speculate on the reason for the longer latency. (Accurate assessment of the latency of the potentials from their published recordings is not possible because the record does not show the time the stimulus was applied.)

Monopolar Recordings

Another consequence of the human auditory nerve being longer than that of animals is that it can be regarded as a long nerve with regard to the sound-evoked potentials that can be recorded from its intracranial portion with a monopolar recording electrode. The length of the nerve between the point where it emerges

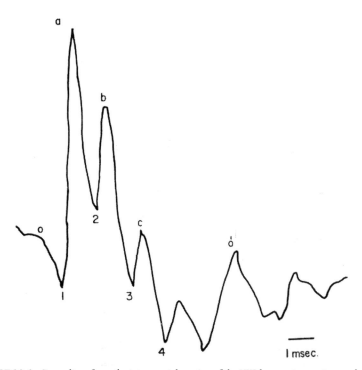

FIGURE 10.6   Recordings from the intracranial portion of the VIIIth nerve in a patient undergoing an operation for Meniere's disease (from Ruben, R. J., and Walker, A. E. (1963). The VIIIth nerve action potential in Ménière's disease. *Laryngoscope* 11:1456–1464).

into the skull cavity (from the porous acousticus) to its entrance into the brain-stem is approximately 1 cm. In individuals with normal hearing a monopolar electrode placed on the exposed intracranial portion of the VIIIth nerve records a triphasic potential in response to click stimulation (Fig. 10.7A), as is typical for recordings with a monopolar electrode from a long nerve. Similar to what is seen in studies in animals the amplitude of the main peak of the CAP recorded from the exposed human auditory nerve increases with increasing stimulus intensity (Fig. 10.7A). The latency of the response decreases with increasing stimulus intensity (Fig. 10.7B) for the same reasons as mentioned above when discussing the responses from the auditory nerve in animals.

> Recording from the intracranial portion of the auditory nerve requires that the VIIIth cranial nerve is exposed in its course in the cerebellopontine angle. This occurs in some operations such as those to treat vascular compression of cranial nerves. Whenever such recordings are done, it must be assured that the auditory nerve is not injured by the surgical dissection necessary to expose the nerve. Therefore, BAEP must be recorded during such dissections to monitor the conduction velocity in the auditory nerve (for details about monitoring neural conduction in the auditory nerve, see [135]).

Recording from a Long Nerve

> A monopolar recording electrode placed on a long nerve along which an area of depolarization propagates will record a characteristic triphasic potential (Fig. 10.8). The initial positive deflection is generated as the area of depolarization approaches the recording electrode. The large negative deflection is generated when the area of depolarization passes directly under the recording electrode. The following small positivity is generated when the area of depolarization leaves the location of the recording electrode. If propagation of neural activity in such a nerve is brought to a halt, for instance by injury to the nerve, a monopolar electrode placed near that location would record a single positive potential. Such a potential is known as the "cut-end" potential and described by Lorente de No [109].

The responses from the exposed intracranial portion of the auditory nerve to short tone bursts have a waveform similar to the responses to click sounds, but the latencies are slightly longer (Fig. 10.9A) [145].

Recordings of auditory evoked potentials from the exposed auditory nerve in humans have helped in understanding some of the differences between the

---

FIGURE 10.7 (A) Typical compound action potentials directly recorded from the exposed intracranial portion of the VIIIth nerve in a patient with normal hearing. Responses to condensation (dashed lines) and rarefaction (solid lines) clicks are shown for different stimulus intensities (given in dB PeSPL). (B) Latency of the negative peak in the CAP shown in A. Reprinted from *Hear. Res.*, Volume 45, Møller, A. R., and Jho, H. D., Late components in the compound action potentials (CAP) recorded from the intracranial portion of the human eighth nerve, pp. 75–86, copyright 1990, with permission from Elsevier Science.

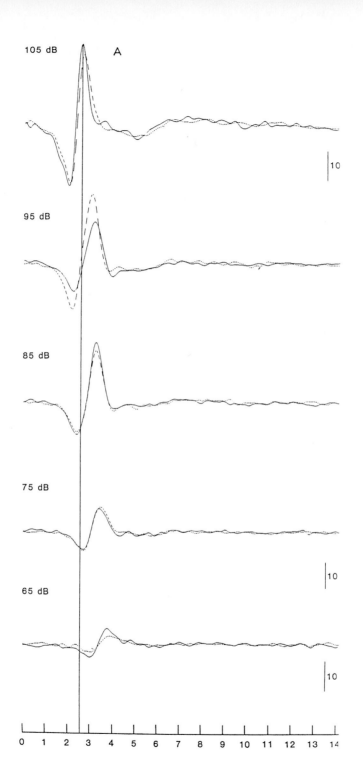

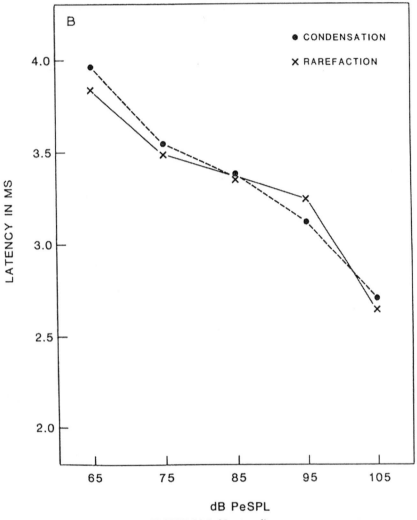

FIGURE 10.7 (*Continued*)

human auditory nervous system and that of small animals often used in studies of the auditory system [139, 141, 152].

Contribution from Other Sources

A recording electrode placed on the intracranial portion of the VIIIth nerve is far from the cochlea and will therefore not record any noticeable cochlear

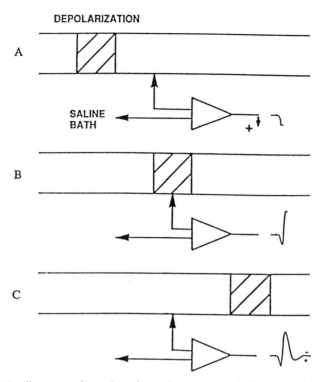

FIGURE 10.8 Illustration of recordings from a long nerve in which an area of depolarization travels from left to right, using a monopolar electrode (from Møller, A. R. (1995). "Intraoperative Neurophysiologic Monitoring." Harwood Academic Publishers, Luxembourg, with permission).

potentials. The total length of the auditory nerve in humans is approximately 2.5 cm and the length of the nerve between the point where it enters into the skull cavity from the porous acousticus to its entrance into the brainstem is approximately 1 cm. If the recording electrode is placed near the porous acousticus it will be approximately 1.5 cm from the cochlea and will therefore not record any noticeable potentials from the cochlea. With the same placement the electrode will be approximately 1 cm from the cochlear nucleus and the potentials generated in the cochlear nucleus will be attenuated before they reach the recording electrode provided that the VIIIth nerve in its intracranial course is submerged in fluid. If the VIIIth nerve is free of fluid in its intracranial course, it will act as an extension of the recording electrode and the recording electrode will record potentials from the cochlear nucleus of noticeable amplitude. Evoked potentials generated in the cochlear nucleus may be noticeable when recording from a location on the auditory nerve that is close to the brainstem, thus near the cochlear nucleus.

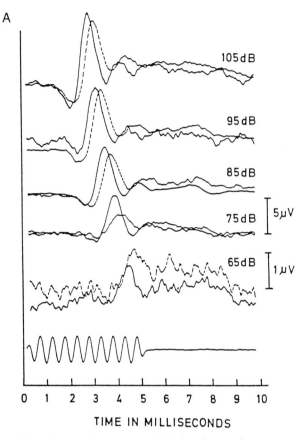

A

105 dB

95 dB

85 dB

75 dB   5 μV

65 dB   1 μV

0 1 2 3 4 5 6 7 8 9 10

TIME IN MILLISECONDS

FIGURE 10.9   (A) Similar recordings as in Fig. 10.7, but showing the responses to tone bursts recorded at two locations along the intracranial portion of the exposed auditory nerve. The solid lines are recordings close to the porous acousticus and the dashed lines are recordings from a location approximately 3 mm more central. The stimuli were short 2-kHz tone bursts. The sound pressure given is in dB PeSPL (from [145]). (B) The latency of the main negative peak of the CAP recorded from two different locations as shown in A (approximately 3 mm apart) on the exposed VIIIth nerve as a function of the stimulus intensity (from [149]).

In practice, it is thus possible to record evoked potentials from the intracranial portion of the auditory nerve in humans without noticeable contamination from potentials generated in the cochlea or the cochlear nucleus using a monopolar recording electrode if the electrode is placed appropriately. The fact that the latency of the response from the auditory nerve to high-intensity clicks increases when the recording electrode is moved from a location near the porous acousticus toward the brainstem (Fig. 10.9B) is an indication that at least the main portion of the

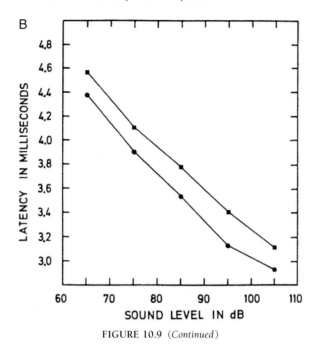

FIGURE 10.9 (*Continued*)

recorded potential results from propagated neural activity in the auditory nerve [145]. The latency of passively conducted potentials would not change when the recording electrode is moved along the auditory nerve, but their amplitude would decrease when moved away from their source. The response from the exposed intracranial portion of the VIIIth nerve to low-intensity click sounds often yields a slow deflection, which is probably generated in the cochlear nucleus and conducted passively in the auditory nerve to the site of recording. This slow component of the response is more pronounced at low stimulus intensities because its amplitude decreases at a slower rate with decreasing stimulus intensity than that of the faster components.

BIPOLAR RECORDINGS

A bipolar recording electrode placed on a nerve with one of its two tips located more peripheral than the other will, under ideal circumstances, only record propagated neural activity. The waveform of the compound action potential recorded from a nerve with a bipolar electrode is different from that recorded by a monopolar electrode and is more difficult to interpret.

The function of a bipolar recording electrode placed on the intracranial portion of the auditory nerve can be understood by assuming that the bipolar electrode consists of two monopolar electrodes, each one recording the potentials at two adjacent locations along the nerve (Fig. 10.10). The electrical potentials generated in a nerve by propagated neural activity appear with a slight time difference at the two tips of such a bipolar recording electrode. The time difference between the activity at the two tips of a bipolar electrode is the time it takes the neural activity to travel the distance between the two tips. The amplifier to which the electrodes are connected senses the difference between the electrical potentials that the two electrodes are recording. Under ideal circumstances, passively conducted potentials will appear equal at the two electrodes and thus not result in any output from the differential amplifier to which the electrodes are connected. To achieve such ideal performance, the two tips of the electrode must have identical recording properties and be placed on a nerve so that they both record from the same population of nerve fibers. While that is rarely achieved in practice, a bipolar electrode is less sensitive to potentials generated by stationary sources than a monopolar recording electrode. If the two tips of the bipolar recording electrode have different recording characteristics or are not placed exactly symmetrically on the nerve, passively conducted potentials will appear different at the two tips and thus generate an output from the amplifier to which the bipolar electrode is connected [139].

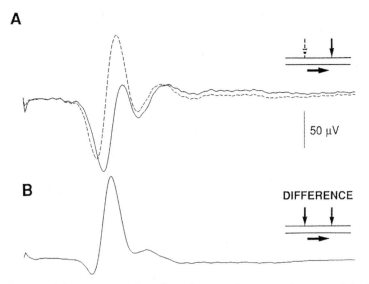

FIGURE 10.10 (A) Separate recordings from the exposed intracranial portion of the VIIIth cranial nerves with two electrodes placed approximately 1 mm apart. (B) The difference between the recordings by the two electrodes in A. Reprinted from *Electroencephalogr. Clin. Neurophysiol.* **92**, Møller, A. R., Colletti, V., and Fiorino, F. G., Neural conduction velocity of the human auditory nerve: Bipolar recordings from the exposed intracranial portion of the eighth nerve during vestibular nerve section, pp. 316–320, copyright (1994), with permission from Elsevier Science.

If no passively conducted potentials reach the recording electrodes, the response recorded by a bipolar recording electrode will be the same as the potentials recorded by a monopolar electrode from which is subtracted a delayed version of the response (Fig. 10.11). The difference between such a simulated bipolar recording and a real bipolar recording is a measure of the passively conducted potentials.

Comparison of bipolar and monopolar recordings from the exposed intracranial portion of the auditory cranial nerve [139] further supports the assumption that click-evoked potentials recorded from the auditory nerve with a monopolar recording electrode, at least at high stimulus intensities, is mainly the result of propagated neural activity.

More space is required for placing a bipolar recording electrode on a nerve compared with using a monopolar recording electrode, but the intracranial portion of the auditory nerve in the humans is sufficiently long to allow the use of bipolar recording electrodes.

CONDUCTION VELOCITY OF THE AUDITORY NERVE

The conduction velocity of the auditory nerve in humans has been determined from bipolar recordings directly from the exposed intracranial portion of the auditory nerve. The difference in the latency of the CAP recorded at two different locations on the exposed intracranial portion of the auditory nerve were used to determine the conduction velocity [139]. The value determined, approximately 20 m/s, is similar to what has been estimated from the fiber diameter of the auditory nerve fibers [103].

Practical Use of Recordings from the Exposed Intracranial Portion of the VIIIth Cranial Nerve

Direct recording of responses from the VIIIth cranial nerve is now in general use in monitoring neural conduction in the auditory nerve in patients undergoing operations in the cerebellopontine angle because such potentials can be interpreted nearly instantaneously [135] due to their large amplitudes. Changes in the function of the nerve from stretching or from slight surgical trauma can therefore be detected almost instantaneously from observation of such direct recordings because only a few responses need to be added (averaged) in order to obtain an interpretable record. (The same monitoring of neural conduction in the auditory nerve can be achieved by recording BAEP, but it takes much longer to obtain an interpretable record because of the small amplitude of the BAEP.)

Click-evoked compound action potentials recorded from the intracranial portion of the VIIIth nerve change systematically when the auditory nerve is injured such as from surgical manipulations or by heat from electrocoagulation [135]. Recorded centrally to the location of the lesion, the latency of the main negative peak of the CAP increases and the amplitude of that peak decreases and it becomes broader, indicating that the prolongation of conduction time is different for different nerve fibers (Fig. 10.12). More severe injury causes the amplitude of the initial positive deflection to increase, indicating neural block in some nerve fibers.

MONOPOLAR

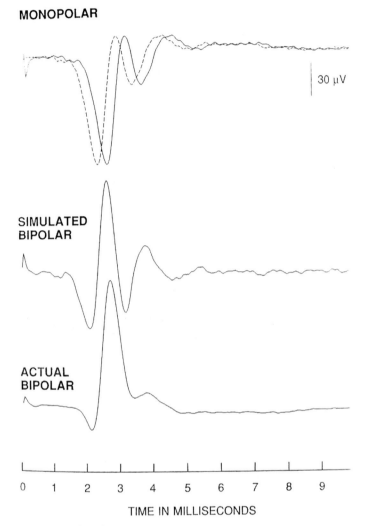

SIMULATED
BIPOLAR

ACTUAL
BIPOLAR

TIME IN MILLISECONDS

FIGURE 10.11   Recordings from the intracranial portion of the auditory nerve in a patient whose vestibular nerve was just cut. Rarefaction clicks presented at 98 dB PeSPL. *Top curves:* Monopolar recordings by the two tips of a bipolar electrode. *Middle curves:* Computed difference between the response recorded by one tip (monopolar recording) and the same response shifted in time with an amount that corresponds to the distance between the two tips of the bipolar electrode. Lower curves are the actual bipolar recording. Reprinted from *Electroencephalogr. Clin. Neurophysiol.* **92**, Møller, A. R., Colletti, V., and Fiorino, F. G., Neural conduction velocity of the human auditory nerve: Bipolar recordings from the exposed intracranial portion of the eighth nerve during vestibular nerve section, pp. 316–320, copyright (1994), with permission from Elsevier Science.

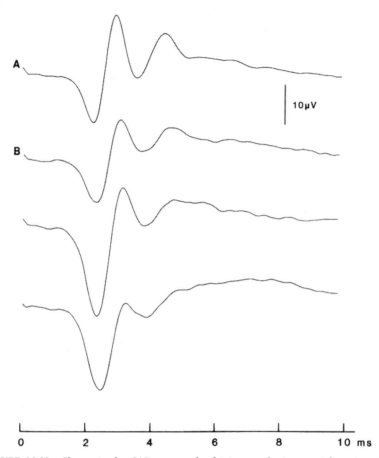

FIGURE 10.12   Change in the CAP as a result of injury to the intracranial portion of the auditory nerve in a patient undergoing an operation where the auditory nerve was heated by electrocoagulation. (A) Before coagulation. (B) Beginning of coagulation.

Frequency Following Response

The frequency following response (FFR), as the name indicates, is a response that follows the waveform of the stimulating sound. The FFR can be demonstrated in the response of the auditory nerve to low-frequency tones and tones that are amplitude modulated at low frequencies. The source of the FFR is phase-locked discharges in nerve fibers and nerve cells. Some investigators have named these potentials the neurophonic response. The FFR has been recorded from the auditory nerve in animals [133, 205, 206] and from the exposed intracranial portion of the auditory nerve in humans [140]. The FFR recorded from the human auditory nerve is similar to that in the cat recorded using bipolar electrodes (cat [205, 206]). When recorded directly from the exposed intracranial portion of the human auditory nerve the FFR (Fig. 10.13) is prominent in the frequency range from 500 to 1500 Hz [140].

> Studies of the FFR to low-frequency pure tones from the auditory nerve in animals are hampered by the contamination from cochlear microphonics that have similar waveforms as the neurophonic potentials. Snyder and Schreiner [206] reduced the contamination of the neural response from potentials generated in the cochlea by using a bipolar recording technique. The fact that the auditory nerve is longer in humans than in the cat makes it possible to record the FFR with a monopolar recording electrode without any noticeable contamination from cochlear potentials. That the FFR recorded from the human auditory nerve with a monopolar electrode is the result of propagated neural activity is supported by the finding that the recorded potentials shifted temporally when the recording electrode was moved along the VIIIth cranial nerve (Fig. 10.14).

The responses to low-frequency tones recorded from the human auditory nerve have two components, a frequency following response and a slow component [140] (Fig. 10.15). When the responses to opposite-phase tones were added, the frequency following response was canceled and only the slow potential was seen. When the responses to opposite-phase tones were subtracted the slow potential was canceled and only the frequency following response remained.

Recordings of the FFR from the auditory nerve in animals and in humans have contributed to the understanding of the function of the cochlea. At high stimulus intensities the frequency following responses are the results of excitation of the basilar membrane at a location that is more basal than the location tuned to the frequency of the stimulation [206]. This is a result of the nonlinearity of the basilar membrane vibration (Chapter 3).

> The waveform of the recorded responses to stimulation with a 500-Hz tone is a distorted sinewave (Fig. 10.16). As a first approximation, the waveform of the responses indicates that auditory nerve fibers are excited by the half-wave rectified stimulus sound, thus a deflection of the basilar membrane in one direction. The waveform of the response to high-intensity tones (104 dB SPL) is more complex than the response to tones of lower intensities and has a high content of second harmonics, thus similar to a full-wave rectified sinewave. This indicates that hair cells respond to deflection of the basilar membrane in both directions, thus supporting the findings in animal experiments that some inner hair cells respond to the condensation phase of a sound while other inner hair cells respond to the rarefaction phase [208, 246]. The distortion of the response to low-frequency pure tones could also be a result of what has been known as "peak splitting" [191, 200]. The distortion of the waveform of the responses from the human auditory nerve seems to be less than it is in the cat at the same sound pressure level. (In the studies of the responses

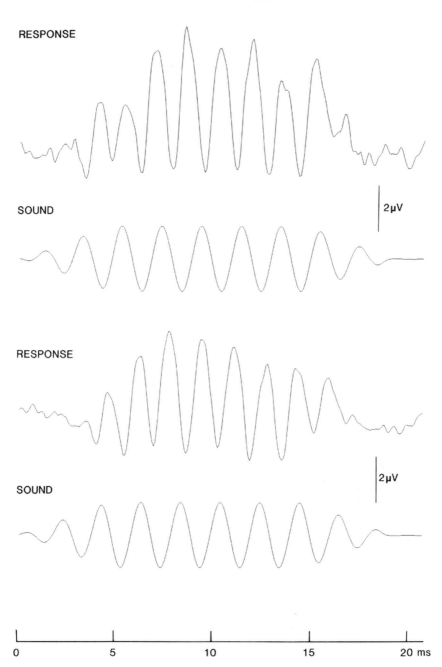

FIGURE 10.13 Responses recorded from the exposed intracranial portion of the auditory nerve to stimulation with 500-Hz tones at 113 dB SPL. Rarefaction of the sound is shown as an upward deflection. Reprinted from *Hear. Res.*, Volume 38, Møller, A. R., and Jho, H. D., Response from the exposed intracranial human auditory nerve to low-frequency tones: Basic characteristics, pp. 163–175, copyright 1989, with permission from Elsevier Science.

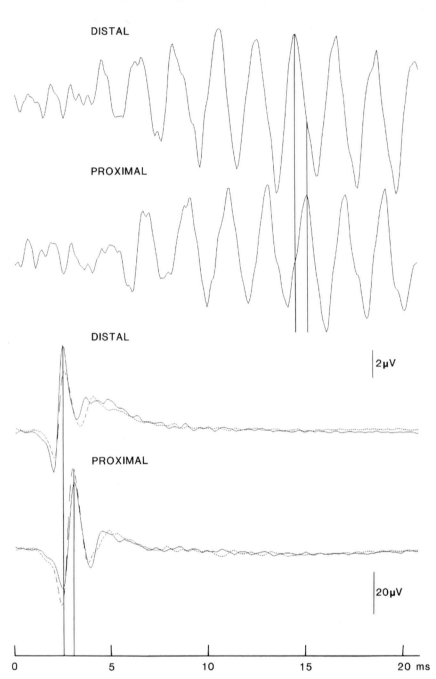

FIGURE 10.14 Comparison of the response to tone bursts and clicks recorded at two different locations along the exposed intracranial portion of the VIIIth cranial nerve. Reprinted from *Hear. Res.*, Volume 38, Møller, A. R., and Jho, H. D., Response from the exposed intracranial human auditory nerve to low-frequency tones: Basic characteristics, pp. 163–175, copyright 1989, with permission from Elsevier Science.

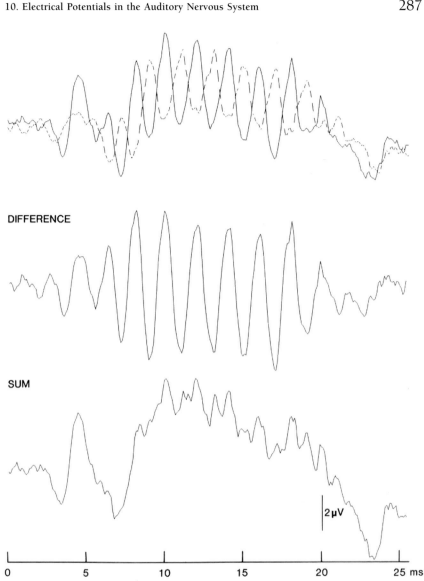

FIGURE 10.15 Similar recordings as in Fig. 10.14 but showing the responses of both polarities of the sound (500-Hz tone at 110 dB SPL) (top traces, solid and dashed lines). The difference between these two responses (middle trace) and the sum (bottom trace) is also shown. Reprinted from *Hear. Res.*, Volume 38, Møller, A. R., and Jho, H. D., Response from the exposed intracranial human auditory nerve to low-frequency tones: Basic characteristics, pp. 163–175, copyright 1989, with permission from Elsevier Science.

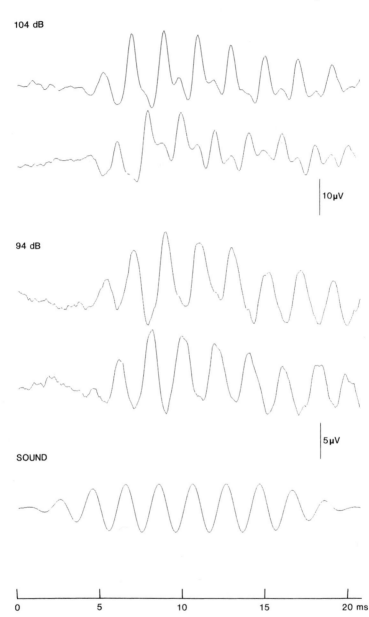

FIGURE 10.16 Examples of responses to 500-Hz tones recorded from the exposed intracranial portion of the VIIIth cranial nerve to show distortion of the waveform. The responses to tones of two different intensities are shown. The two curves at each intensity are the responses to stimulation of opposite polarity. Reprinted from *Hear. Res.*, Volume 38, Møller, A. R., and Jho, H. D., Response from the exposed intracranial human auditory nerve to low-frequency tones: Basic characteristics, pp. 163–175, copyright 1989, with permission from Elsevier Science.

from the exposed VIIIth nerve in humans, the BAEP was monitored during surgical exposure to ensure that the surgical manipulations of the auditory nerve did not cause noticeable change in the neural conduction in the auditory nerve; for details about intraoperative monitoring of neural conduction in the auditory nerve see Moller, 1995.)

RECORDINGS FROM THE COCHLEAR NUCLEUS

Recordings of the responses from the exposed cochlear nucleus to various kinds of sound stimuli have been done in both humans and animals. When a monopolar recording electrode is placed directly on the surface of the cochlear nucleus in humans it records an initial positive–negative deflection in response to a transient sound (Fig. 10.17). That component represents the arrival of the neural volley from the auditory nerve and it is followed by a slower deflection on which peaks are often riding. The slow potential is assumed to be generated by dendrites in the nucleus and its polarity depends on the placement of the recording electrode. The source of the slow potential can be described by a dipole with a certain orientation. The peaks that are seen riding on this slow wave are assumed to be generated by discharges of cells of the nucleus (somaspikes). The latency of the sharp negative peak (N_1) that follows the initial positive deflection (P_1) is approximately 1 ms longer than that of the positive deflection (Fig. 10.17).

The cochlear nucleus consists of three major subdivisions with different response characteristics as judged from recordings from single nerve cells. Therefore, the evoked responses recorded from the surface of the cochlear nucleus are likely to be different depending on which subdivision they are recorded from.

Interpreting the sources of the different components of the response waveform from a nucleus is based on studies of nuclei of the somatosensory system, which was done early in the history of neurophysiology. The oscilloscope was just developed at that time, making it possible to view such potentials (Fig. 10.18) and Gelfan and Tarlov [54] confirmed what Gasser and Graham 2 decades earlier had predicted [50] namely that the initial positive–negative deflection was generated when the incoming volley reached a nucleus. Gelfan and Tarlov [54] in experiments in a dog, showed how the different waves gradually disappeared during anoxia. The slow negative wave was affected first when the animal was exposed to anoxia, and the initial positive–negative (A) complex was only affected after long exposure to anoxia, thus indicating that the main (negative) component that they labeled N_1 was dependent on synaptic transmission, while the initial deflections were generated in a nerve or a fiber tract (Fig. 10.19). It is known that synaptic transmission is more insensitive to anoxia than propagation of neural activity in nerves

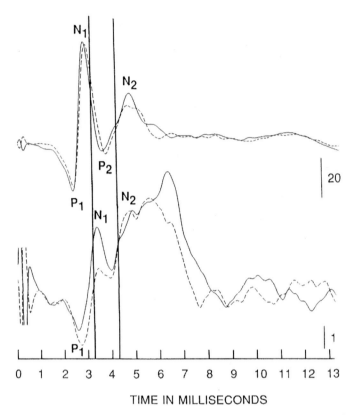

TIME IN MILLISECONDS

FIGURE 10.17   Recordings from the exposed VIIIth nerve (top tracings) and the surface of the cochlear nucleus in a human (bottom tracings). The response from the cochlear nucleus was obtained by placing an electrode in the lateral recess of the fourth ventricle. Solid lines are the responses to rarefaction clicks and the dashed lines are the responses to condensation clicks. Reprinted from *Electroencephalogr. Clin. Neurophysiol.* **92**, Møller, A. R., Jannetta, P. J., and Jho, H. D., Click-evoked responses from the cochlear nucleus: A study in human, pp. 215–224, copyright (1994), with permission from Elsevier Science.

and fiber tracts. Since the response shown in Figs. 10.18 and 10.19 were recorded from the spinal cord it was concluded that the negative component, labeled N in Fig. 10.18, was generated by interneurons. Since the neural activity of nerve cells may be regarded as a dipole source, the N potential in Fig. 10.18 may be recorded as a positive potential at a different location on the nucleus. A reversal of polarity can be observed by passing a recording electrode trough a nucleus (Fig. 10.20) [5].

The responses from the cochlear nucleus, shown in Fig. 10.17, were obtained from monopolar electrodes placed in the lateral recess of the fourth ventricle (Fig. 10.21) [96, 148, 150].

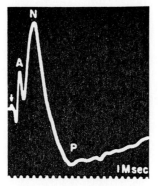

FIGURE 10.18   Evoked potentials recorded from the surface of the spinal cord in a dog to electrical stimulation of a dorsal root (from Gelfan, W. R., and Tarlov, I. M. (1955). Differential vulnerability of spinal cord structures to anoxia. *J. Neurophysiol.* **18**:170–188).

The caudal portion of the floor of the lateral recess of the fourth ventricle is the (dorsal) surface of the dorsal cochlear nucleus and the rostral portion of the floor of the lateral recess is the dorsal surface of the ventral cochlear nucleus [96]. When the lateral side of the brainstem is viewed in operations using a retromastoid craniectomy, the foramen of Luschka, which leads to the lateral recess of the fourth ventricle, is found dorsally to the exit of the IXth and Xth cranial nerves. Often a smaller or larger portion of the choroid plexus protrudes from the foramen of Luschka and may have to be reduced by coagulation in order to place a recording electrode in the lateral recess of the fourth ventricle.

Recordings from the surface of the cochlear nucleus are used clinically in intraoperative neurophysiologic monitoring because it offers a more stable electrode position than the exposed VIIIth cranial nerve. Both methods provide auditory evoked potentials of high amplitude [135, 143].

RECORDINGS FROM MORE CENTRAL PARTS OF THE ASCENDING AUDITORY PATHWAY

Reports on recordings from more central brainstem structures of the ascending auditory pathway have been few compared with recordings from the auditory nerve and the cochlear nucleus. While such recordings have been done in humans, they are not yet used in intraoperative monitoring. Nevertheless, such recordings have been important in identifying the neural generators of the BAEP (to be discussed in Chapter 11).

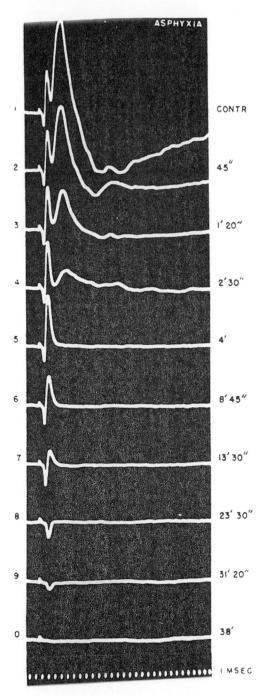

FIGURE 10.19 Effect of asphyxia on spinal cord potentials such as those seen in Fig. 10.18. The time after the respirator was stopped is given to the right (from Gelfan, W. R., and Tarlov, I. M. (1955). Differential vulnerability of spinal cord structures to anoxia. *J. Neurophysiol.* **18:**170–188).

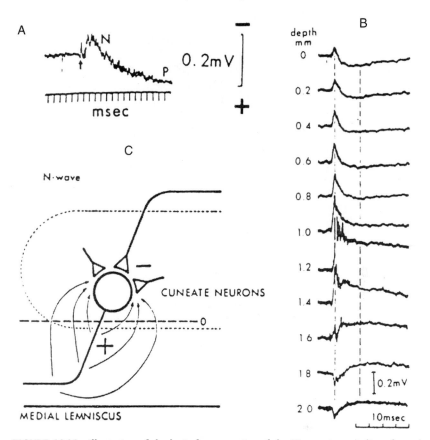

FIGURE 10.20 Illustration of the basis for generation of the N wave in recordings from the cuneate nucleus. (A) Response from the cuneate nucleus. (B) Responses recorded by an electrode passed through the nucleus. (C) The generation of the N potential. When relay neurons are depolarized, current will flow as indicated by arrows (from Andersen, P., Eccles, J. C., Schmidt, R. F., and Yokota, T. (1964). Slow potential wave produced in the cunate nucleus by cutaneous volleys and by cortical stimulation. *J. Neurophysiol.* 27:78–91).

FREQUENCY FOLLOWING RESPONSES FROM THE SUPERIOR OLIVARY NUCLEI

Recordings of the FFR from nuclei of the superior olivary complex have been used to study binaural interactions that depend on the phase shift between tones applied to the two ears. Thus, Moushegian et al [159] has studied the FFR recorded from the superior olivary complex of the cat in response to tones applied to both ears (Fig. 10.22) and found that the response depends on the phase angle between the sounds applied to the two ears (see Chapter 9).

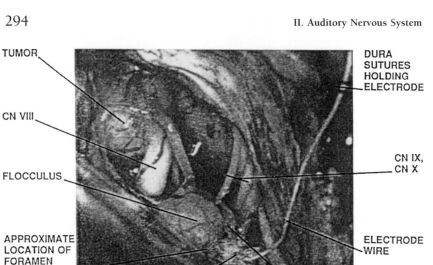

TUMOR

CN VIII

FLOCCULUS

APPROXIMATE
LOCATION OF
FORAMEN
LUSCHKA

COTTON
WICK
ELECTRODE

DURA
SUTURES
HOLDING
ELECTRODE

CN IX,
CN X

ELECTRODE
WIRE

CHOROID
PLEXUS

FIGURE 10.21   Placement of recording electrode in the lateral recess of the fourth ventricle (from Moller, *et al.* (1994) with permission).

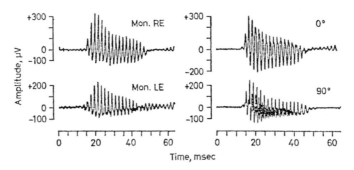

FIGURE 10.22   Frequency following responses (FFR) recorded from the superior olivary nucleus of a cat in response to 30-ms bursts of pure tones (539Hz) presented to the right ear (RE), left ear (LE), and both ears in phase (0°) or 90° out of phase (from Moushegian, G., Rupert, A. L., and Gidda, J. S. (1975). Functional characteristics of superior olivary neurons to binaural stimuli. *J. Neurophysiol.* 38:1037–1048).

RECORDINGS FROM THE MIDBRAIN

As is the case for recordings from the floor of the fourth ventricle, recordings directly from the exposed surface of the midbrain have been relatively few both from animals and from humans where access has been gained in specific but rare operations. Recordings from the surface of the inferior colliculus in the monkey resemble that of other nuclei in the sensory nervous system by having an initial positive–negative wave followed by a broader negative wave [138]. The slow late negative wave has a large amplitude when recorded by a monopolar electrode but a much smaller amplitude in a bipolar recording, which indicates that the generators of the slow potentials are more widespread in the nucleus. The initial positive–negative wave is generated by the incoming

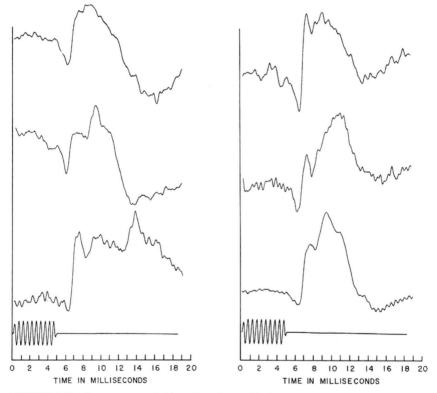

FIGURE 10.23 Responses recorded from the inferior colliculus in patients undergoing operations where the inferior colliculus was exposed or responses recorded from an electrode placed along the path of the fourth cranial nerve (from [146]).

fiber tract, i.e., the lateral lemniscus. The slow negative wave is assumed to reflect dendritic potentials and the sharp peaks riding on this slow potential are probably soma spikes.

The waveform of the response to short tone bursts recorded from the surface of the contralateral inferior colliculus in humans (Fig. 10.23) [69, 146] is typical of a nucleus. The earliest positive deflection is presumably generated when the volley of neural activity in the lateral lemniscus reaches the termination of the lateral lemniscus in the inferior colliculus. The slow negative deflection is most likely a result of dendritic activity. The response from the inferior colliculus to ipsilateral stimulation has a much smaller amplitude and a different waveform indicating that the number of uncrossed fibers that reach in inferior colliculus is small. Recordings from the inferior colliculus and its vicinity using chronically implanted electrodes have been done recently [241].

Far-Field Auditory Evoked Potentials

ABSTRACT

1. Far-field evoked potentials are the potentials that can be recorded from locations that lie far from the anatomical location of their generators.
2. Neural activity in many of the structures of the classical ascending auditory pathways, but not all, give rise to far-field evoked potentials that can be recorded from electrodes placed on the scalp.
3. Propagated neural activity in a nerve or fiber tract in the brain may generate stationary peaks in the far-field potentials when the propagation is halted, when the electrical conductivity of the medium surrounding the nerve changes, or when the nerve or fiber tract bends.
4. The far-field potentials of nuclei depend on their internal organization.
5. Brainstem auditory evoked potentials (BAEP) and the middle latency responses (MLR) are far-field responses from the auditory nervous system.
6. The BAEP is commonly recorded from electrodes placed on the scalp. Normal click evoked BAEP consists of five prominent and constant vertex

positive peaks that occur during the first 10 ms after the click sound. They are labeled by Roman numerals (I–V).

7. Most studies of the neural generators of the BAEP have concentrated on the generators of the vertex positive peaks.

8. Peaks I and II of the human BAEP are generated by the auditory nerve (distal respective proximal portions).

9. All other constant peaks (III, IV, V) have contributions from more than one anatomical structure and each anatomical structure of the ascending auditory pathway contributes to more than one peak.

10. Peak III is mainly generated by the cochlear nucleus and Peak IV is most likely generated by structures that are located close to the midline, such as the superior olivary complex.

11. The sharp tip of Peak V is generated by the lateral lemniscus, most likely where it terminates in the inferior colliculus on the side contralateral to the ear from which the response is elicited.

12. The individually variable slow negative potential following Peak V (SN_{10}) is generated by (dendritic) potentials in the contralateral inferior colliculus.

13. The middle latency response (MLR) is composed of the potentials that occur during in the time interval of 10–80 ms or 10–100 ms.

14. The neural generators of the MLR are less well understood than those of the BAEP. Cortical potentials contribute to the MLR and muscle (myogenic) responses may also contribute to the MLR.

15. The 40-Hz response is a far-field response that results from summation of components of the evoked potentials that repeat every 25 ms.

16. The frequency following response (FFR) can be recorded in response to low-frequency tones from the electrodes placed on the scalp.

17. Sounds may evoke electrical activity from muscles of the head, which may be recorded as myogenic auditory evoked potentials.

INTRODUCTION

Far-field evoked potentials are the responses that can be recorded at a long distance from their source. Far-field potentials therefore have much smaller amplitudes than near-field potentials and it is necessary to add the responses to many stimuli in order to discern the various components of far-field potentials from the background of other biologic signals such as EEG activity, potentials from muscles, and electrical interference signals. Far-field evoked potentials could therefore not be studied before the development of the signal averager.

While evoked potentials can always be recorded from electrodes placed directly on nerves, fiber tracts, and nuclei, these structures generate far-field potentials only when certain criteria are fulfilled. Thus, neural activity that propagates in a nerve or a fiber tract generates stationary peaks in the far field when the electrical conductivity of the surrounding medium changes or when the nerve or fiber tract is bent. Neural activity that propagates in a straight nerve, the surrounding medium of which has uniform electrical conductivity, generates very weak far-field potentials. A nucleus generates strong far-field potentials when its dendrites are organized uniformly, whereas a nucleus where the dendrites are randomly organized and points in all directions generates only weak far-field potentials. These two different types of nuclei are known to have an open and a closed field, respectively.

Far-field potentials are more complex than near-field potentials because they are likely to have contributions from many different sources, located at anatomically different places. Neural structures activated sequentially by transient stimulation may generate a sequence of components, each of which occurs with different latencies. Brainstem auditory evoked potentials (BAEP) (Fig. 11.1) are an example of far-field evoked potentials that are commonly used for clinical diagnosis and for intraoperative neurophysiologic monitoring. It is the most important functional test for detecting acoustic tumors. The middle latency responses (MLR) are another kind of far-field auditory evoked potential that are used clinically for diagnostic purposes, but to a lesser extent. Proper interpretation of these auditory evoked potentials for diagnostic purposes depends on knowledge about the anatomical origin of the different components of these potentials and how they are affected by pathologies. During the past 2 decades much knowledge about the neural generators of the BAEP has accumulated, but the neural generators of the MLR are not as well known, which has hampered the use of the MLR in diagnosis of neurologic disorders. The MLR is considerably more variable than the BAEP. The MLR is mainly used as an objective test of hearing threshold but the MLR has a potentially important role for diagnosis of disorders of the auditory nervous system, but the individual variability of the MLR and insufficient knowledge about the origin of these potentials have prevented such use so far.

The far-field response to periodic sounds such as pure tones (frequency following responses, FFR) and the modulation waveform of amplitude-modulated sounds can also be recorded from electrodes placed on the scalp. Recording electrodes placed on the scalp also pick up responses from muscles that are elicited by sound stimulation (myogenic evoked potentials), which complicate interpretation of these responses.

BRAINSTEM AUDITORY EVOKED POTENTIALS

The human brainstem auditory evoked potentials (BAEP) consist of far-field evoked potentials from the auditory nervous system that occur during the first

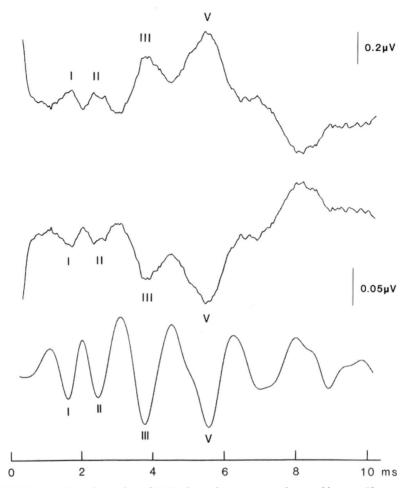

FIGURE 11.1 Typical recording of BAEP obtained in a person with normal hearing. The curves are the average of 4096 responses to rarefaction clicks recorded from electrodes placed on the forehead at the hairline and the mastoid on the side where the stimuli were applied. The upper curve is shown with vertex positivity as an upward deflection, the middle curve is the same recording shown with positivity downward. These two curves are recordings that were filtered electronically with a band-pass of 10–3000 Hz. The bottom curve is the same recording after digital filtering designed to enhance all five peaks of the BAEP.

10 ms after the presentation of a transient sound such as a click. The amplitudes of the BAEP are small, less than 1 μV, and thus much smaller than the ongoing spontaneous activity of the brain (EEG). The responses to many stimuli must be added to obtain a record where the individual components can be discerned.

The different components of the BAEP are generated by neural activity in the ear, the auditory nerve and the nuclei, and fiber tracts of the ascending auditory pathway. When recorded differentially between two electrodes, one placed at the vertex and one at the mastoid or earlobe on the side where the stimulus sounds are presented, the BAEP typically is characterized by five to seven vertex-positive waves (Fig. 11.1). These peaks are traditionally labeled with Roman numerals. The first five of these peaks of the human BAEP (maybe except Peak IV) are constant and can usually be discerned in individuals with normal hearing. The labeling of (only) the vertex-positive waves with Roman numerals, which Jewett and Williston [81] originally suggested, is still the most common way to label the components of the BAEP. This labeling is different from the way different components of other sensory evoked potentials are labeled. Usually, both positive and negative components of evoked potentials are labeled, with the letter P and N respectively, followed by a number that gives the normal value of the latency of the respective peak.

History of the BAEP

It was probably Kiang and his colleagues [87] at the Eaton Peabody Laboratory in Boston who first demonstrated these potentials. Dr. Kiang was a member of the group at MIT assembled by Professor Walter Rosenblith, who pioneered signal analysis of neuroelectric potentials and was at the forefront of developing the signal averaging technique into a routine method for studies of neuroelectric potentials. Dr. Kiang also predicted that these potentials might be useful in diagnosing disorders of the auditory system and in intraoperative monitoring [87, see 136]. However, systematic studies of the BAEP were not published until 10–15 years later, at which time Jewett and his collaborators identified and described the different components of the BAEP and introduced the placement of the recording electrodes that produced potentials of larger amplitudes. This electrode montage, one electrode placed at the vertex and the other placed at the mastoid on the side where the stimuli are applied [81], remains the most widely used way of recording the BAEP for clinical purposes.

One of the consequences of only labeling the vertex positive peaks of the BAEP has been that only the positive peaks have been used for diagnostic purposes and most studies of neural generators of the BAEP have ignored the vertex-negative peaks. At the time when this labeling was introduced it was not known which of the different components of the BAEP were most important for diagnostic purposes. It would seem likely that the vertex-negative waves would also be of diagnostic value, as these negative peaks also have distinct neural generators.

Recording of BAEP

When evoked potentials are recorded differentially between two electrodes, one electrode is usually placed at a location where the potential to be recorded is large and the other electrode (reference electrode) is placed on a location where it records as little as possible of the evoked potentials that are studied. With the electrode placement commonly used for recording BAEP both recording electrodes record auditory evoked potentials that contribute to the BAEP. Although the amplitude of the different peaks recorded by the two electrodes are slightly different, it is not possible to distinguish between an active electrode and a reference electrode. Peak

V has a larger amplitude in the vertex recording than in the recording from the ear while Peaks I–III have larger amplitudes in the recordings from the ear than from the vertex. Some authors prefer to show the BAEP with the vertex positivity as an upward deflection, whereas others display the BAEP with the vertex positivity as a downward deflection (Fig. 1.11), the latter being in accordance with the common convention of displaying negative potentials as an upward deflection, assuming the vertex electrode to be the most active electrode. In this book, BAEPs are usually shown with vertex positivity as a downward deflection.

Effect of Filtering

The main reason for filtering the BAEP is to reduce the number of responses that must be averaged in order to obtain an interpretable record. However, the appearance of the BAEP depends on the way that the potentials are filtered (Fig. 11.2). It is the latencies of the different peaks that are important for diagnostic purposes and the filters used should therefore enhance the peaks without shifting the peaks in time. When recorded with an open band-pass filter (10–3000 Hz), the BAEP has the appearance of a series of three clear positive peaks followed by Peak V, which consists of a sharp positive peak followed by a broad negative peak known as the SN_{10} (Fig. 11.3, top records). When low-frequency components of the BAEP are attenuated, the SN_{10} component may not be noticeable but the first five peaks will appear clearly (Fig. 11.3). Electronic filters shift the peaks in time to an extent that depends on the spectrum of the peaks, the type of filters used, and their settings. The different peaks of the BAEP may be shifted differently by electronic filters. It is possible to design electronic filters with linear phase shift and such filters (known as Bessel filters) will shift all components of the BAEP by the same amount but such filters are rarely available. The problems related to phase shift in filters can best be solved by using so-called digital filters that can be designed to have no phase shift at all (zero-phase digital filters). Such filters thus do not shift any component of the waveform that is filtered. Digital filters are computer programs that operate on the averaged waveform of the BAEP. Digital filters are more flexible than electronic filters and optimal filtering can be obtained using digital filters, whereas electronic filters have considerable limitations. Aggressive filtering that is made possible using digital filters can enhance specific components of the BAEP that are of interest (peaks). Peak II and Peak IV of the BAEP are often difficult to identify, but appropriate (digital) filtering can make these peaks appear clearly (Figs. 11.2, 11.3) [134]. Such filtering also makes it possible to have computer programs identify the individual peaks and print their latencies automatically (Fig. 11.2).

Effect of Stimulation Parameters

The amplitude of the BAEP decreases with decreasing stimulus intensity and the latencies increase but the different components of the BAEP are affected differently.[1] A decease in stimulus intensity causes the amplitude of early peaks (I,

---

[1] The intensity of clicks used for recording BAEP is usually given in peak equivalent SPL (PeSPL), which is the sound pressure of a pure tone with the same peak sound pressure as the clicks. It is also common to give the intensity of click stimulation in decibels above threshold, measured in individuals with normal hearing. While the physical measure of click intensity (PeSPL) is independent of the rate at which the clicks are presented, the behavioral threshold

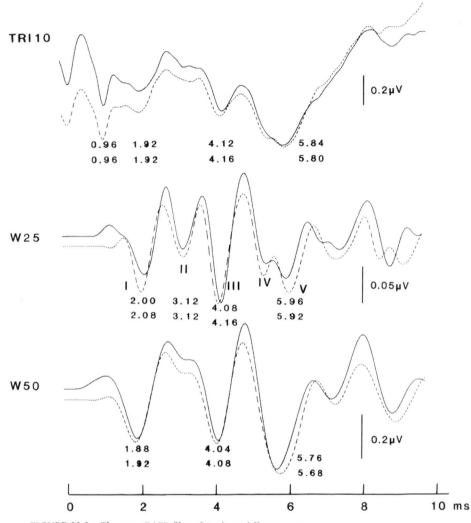

FIGURE 11.2 The same BAEP filtered in three different ways.
Top traces: Low-pass filter with a digital filter with a triangular shape and a base of 0.4 ms. *Middle trace:* digital band-pass filter with a base of approximately 1 ms. *Lower trace:* digital filter with a base of approximately 2 ms.

decreases with increasing repetition rate because of temporal integration. Thus clicks presented to one ear at a rate of 5 pps in an individual with normal hearing has threshold of approximately 37 dB PeSPL, at 20 pps it is 35 dB PeSPL, and at a rate of 80 pps, the threshold is 32 dB PeSPL [211]. That means that clicks of an intensity of 105 dB PeSPL have a hearing level (HL) of 70 dB when presented at 20 pps.

exposed. When the data from animals are compared with data from humans, it is important to recognize that there are specific anatomical differences between animals commonly used in such studies and humans. The responses recorded from structures exposed surgically in animals and humans may be affected by changes in the function of the classical auditory nervous system caused by surgical manipulations. In humans, the recorded responses may be affected by preexisting hearing loss and disease processes that affect the auditory nerve and other parts of the classical auditory nervous system. On the other hand, the BAEP is remarkably little affected by common anesthetics, which makes it possible to study these potentials and their neural generators during general anesthesia.

At a first approximation, the different components of the BAEP are generated by neural activity in sequentially activated structures of the ascending auditory pathway. However, the classical auditory nervous system is not just a string of nuclei connected with fiber tracts, but rather a complex series of nuclei with many interconnections, including a high degree of parallel processing, which complicates the interpretation of the BAEP.

Much of our understanding of evoked potentials recorded from the classical ascending auditory nervous system and their contributions to the BAEP has been gained from studies in animals but the large anatomical differences between the auditory nerve in humans and the commonly used experimental animals including the monkey have at times contributed to misinterpretation of the contribution to the BAEP in humans. During the past 15–20 years extensive studies of the responses recorded directly from exposed parts of the human auditory nervous system during neurosurgical operations have contributed to the understanding of the generation of the human BAEP.

The abnormalities of the BAEP in patients with known pathologies, such as tumors, have been used to identify the neural generators of the BAEP. This presumes that the pathology affects specific parts of the ascending auditory pathway and that its location is known. The disadvantages of such methods are related to difficulties in assessment of injuries by imaging techniques. Imaging studies such as the magnetic resonance imaging (MRI) scans can only detect changes in structure, not in function. It is also an obstacle to such studies that it is unknown how specific morphological abnormalities, as they appear in the different imaging studies, relate to functional abnormalities.

Reconstruction of the dipoles of the generators of sensory evoked potentials based on recordings of three-dimensional evoked potentials, accomplished by placing three pairs of recording electrodes orthogonally on the scalp (three-dimensional Lissajous trajectories, 3-CLT) make it possible to obtain information about the anatomical location of generators of sensory evoked potentials, but the spatial resolution limits the use of this method.

Most studies of the generators of the BAEP have focused on generators of the vertex positive peaks in the BAEP and only a few studies have concerned the vertex negative peaks of the BAEP. It is also the vertex positive peaks that are used for clinical diagnostic purposes. The choice of the vertex positive peaks is not a result of any evidence that the vertex positive peaks are more important than the vertex negative peaks but rather a result of the fact that only the vertex positive peaks have been labeled.

Comparison between the Brainstem Auditory Evoked Potentials and Directly Recorded Potentials from Specific Structures of the Ascending Auditory Pathway

Perhaps the most successful method for identifying the neural generators of the BAEP is the one that makes use of comparisons between the BAEP and evoked potentials recorded directly from specific structures of the ascending auditory pathway. Coincidence in time between main components of the directly recorded potentials and the different (vertex-positive) peaks of the BAEP is taken as an indication, but not proof, that a specific structure is the generator of a certain peak of the BAEP. Such studies are possible in selected neurosurgical operations where it becomes possible to place a recording electrode directly on the intracranial portion of the auditory nerve, the cochlear nucleus, or other structures of the ascending auditory pathway or in their immediate vicinity. In such studies the BAEP is recorded simultaneously before and during intracranially recordings to ensure that the surgical manipulations have not affected the function of the structures that contribute to the BAEP [139–152]. Some investigators [67–69] have recorded directly from the auditory nervous system by inserting electrodes through burr holes in the skull and passing them through the brain to reach the desired location. Clicks have been the most commonly used stimuli in such studies, but some investigators have used tone bursts.

Neural Generators of Peaks I and II

Recordings from the intracranial portion of the VIIIth cranial nerve in operations where that portion of the nerve became exposed revealed that the negative peak of the response (CAP) occurs with about the same latency as the second vertex positive peak in the BAEP (Peak II) [68, 69, 145, 151, 209] (Figs. 11.5 and 11.6). This has been taken to indicate that Peak II is generated by the central portion of the auditory nerve. The relationship between the negative peak in the CAP recorded directly from the intracranial portion of the VIIIth cranial nerve becomes even more convincing when the latencies of

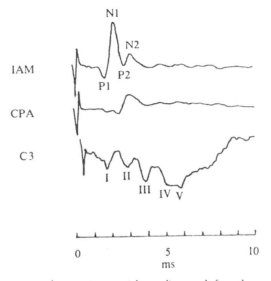

FIGURE 11.5   Comparison between intracranial recordings made from the exposed VIIIth nerve and the BAEP.

IAM, recording of the CAP from the intracranial portion of the VIIIth nerve where it exits the bony canal (porus acousticus); CPA, recordings of the CAP directly from the exposed VIIIth nerve in the cerebellopontine angle; C3, BAEP recorded between the ipsilateral earlobe and the C3 on the parietal ipsilateral scalp (international EEG recording nomenclature) (from Hashimoto, I., Ishiyama, Y., Yoshimoto, T., and Nemoto, S. (1981). Brainstem auditory evoked potentials recorded directly from human brain-stem and thalamus. *Brain* **104**:841–859).

Peak II of the BAEP and the main negative peak of the CAP were compared over a large range of stimulus intensities [149] (Fig. 11.6).

Before these studies were published, comparison of the electrocochleographic (ECoG) potentials with the BAEP had shown that Peak I of the BAEP occurs with the same latency as the negative peak (N_1) in the ECoG [175]. The N_1 peak of the ECoG is generated by the most peripheral portion of the auditory nerve and therefore Peak I of the BAEP is also assumed to be generated in the most peripheral portion of the auditory nerve (in the ear). That means that in humans Peak I is generated by the distal portion of the auditory nerve and Peak II is generated by the proximal (intracranial) portion of the auditory nerve.

Neural Generators of Peak III

The responses recorded directly from the surface of the cochlear nucleus have a less clear relationship with the simultaneously recorded BAEP than the CAP recorded directly from the auditory nerve. Comparison between the

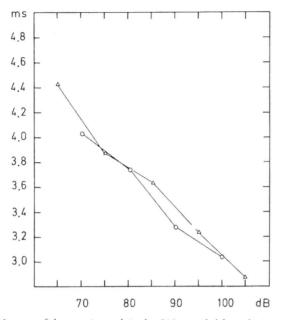

FIGURE 11.6   Latency of the negative peak in the CAP recorded from the intracranial portion of the VIIIth nerve as a function of the stimulus intensity (open triangles) and the latency of peak II of the BAEP postoperatively (open circles). The sound stimuli were 5-ms-long 2-kHz tone bursts (from [145]).

responses recorded from the exposed VIIIth cranial nerve, the cochlear nucleus, and the BAEP (Fig. 11.7) [149] show that the initial positive–negative deflection has the same latency as Peak II of the BAEP and the sharp negative peak that dominates the response from the cochlear nucleus has the same latency as Peak III of the BAEP. This large negative peak in the cochlear nucleus response is probably a result of firings of nerve cells in the cochlear nucleus, or it may be the fiber tract that leaves the cochlear nucleus that generates Peak III (see Chapter 10).

Peak III is thus the earliest manifestation of neural activity in secondary neurons and the cochlear nucleus is the main generator of Peak III. Peak III may, in addition, receive contributions from (late) firings of auditory nerve fibers, but Peak III probably does not receive input from neurons of a higher order than the cochlear nucleus. It is likely that the vertex-negative peak that follows after Peak III also receives contributions from the cochlear nucleus, although perhaps different divisions from those contributing to Peak III.

The timing of the neural activities in the three different divisions of the cochlear nucleus may be different and the different divisions may contribute

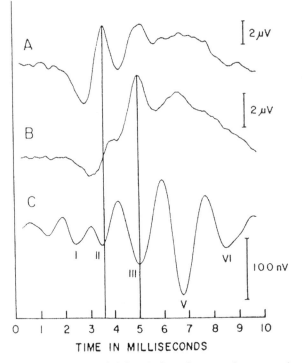

FIGURE 11.7 (A) Responses recorded directly from the exposed intracranial portion of the eighth nerve.

(B) Responses recorded from an electrode placed in the lateral recess of the fourth ventricle to record evoked potentials from the cochlear nucleus.

(C) BAEP recorded during the same operation.

The stimuli were 5-ms-long 2000-Hz tone bursts at 95 dB SPL. (from Møller, A. R., and Jannetta, P. J. (1983). Auditory evoked potentials recorded from the cochlear nucleus and its vicinity in man. *J. Neurosurg.* 59:1013–1018).

to different peaks of the BAEP. The far-field potentials generated by the three striae when they merge to form the lateral lemniscus near the contralateral cochlear nucleus may also contribute to Peak III and the following negative wave.

Neural Generators of Peak IV

The anatomical locations of the neural generators of Peak IV are poorly understood and only few published studies have addressed the sources of Peak IV of the BAEP. Recordings of evoked potentials directly from the exposed lateral brainstem, rostral to the entrance of the VIIIth cranial nerve near the Vth cranial nerve, have revealed a distinct component with a latency similar to that of Peak IV of the BAEP [144, 147, 148]. This recording location is anatomically close to the superior olivary complex, indicating that the superior olivary complex might generate these near-field potentials and thus suggesting that Peak IV of the BAEP might be generated by the nuclei of the superior olivary complex [147]. Recent studies (144) indicate that the sources of Peak IV are located close to the midline.

Neural Generators of Peak V

Responses recorded from the floor of the fourth ventricle near the inferior colliculus to contralateral stimulation [67, 69, 144, 146] reveal a sharp positive deflection that is followed by a broad negative wave. The sharp positive deflection occurs with nearly the same latency as Peak V of the BAEP (Fig. 11.8) [67, 69]. The results in Fig. 11.8 were obtained using binaural stimulation, which makes it impossible to determine which of these responses originate in the crossed pathway and which originate in the uncrossed pathway. Studies using monaural stimulation [144, 146] support the hypothesis that the initial sharp positive deflection is generated by the termination of the lateral lemniscus in the inferior colliculus (Fig. 11.9), and other studies show that the response is larger and more distinct in response to contralateral stimulation indicating that Peak V is generated mainly by structures that are activated by contralateral stimulation.

The slow deflection in the response from the inferior colliculus (Fig. 11.9) has a similar latency as the broad negative deflection seen to follow Peak V in the BAEP. That peak (SN_{10} [22]) is variable in humans and usually attenuated by the commonly used filtering of the BAEP. When the responses recorded directly from the inferior colliculus or its close vicinity are filtered so that low-frequency components are attenuated, a series of sharp peaks appear after the initial positive peaks. These peaks are probably firings of nerve cells in the inferior colliculus (somaspikes). Comparison between the BAEP and such filtered recordings from the inferior colliculus indicate that these components may be the generators of Peaks VI and VII of the BAEP (Fig. 11.9).

Use of Dipole Identification to Determine the Neural Generators of the Brainstem Auditory Evoked Potentials Using Lissajous' Trajectories

Identifying the anatomical location of the generators of the BAEP has been attempted by recording the BAEP in three orthogonal planes [176, 198, 230]. Such three-dimensional recordings, known as the three-channel Lissajous' trajectory (3CLT), display evoked potentials as a line, each point of which represents the voltage at any given time after the stimulus (Fig. 11.10). The use of the 3CLT recordings to identify the anatomical location of the generators

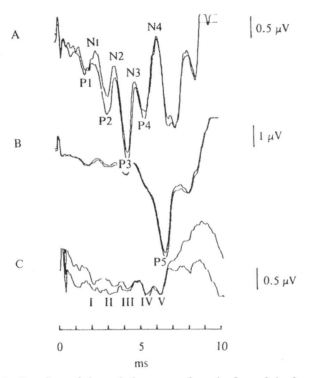

FIGURE 11.8   Recordings of the evoked responses from the floor of the fourth ventricle.
(A) The midline of the dorsal pons at the level of the facial colliculus. (B) The surface of the
inferior colliculus. (C) Recording of the BAEP from electrodes placed at C3 (left parietal scalp)
and the left earlobe. Note that positive peaks are labeled by the letter P and negative peaks with
the letter N, followed by a number. For the BAEP the peaks are labeled in the traditional way.
The stimuli (clicks at 15 pps and 65 dB SL) were applied to both ears (binaural stimulation)
(from Hashimoto, I., Ishiyama, Y., Yoshimoto, T., and Nemoto, S. (1981). Brainstem auditory
evoked potentials recorded directly from human brain-stem and thalamus. *Brain* 104:841–859).

of evoked potentials such as the BAEP is based on the fact that the voltage
generated on any point of the surface of a sphere by an electrical source
(dipole) placed at a known location inside the sphere can be accurately computed. However, the opposite, namely determining the location of generators
when the potentials on the surface are known, can only be done when certain
conditions are assumed because a certain voltage distribution on the scalp of
an individual can be caused by more than one location of dipoles. Thus while
such 3CLT recordings provide a complete description of the potentials on the
surface of the head, attempts to determine the location of a source of the
potentials on the basis of distribution of the electrical voltage on the surface
of the head does not have a unique solution. Despite this deficiency, the

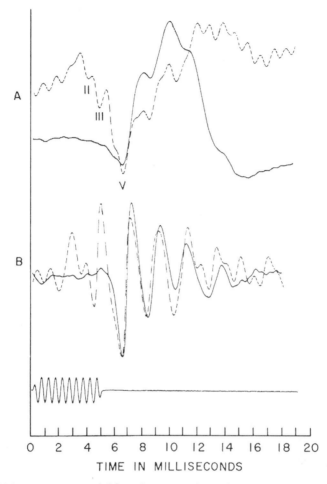

FIGURE 11.9 Responses recorded from the vicinity of the inferior colliculus with the reference electrode on the clavicle (solid lines) compared with the BAEP recorded in the same operation between the vertex and a position immediately above the ipsilateral pinna (dashed lines).

Top tracings: Recordings with electronic filtering 3–10,000 Hz and digital low-pass filtering by a triangular weighting function with a 0.8-ms base.

Bottom tracings: The same recordings after digital filtering with a triangular weighting function that has a band-pass characteristic and attenuates slow potentials. The stimuli were 5-ms-long, 2-kHz tone bursts at 95dB SPL, presented at a rate of 7 pps (from Møller, A. R., and Jannetta, P. J. (1983). Interpretation of brainstem auditory evoked potentials: Results from intracranial recordings in humans. *Scand. Audiol.* **12**:125–133).

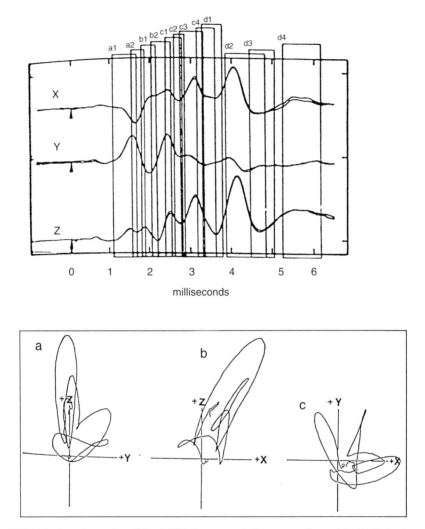

FIGURE 11.10   Illustration of the 3 CLT. Upper graph shows recordings of the BAEP in three orthogonal planes. Lower graphs are two-dimensional plots (Lissajous' trajectories) (from Pratt, H., Bleich, N., and Martin, W. H. (1985). Three-channel Lissajous' trajectory of human auditory brainstem evoked potentials. I. Normative measures. *Electroencephalogr. Clin. Neurophysiol.* **61**:530–538).

3CLT method has yielded valuable results regarding identifying the anatomical location of individual generators of the components of the BAEP [111, 176, 198, 230]. Scherg and von Cramon [198] showed that the BAEP can be synthesized by six dipoles (Fig. 11.11A), approximately located in the coronal

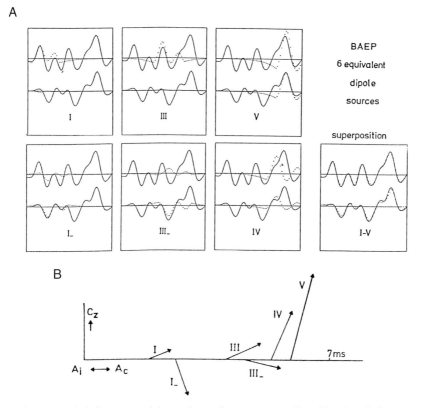

FIGURE 11.11 (A) Illustration of the synthesis of source activity (dotted lines) to fit the BAEP (solid lines). The BAEP was recorded between the vertex and the ipsilateral ear (upper tracings) and between the vertex and the contralateral ear (lower tracings). Right-hand insert: The sum of all synthesized dipoles (dotted lines) and the BAEP. (B) Orientation and strength of the six dipoles identified from recordings from electrodes placed in three planes. The horizontal line is a line between the two ears and is also the time axis. The vertical axis is a line between the middle of that line and the vertex. The origin of the vectors is the latency of the first peak in the dipole and the length is the relative strength of the dipoles. Note that the short distance between the two first dipoles (Peaks I and II of the BAEP) and the third (peak III). Reprinted from *Electroencephalogr. Clin. Neurophysiol.* **62**, Scherg, M., and von Cramon, D., A new interpretation of the generators of BAEP waves I–V: Results of a spatio-temporal dipole, pp. 290–299, copyright (1985), with permission from Elsevier Science.

plane (a vertical plane that is perpendicular to the sagittal plane). Dipole I- and I are horizontally oriented and represent the auditory nerve. Dipole III and III- are also horizontally oriented toward the contralateral ear and located in the lower brainstem on the ipsilateral side at approximately the same distance from the midline as the cochlear nucleus. The fifth and sixth dipole, representing Peaks IV and V, are orientated vertically but the resolution of these vertical components did not allow determination of their exact location, nor was it possible to determine whether they were located ipsilaterally or contralaterally to the stimulated ear.

Studies of such 3CLT recordings confirmed results obtained by other methods, particularly regarding the generator of Peaks I and II, and helped toward understanding how the recorded BAEP depends on the electrode positions. Since the orientation of the dipoles of Peaks I, II, and III are mostly along a line between the two ears, these peaks will appear with their highest amplitudes in recordings with electrodes placed at the earlobes (or mastoids), whereas Peaks IV and V are best recorded from electrodes on the vertex and one earlobe or with a noncephalic reference. This is in good agreement with common experience from recordings of the BAEP.

Use of Pathologies in Studies of Neural Generators of the Brainstem Auditory Evoked Potentials

Studies of pathologies that affect the auditory nervous system have confirmed that Peak II is generated by the intracranial portion of the auditory nerve. Studies of patients with a certain kind of hereditary motor-sensory neuropathy (type I) showed evidence that both Peak I and Peak II of the BAEP were generated by the auditory nerve. That Peak III is generated in the pontine region of the ipsilateral brainstem was supported by studies of the abnormalities of the BAEP in patients with discrete lesions in the brainstem. However, some studies of changes in the BAEP in individuals with known pathologies have produced results regarding the neural generators of Peak V that, at a glance, contradict other studies of the neural generators of the BAEP.

Determining the anatomical location of the neural generators of the BAEP on the basis of abnormalities in the BAEP in patients with lesions that affect the ascending auditory pathway has pitfalls that are often overlooked. The results of such studies are also more difficult to interpret than results from intracranial recordings. Nevertheless, such methods have the advantage that the results are directly related to the use of the BAEP in diagnosis of disease processes that may affect the auditory nervous system.

Question about Laterality of the Generators of the Brainstem Auditory Evoked Potentials

While most investigators agree that Peaks I, II, and III are generated on the ipsilateral side of the brainstem, some investigators disagree about the anatomical location of the generators of the later peaks. Thus, Markand and co-workers [110] have interpreted available data to show that Peak V is generated by brainstem structures on the side from which the BAEP is elicited. Other investigators find it more plausible that Peak V is mainly generated by structures located on the contralateral side. The main crossing of the ascending auditory pathway occurs at the level of the superior olivary complex (see Chapter 5), but anatomical studies show an uncrossed pathway as well. The anatomical studies, however, do not provide precise information about exactly how many fibers cross and how many continue on the same side and the importance of the uncrossed pathway for hearing is unknown.

The results of studies of the abnormalities of the BAEP in patients with discrete intrinsic lesions of the brainstem [110, 163, 238] have been summarized by the widely cited statement: "When BAEP abnormalities either occur exclusively on stimulation of one ear or are asymmetric on right and left ear stimulation, the responsible lesion in the brain stem is on the side of the ear eliciting maximal abnormality" [110]. These findings have been interpreted to show that the neural generators of all peaks, including Peak V, are located on the side from which the BAEP is elicited, thus contradicting results of electrophysiologic studies that have consistently shown that Peak V of the BAEP is generated by contralateral structures.

Chiappa [15], in a recognized handbook on evoked potentials, has extended the conclusions from lesion studies on the BAEP to mean that the BAEP does not reflect the parts of the ascending auditory pathway that is normally associated with hearing (the crossed pathway). This seems too strong a statement and the results from studies of pathologies can be explained in a more plausible way. The findings by some studies that lesions (tumors, bleeding, etc.) verified by imaging techniques (such as MRI) affect components of the BAEP elicited from the side of the lesion more than they affect Peak V of the BAEP elicited from the contralateral ear [110] does not need to be a contradiction to results of electrophysiologic studies that show that Peak V of the BAEP is generated mainly by structures located on the opposite side from which the BAEP is elicited. These findings can be explained without resorting to such a drastic assumption that the uncrossed pathway is the (main) generator of the BAEP [15].

In view of the great clinical importance of the question of laterality of the BAEP, let us examine the available results that claim to show that all five BAEP peaks are generated in structures that are located ipsilaterally to the stimulated ear.

First, recognize that changes at the peripheral levels are imposed on more centrally located structures. Lesions of peripheral structures will therefore also affect components of the BAEP that are generated by more centrally located structures. Second, assume that Peak V is generated when propagated neural activity in the lateral lemniscus halts, which normally occurs where the lateral lemniscus terminates in the inferior colliculus. If the lateral lemniscus is affected by disease processes such as a tumor, it may cause a total or partial arrest of propagation of neural activity at the location of the lesion. Such a halt in propagation of neural activity in the lateral lemniscus can generate a stationary peak in the far field. This means that a lesion of the lateral lemniscus that causes an arrest of the propagated neural activity may generate a component in the far-field potential that is indistinguishable from a normal Peak V.

A lesion located at the lateral lemniscus anywhere between the (ipsilateral) cochlear nucleus and the inferior colliculus will have the same effect. The only obvious abnormality in the BAEP would be a slightly shorter latency of Peak V, but that is not likely to be noticeable. The SN_{10} would be abolished by interruption of the neural transmission in the lateral lemniscus but the SN_{10} is normally not detectable because the high-pass filter commonly used eliminates the SN_{10}. That means that a lesion of the lateral lemniscus in its contralateral course may not produce any noticeable change in the BAEP. Tumors or other lesions that affect the inferior colliculus may not change the BAEP noticeably either because the normal contribution to the BAEP from the inferior colliculus is small except for the SN_{10}, but that component is rarely noticed because low frequencies of the BAEP are usually attenuated by filtering.

> That a lesion of a fiber tract can create a generator of far-field potentials was supported by studies by Hatayama *et al.* [70] in experiments in dogs. These investigators presented evidence that interruption of neural conduction by injury to the auditory nerve may result in the generation of a stationary peak in the far-field response (BAEP). After compression injury to the auditory nerve, the BAEP contained a peak (1b, in Fig. 11.12) that was not present before the injury. That peak is thus the far field of the "cut end" potential of the injured auditory nerve. That arresting neural conduction in the auditory nerve can create a stationary peak in the far field supports the hypothesis that interruption of neural conduction in the lateral lemniscus can also create a stationary far-field peak.

Lesions of the brainstem are thus unlikely to produce changes in the BAEP elicited from the contralateral ear, which may explain why Markand et. al. [110] found that abnormalities in the BAEP elicited from the side contralateral to a lesion were smaller than those observed in the BAEP elicited from the side of the lesion. But why did they find changes in the BAEP elicited from the ipsilateral side? The answer to that question may be found in the fact that the anatomical extension of the type of lesions they studied is often poorly defined. After all, imagining techniques only show changes in structure but changes in function are likely to be more extensive. That means that lesions that are located at the midbrain, as determined by examining MRI scans, may extend further caudally and affect the cochlear nucleus and superior olivary complex. Space-occupying lesions may affect the function of nuclei more than that of fiber tracts, but nuclei may not contribute noticeably to the BAEP.

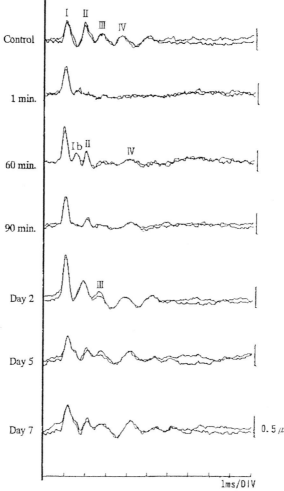

FIGURE 11.12   Serial recordings of BAEPs in a dog to show changes in the BAEP after experimentally induced compression of the auditory nerve. Peak II and all subsequent peaks are obliterated immediately after compression of the auditory nerve and a peak (Ib) appears between Peak I and Peak II. Reprinted from *Neurological Research* 21, Hatayama, T., Sekiya, T., Suzuki, S., and Iwabuchi, T., Effect of compression on the cochlear nerve: A short- and long-term electrophysiological and histological study, pp. 559–610, copyright (1999), with permission from Elsevier Science.

This means the effect of compression of a nucleus will produce the greatest abnormality in the components of the BAEP that are generated by the fiber tract that leaves the nucleus in question. Thus, compression of auditory nuclei located in the pons may affect Peak V because the lesion causes the neural activity in the lateral lemniscus to become abnormal. Space-occupying lesions do not necessarily cause their greatest change in the components of the BAEP that are generated by structures that are affected by the lesion. Rather, the greatest changes may occur in components of the BAEP that are generated by structures that are located centrally to the lesion. This may explain why the greatest change occurs in the BAEP elicited from the side of the lesion.

Widely cited studies may become overinterpreted, or misinterpreted, while the results of other studies become ignored. Thus, studies that show that lesions that affect the midbrain may cause changes in Peak V of the BAEP elicited from the contralateral ear have often been overlooked. Thus, Zanette *et al.* [240] found that Peak V was absent in the BAEP of certain patients with brainstem hemorrhage when elicited from the ear contralateral to the bleeding. Fischer *et al.* [39] showed that wave V of the BAEP was delayed and had a reduced amplitude when elicited from the opposite side in a patient with a lesion involving the inferior colliculus. It should also be noted that the study by Markand indeed found changes in the BAEP elicited from the side opposite to the lesion, but the changes were not noticeably larger than those in the BAEP elicited from the side of the lesion [110].

The discussions above emphasize that it is not only the lesions that affect the generator of a certain component of the BAEP that can cause changes in that component but also lesions of more peripheral location will result in changes in such components of the BAEP.

Comparison between Brainstem Auditory Evoked Potentials Obtained in Humans and in Animals

The BAEP obtained from animals that have been studied, including the monkey, consist of only four constant vertex-positive peaks. The reason is that the auditory nerve is much shorter in animals used in auditory experiments than it is in humans. This causes the travel time in the auditory nerve in animals to be too short to generate two clearly separate peaks. There may be other differences attributable to the differences in the ascending auditory pathway, mainly concerning the superior olivary complex [153, 154].

Animal experiments make it possible to study the effect of inactivation (ablation) of specific neural structures on the BAEP in addition to comparing the potentials recorded from specific structures of the ascending auditory pathway with the BAEP. The results of early studies of the neural generators of the BAEP in animals [11] which showed that Peak II of the cat BAEP was generated in the cochlear nucleus, resulted in the erroneous assumption that Peak II of the human BAEP was also generated in the cochlear nucleus. This misinterpretation of animal data was a result of anatomical differences between the auditory nerve in humans and in the animals used in such experiments.

The auditory nerve in humans is approximately 2.5 cm long [97, 98] compared with 0.5–0.8 cm in the cat [45]. The auditory nerve in humans is therefore sufficiently long to generate two well-separated peaks in the BAEP (I and II), while it generates only one peak in the BAEP in animals. The auditory nerve in humans is longer than in small animals because humans have a larger head and a much larger subarachnoidal space in the cerebellopontine angle than animals commonly used for studies of the auditory system.

Later it was shown that under favorable circumstances two separate peaks generated by the auditory nerve could be identified in the BAEP of small animals. The latency of the peak that follows the initial vertex-positive peak (Peak I) of the BAEP in small animals is approximately 0.4 ms longer than that of Peak I [1, 212] and this peak may correspond to Peak II in humans.

CONCLUSION ABOUT THE NEURAL GENERATORS OF THE BRAINSTEM AUDITORY EVOKED POTENTIALS

A synthesis of the results, using several different methods, provides the following general description of the neural generators of vertex-positive peaks of the click-evoked BAEP recorded between electrodes placed on the vertex and the earlobe or mastoid on the side stimulated:

Peak I: Distal (peripheral) portion of the auditory nerve.
Peak II: Proximal (central) portion of the auditory nerve.
Peak III: Cochlear nucleus.
Peak IV: Probably structures that are close to the midline (superior olivary complex?)
Peak V: Sharp vertex positive peak: the termination of the lateral lemniscus in the inferior colliculus on the contralateral side. The slow negative peak (SN_{10}): dendritic potentials from the inferior colliculus.

It seems unlikely that any other structure of the auditory system than the auditory nerve can contribute to Peaks I and II because that cochlear nucleus cells would not fire sooner than 0.5 to 0.7 ms after the arrival of the neural volley in the auditory nerve. Peaks I and II are thus the only components of the BAEP that are generated by a single anatomical structure of the auditory nervous system (the auditory nerve). Later components of the BAEP are likely to receive contributions from several structures, including the auditory nerve. In fact, all structures of the ascending auditory nervous system other than the auditory nerve are likely to contribute to more than one peak of the BAEP. It is incompletely known to what extend the uncrossed pathway above the cochlear nucleus contributes to the BAEP. It is also important to consider that not only the vertex-positive peaks are generated by specific structures of the auditory nervous system, but also the vertex-negative peaks have more or less specific neural generators [144, 150, 241]. It is interesting to speculate what would have happened if it had been the vertex-negative peaks that were labeled and attention therefore drawn to the vertex-negative peaks instead of the vertex-positive peaks.

Despite all these studies of the neural generators of the BAEP it is not known with certainty whether the different components of the BAEP are generated by fiber tracts (white matter) or cell bodies in nuclei (gray matter). It seems as both of those two kinds of structures may contribute to far-field potentials. The fact that the auditory nerve is the sole generator of Peaks I and II shows that a nerve can contribute to the BAEP. This means that it is also reasonable to assume that fiber tracts can generate stationary peaks in the far field and thus contribute to the BAEP. The contribution to the BAEP that has been ascribed as coming from nuclei may in fact be generated by fiber tracts that lead from the nuclei. Whatever structures generate the sharp peaks in the far-field auditory potentials, their amplitude seems to depend on how well synchronized the neural activity is in the structure in question (nerve, fiber tracts, or nuclei) and whether activation of nuclei generates a dipole or if it has a closed field.

MIDDLE LATENCY RESPONSES

The middle latency responses (MLR) consists of evoked potentials that occur between 10 and 80 ms (or 10–100 ms) after a sound stimulus. The MLR is commonly recorded in a similar way as the BAEP, thus differentially between electrodes on the vertex and the earlobe contralateral to the ear to which the sound stimuli are applied. It is generated by more central neural structures than the BAEP, including the auditory cortex. These potentials were first described by Geisler et al. [52] and were later studied by many investigators such as Picton et al. [174] and more recently by Nina Kraus and her co-workers [95] (see [64]).

The labeling of the components of the MLR is maybe even more confusing than is the case for the BAEP. The most prominent components are labeled Na, Pa, Nb, Pb, Nc, Pc, and Nd with N for negative and P for positive waves (Fig. 11.13 [47]). The slow (SN_{10}) component of Peak V of the BAEP is usually visible in the MLR (as Na) because these responses are recorded with preservation of low frequencies (low settings of the high-pass filter in the amplifiers used). The Na component is followed by a large negative peak, Nb,

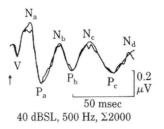

40 dBSL, 500 Hz, Σ2000

FIGURE 11.13 Middle latency responses (MLR) shown on a 100-ms time scale. The responses were elicited by click stimulation, presented at a rate of 10 pps (from Galambos, R., Makeig, S., and Talmachoff, P. J. A. (1981). 40 Hz auditory potentials recorded from the human scalp. *Proc. Natl. Acad. Sci. USA* 78:2643–2647).

with a latency of approximately 35 ms. Two more negative peaks, Nc and Nd, can usually be identified.

NEURAL GENERATORS OF THE MIDDLE LATENCY RESPONSES

Our knowledge about the origin of these potentials is sketchy at best. The fact that myogenic auditory evoked potentials occur in the time window of the MLR has added to the difficulties in determining the neural generators of the MLR. Other factors such as a much greater variability than that of the BAEP make interpretation of studies of the neural generators of the MLR more difficult than that of the BAEP. The degree of wakefulness affects the MLR and intraoperative studies are difficult to do because the MLR is affected by anesthesia. Thus, only a few studies have addressed the neural generators of the MLR in humans.

As early as 1958, Geisler *et al.* [52] suggested that these potentials might be generated by the auditory cortex and later Lee *et al.* [104] found evidence in intracranial recordings that the Pa component with normal latencies of 24–30 ms was generated by the auditory cortex. Some investigators have claimed that these potentials are generated by muscle activity elicited by sound stimulation.

Recent studies in animals have revealed that some of the components of the MLR may be generated by the nonclassical ascending auditory system [115]. The ventral and the caudo-medial portions of the medial geniculate body in the guinea pig give specific contributions to the MLR. The ventral portion of the medial geniculate body is associated with the classical (lemniscal) auditory pathway and relays information to the primary cerebral auditory cortex, where all information passes (see Chapter 5). The caudo-medial portion of the medial geniculate body contains relay neurons for the nonclassical ascending auditory system. McGee *et al.* [115] studied the contribution to the MLR recorded from these two parts of the medial geniculate body by injecting lidocaine in specific areas of the geniculate body. The MLR recorded from the scalp in guinea pigs overlying the temporal lobe was different from that recorded from a midline electrode position. Some of the components of the MLR recorded from a midline position may thus be generated by the nonclassical system, whereas those recorded from the skull over the temporal lobe are (mostly) generated by the classical system. The neural generators of the MLR in humans may be different from those in animals and it is uncertain how the different components of the MLR in animals correspond to the components of the MLR in humans.

THE 40-Hz RESPONSE

It is a general rule that the amplitude of sensory evoked potentials such as auditory evoked potentials decreases with increasing repetition rate of the stimuli. However, Galambos and his co-workers [47] have shown that the amplitude of the response to repetitive sound stimuli has its highest value at stimulus repetition rates of approximately 40 Hz (or pps) when elicited by clicks or short tone bursts (Fig. 11.14).

The reason for the increased amplitude at a stimulus repetition rate of 40 pps is that several peaks in the MLR occur with intervals of approximately 25 ms and therefore add to each other when the interval between individual stimuli is 25 ms, thus a repetition rate of 40 pps (or Hz). The 40-Hz response elicited by bursts of pure tones seems to be useful for determining the hearing threshold in noncooperative individuals.

OTHER SOUND-EVOKED POTENTIALS

FAR-FIELD FREQUENCY FOLLOWING RESPONSES IN HUMANS

Moushegian and co-workers showed [160, 161] that FFR to low-frequency tones could be recorded from electrodes placed on the scalp of human volunteers (Fig. 11.15). Masking studies confirmed that these potentials were of neural origin and not cochlear microphonics (CM) (Fig. 11.15A). The FFR recorded from electrodes placed on the scalp (vertex and mastoid) can be recorded in the frequency range from 250 to 2000 Hz and is most pronounced at the lowest frequencies (250 and 500 Hz) (Fig. 11.15A). The amplitude of

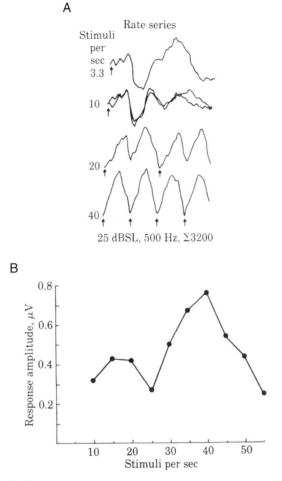

FIGURE 11.14 (A) The 40-Hz response. The change in the response when the repetition rate of the stimulation is changed. (B) Peak-to-peak amplitude of the response as a function of the repetition rate of the stimulation (A and B from Galambos, R., Makeig, S., and Talmachoff, P. J. A. (1981). 40 Hz auditory potentials recorded from the human scalp. *Proc. Natl. Acad. Sci. USA* 78:2643–2647).

these potentials increases with increasing stimulus intensity (Fig. 11.15B) [160]. The latency of these responses is approximately 6 ms, indicating that the FFR is generated by structures that are rostral to the auditory nerve. It is probably the part of the cochlea that has best frequencies below 2 kHz, the more apical portion, that generates the FFR [160, 161].

A

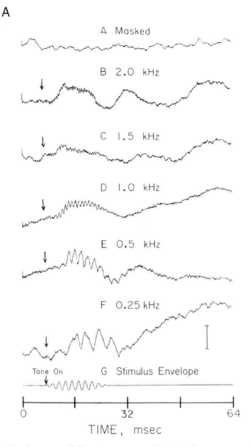

FIGURE 11.15   (A) The frequency following response (FFR). The responses to tones of different frequency presented in bursts of 18-ms duration and 5-ms rise and fall time. The stimulus envelope is indicated below by the stimulus artifact of a 500-Hz tone burst [160]. (B) The frequency following response (FFR). The response to 500-Hz tone bursts of 14-ms duration and 5-ms rise and fall time at different sound intensities (in dB SL) recorded from a human subjects from electrodes placed on the vertex and the earlobes. Reprinted from *Electroencephalogr. Clin. Neurophysiol.* **35**(6), Moushegian, G., Rupert, A. L., and Stillman, R. D., Laboratory note: Scalp-recorded early responses in man to frequencies in the speech range, pp. 665–667, copyright (1973), with permission from Elsevier Science.

MYOGENIC AUDITORY EVOKED POTENTIALS

When sensory evoked potentials are concerned, it is usually assumed that the recorded potentials are generated in the nervous system. Sensory stimuli can, however, also evoke motor responses that can be recorded as electromyographic

B

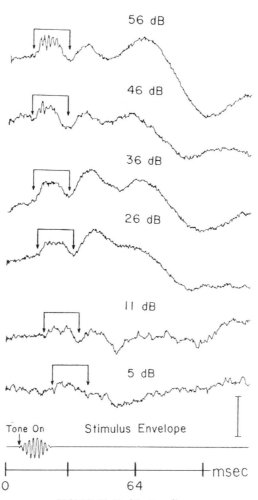

FIGURE 11.15 (*Continued*)

(EMG) potentials. Several investigators have described acoustically evoked muscle responses that occur with latencies between 10 and 30 ms in response to loud transient sounds [17, 52, 89, 112]. Auditory myogenic potentials can be recorded from an electrode placed behind the ear [99] and from electrodes placed on the parietal region of the scalp (Fig. 11.16). The responses are very variable and affected by muscle tension (Fig. 11.16). It is not only muscles on the head that respond to sound stimulation; extracranial muscles also respond to strong click sounds [52].

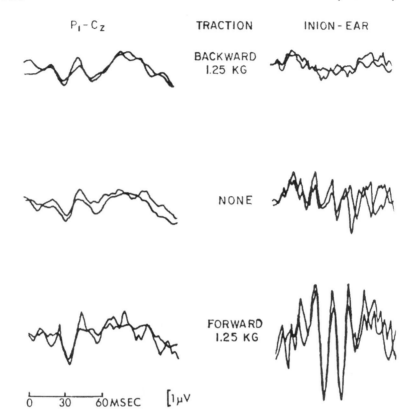

FIGURE 11.16 Myogenic potentials recorded from an electrode placed on P1 and Cz (left column) and on Pz (right column). The different rows of records were obtained with different tension of neck muscles (from Mast, T. E. (1965). Short latency human evoked responses to clicks. *J. Appl. Physiol.* **20**:725–730).

The myogenic responses are not associated with visible contractions of muscles, such as the general sound-evoked startle response, which may occur in response to unanticipated loud sounds involving a large number of muscles. Myogenic auditory evoked potentials are affected by attention, arousal, voluntary and involuntary muscle tension, and other factors (Fig. 11.16). A distinct component with a latency of 30 ms can be recorded from various positions on the scalp in response to click stimulation. The myogenic response is enhanced by voluntary muscle contraction. The variability of the responses makes it difficult to interpret the results of such recordings and this is why myogenic evoked responses never gained clinical use [24]. The latency of the earliest components of these myogenic potentials are between 10 and 30 ms, which

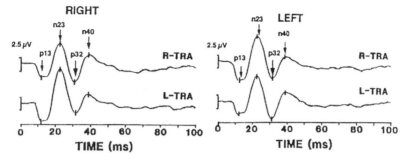

FIGURE 11.17   Responses from the trapezius muscle to click stimulation. The grand average of the responses to monaural stimulation (clicks, 100-dB hearing level presented at 3 pps) in 12 subjects is shown. Responses from the right (R-TRA) and left (L-TRA) trapezius muscles to stimulation of the right ear is shown (from Ferber-Viart, C., Soulier, N., Dubreuil, C., and Duclaux, R. (1998). Cochleovestibular afferent pathways of trapezius muscle responses to clicks in human. *Acta Otolaryngol.* (*Stockholm*) **118:**6–10).

makes such potentials sometimes occur at the end of the 10-ms recording window of the BAEP and thus easily distinguishable from the components of the BAEP that originate from auditory brainstem structures. Myogenic responses, however, may interfere with the neural components of the MLR, responses which has been an obstacle in the clinical use of the MLR.

Some of these responses recorded from extracranial muscles have been attributed to activation of the vestibular system and thus as part of the vestibular spinal reflexes (Fig. 11.17) [37].

SECTION II REFERENCES

1. Achor, L., and Starr, A. (1980a). Auditory brain stem responses in the cat. I. Intracranial and extracranial recordings. *Electroenceph. Clin. Neurophysiol.* **48**:154–173.
2. Achor, L., and Starr, A. (1980b). Auditory brain stem responses in the cat. II. Effects of lesions. *Electroenceph. Clin. Neurophysiol.* **48**:174–190.
3. Aitkin, L. (1986). "The Auditory Midbrain." Humana Press, Clifton, NJ.
4. Allen, J. B. (1996). Harvey Fletcher's role in the creation of communication acoustics. *J. Acoust. Soc. Am.* **99**:1825–1839.
5. Andersen, P., Eccles, J. C., Schmidt, R. F., and Yokota, T. (1964). Slow potential wave produced in the cunate nucleus by cutaneous volleys and by cortical stimulation. *J. Neurophysiol.* **27**:78–91.
6. Arthur, R. M., Pfeiffer, R. R., and Suga, N. (1971). Properties of "two tone inhibition" in primary auditory neurons. *J. Physiol. (Lond.)* **212**:593–609.
7. Blackburn, C. C., and Sachs, M. B. (1989). Classification of unit types in the anteroventral cochlear nucleus: PST histograms and regularity analysis. *J. Neurophysiol.* **62**:1303–1329.
8. Blackburn, C. C., and Sachs, M. B. (1990). The representations of steady-state vowel sound /ɛ/ in the discharge patterns of cat anteroventral cochlear nucleus neurons. *J. Neurophysiol.* **63**:1191–1212.
9. deBoer, E. (1967). Correlation studies applied to the frequency resolution of the cochlea. *J. Aud. Res.* **7**:209–217.
10. Britt, R., and Starr, A. (1978). Synaptic events and discharge patterns of cochlear nucleus cells. II. Frequency-modulated tones. *J. Neurophysiol.* **39**:179–194.
11. Buchwald, J., and Huang, C. (1975). Far-field acoustic response: Origins in the cat. *Science* **189**:382–384.
12. Carr, C. E., and Konishi, M. (1990). A circuit for detection of interaural time differences in the brainstem of the barn owl. *J. Neurosci.* **10**:3227–3246.
13. Casseday, J. H., Ehrlich, D., and Covey, E. (1994). Neural tuning for sound duration: Role of inhibitory mechanisms in the inferior colliculus. *Science* **264**:847–850.
14. Chatterjee, M., and Zwislocki, J. J. (1997). Cochlear mechanisms of frequency and intensity coding. I. The place code for pitch. *Hear Res.* **111**:65–75.
15. Chiappa, K. (1997). "Evoked Potentials in Clinical Medicine" (3rd ed.). Lippincott–Raven, Philadelphia.
16. Chiappa, K. H., and Hill, R. A. (1997). Brain stem auditory evoked potentials: Interpretation. *In* K. H. Chiappa (Ed.), "Evoked Potentials in Clinical Medicine" (3rd ed.). Lippincott–Raven, Philadelphia.

17. Cody, D. T. and Bickford, R. G. (1969). Averaged evoked myogenic responses in normal man. *Laryngoscope* **79**(3):400–416.

18. Cooper, N. P., Roberston, D., and Yates, G. K. (1993). Cochlear nerve fiber responses to amplitude-modulated stimuli: Variations with spontaneous rate and other response characteristics. *J. Neurophysiol.* **70**:370–386.

19. Covey, E., and Casseday, J. H. (1991). The monaural nuclei of the lateral lemniscus in an echolocating bat: Parallel pathways for analyzing temporal features of sound. *J. Neurosci.* **11**(11):3456–3470.

20. Covey, E., Vater, M., and Casseday, J. H. (1991). Binaural properties of single units in the superior olivary complex of the mustached bat. *J. Neurophysiol.* **66**(3):1080–1094.

21. Dallos, P. J. (1964). Study of the acoustic reflex feedback loop. *IEEE Trans. Bio-Med. Eng.* **11**:1–7.

22. Davis, H., and Hirsh, S. (1976). The audiometric utility of brain stem responses to low-frequency sounds. *Audiology* **15**:181–195.

23. Doucet, J. R., and Relkin, E. M. (1995). The perstimulus compound action potential: A new method for recording a compound potential from the chinchilla auditory nerve. *Aud. Neurosci.* **1**:151–168.

24. Douek, E. E., Ashcroft, P. B., and Humphries, K. N. (1976). The clinical value of the postauricular myogenic (crossed acoustic) response in neuro-otology. *In* S. D. G. Stephens (Ed.), "Disorders of Auditory Function" (pp. 139–144). Academic Press, London.

25. Eggermont, J. J., (1993). Wiener and Volterra analyses applied to the auditory system. *Hear. Res.* **66**:177–201.

26. Eggermont, J. J. (1994). Temporal modulation transfer functions for AM and FM stimuli in cat auditory cortex: Effects of carrier type, modulating waveform, and intensity. *Hear. Res.* **74**:51–66.

27. Eggermont, J. J., Johannesma, P. I. M., and Aertsen, A. M. H. (1983). Reverse-correlation methods in auditory research. *Q. Rev. Biophysiol.* **16**:341–414.

28. Ehret, G. and Romand, R. (Eds.). (1997). "The Central Auditory Pathway." Oxford Univ. Press, New York.

29. Erulkar, S. D. (1959). The responses of single units of the inferior colliculus of the cat to acoustic stimulation. *Proc. R. Soc. Ser. B.* **150**:336–355.

30. Erulkar, S. D., Nelson, P. G., and Bryan, J. S. (1968). Experimental and theoretical approaches to neural processing in the central auditory pathway. *In* "Contributions to Sensory Physiology" (Vol. 3). Academic Press, New York.

31. Evans, E. F. (1972). The frequency response and other properties of single fibers in the guinea-pig cochlear nerve. *J. Physiol.* **226**:263–287.

32. Evans, E. F. (1975). Normal and abnormal functioning of the cochlear nerve. *Symp. Zool. Soc. Lond.* **37**:133–165.

33. Evans, E. F. (1977). Frequency selectivity at high signal levels of single units in cochlear nerve nucleus. *In* E. F. Evans and J. P. Wilson (Eds.), "Psychophysics and Physiology of Hearing" (pp. 185–192). Academic Press, New York.

34. Evans, E. F. (1992). Auditory processing of complex sounds: an overview. *Phil. Trans. R. Soc. Lond. B.* **336**:295–306.

35. Evans, E. F., and Nelson, P. G. (1973). The responses of single neurons in the cochlear nucleus of the cat as a function of their location and the anaesthetic state. *Exp. Brain Res.* **17**:402–427.

36. Feddersen, W. E., Sandel, T. T., Teas, D. C., and Jeffress, L. A. (1957). Localization of high frequency tones. *J. Acoust. Soc. Am.* **29**:988–991.

37. Ferber-Viart, C., Soulier N., Dubreuil, C. and Duclaux, R. (1998). Cochleovestibular afferent pathways of trapezius muscle responses to clicks in human. *Acta Otolaryngol. (Stockholm)* **118**:6–10.

38. Fletcher, H. (1953). "Speech and Hearing in Communication." Krieger, Huntington, New York.

39. Fischer, C., Bognar, L., Turjman, F., and Lapras, C. (1994). Auditory early- and middle-latency evoked potentials in patients with quadrigeminal plate tumors. *Neurosurgery* **35**:45–51.

40. Frisina, R. D., Karcich, K. J., Tracy, T. C., Sullivan, D. M., Walton, J. P. and Colombo, J. (1996). Preservation of amplitude modulation coding in the presence of background noise by chinchilla auditory-nerve fibers. *J. Acoust. Soc. Am.* **99**(1):475–490.

41. Frisina, R. D., Smith, R. L., and Chamberlain, S. C. (1985). Differential encoding of rapid changes in sound amplitude by second-order auditory neurons. *Exp. Brain Res.* **60**:417–422.

42. Frisina, R. D., Smith, R. L., and Chamberlain, S. C. (1990a). Encoding of amplitude modulation in the gerbil cochlear nucleus. I. A hierarchy of enhancement. *Hear. Res.* **44**: 99–122.

43. Frisina, R. D., Smith, R. L., and Chamberlain, S. C. (1990b). Encoding of amplitude modulation in the gerbil cochlear nucleus. II. Possible neural mechanisms. *Hear. Res.* **44**: 123–142.

44. Frisina, R. D., Walton, J. P., and Karcich, K. J. (1994). Dorsal cochlear nucleus single neurons can enhance temporal processing capabilities in background noise. *Exp. Brain Res.* **102**(1):160–164.

45. Fullerton, B. C., Levine, R. A., Hosford-Dunn, H. L., Kiang, N. Y.-S. (1987). Comparison of cat and human brainstem auditory evoked potentials. *Electroencephalography and Clinical Neurophysiology.* **66**:547–570.

46. Galambos, R., and Hecox, K. (1977). Clinical application of the brainstem auditory evoked potentials. *In* J. E. Desmedt "Progress in Clinical Neurophysiology: Auditory Evoked Potentials in Man: Psychopharmacology Correlates of EPs." (Vol. 2, pp. 1–19). Karger, Basel.

47. Galambos, R., Makeig, S., and Talmachoff, P. J. A. (1981). 40 Hz auditory potentials recorded from the humans scalp. *Proc. Natl. Acad. Sci. USA* **78**:2643–2647.

48. Galambos, R., Myers, R., and Sheatz, G. (1961). Extralemniscal activation of auditory cortex in cats. *Am. J. Physiol.* **200**:23–28.

49. Gasser, H. S. (1941). The classification of nerve fibers. *Ohio. J. Sci.* **41**:145–159.

50. Gasser, H. S., and Graham, H. T. (1933). Potentials produced in the spinal cord by stimulation of dorsal roots. *Am. J. Physiol.* **103**:303–320.

51. Geisler, C. D. (1998). "From Sound to Synapse." Oxford Univ. Press, New York.

52. Geisler, C. D., Frishkopf, L. S., and Rosenblith, W. A. (1958). Extracranial responses to acoustic clicks in man. *Science* **128**:1210–1211.

53. Geisler, C. D. Rhode, W. S. and Kennedy, D. T. (1974). The responses to tonal stimuli of single auditory nerve fibers and their relationship to basilar membrane motion in the squirrel monkey. *J. Neurophysiol.* **37**:1156–1172.

54. Gelfan, W. R., and Tarlov, I. M. (1955). Differential vulnerability of spinal cord structures to anoxia. *J. Neurophysiol.* **18**:170–188.

55. Ghoshal, S., Kim, D. O., and Northrop, R. B. (1992). Amplitude-modulated tone encoding behavior of cochlear nucleus neurons: Modeling study. *Hear. Res.* **58**:153–165.

56. Goldberg, J. M., and Brown, P. B. (1968). Functional organization of the dog superior olivary complex: An anatomical and electrophysiological study. *J. Neurophysiol.* **31**:639–656.

57. Goldberg, J. M., and Brown P. B. (1969). Response of binaural neurons of dog superior olivary complex to dichotic tonal stimuli: Some physiological mechanisms of sound localization. *J. Neurophysiol.* **32**:613–636.

58. Goldstein, J. (1978). Mechanisms of signal analysis and pattern perception in periodicity pitch. *Audiology* **17**:421–445.

59. Goldstein, M. H., Jr. (1960). A statistical model for interpreting neuroelectric responses. *Informat. Control* **3**:1–17.

60. Graybiel, A. M. (1972). Some fiber pathways related to the posterior thalamic region in the cat. *Brain Behav. Evol.* **6**:363–393.

61. Guinan, J. J., Guinan, S. S., and Norris, B. E. (1972). Single auditory units in the superior olivary complex. I. Responses of sounds and classifications based on physiological properties. *Int. J. Neurosci.* **4**:101–120.

62. Guinan, J. J., Jr., Warr, W. B. and Norris, B. E. (1984). Topographic organization of the olivocochlear projections from the lateral and medial zones of the superior olivary complex. *J. Comp. Neurol.* **226**(1):21–27.

63. Gummer, M., Yates, G. K., and Johnstone, B. M. (1988). Modulation transfer function of efferent neurons in the guinea pig cochlea. *Hear. Res.* **36**:41–52.

64. Hall, J. W. (1992). "Handbook of Auditory Evoked Responses." Allyn and Bacon, Boston.

65. Harrison, J. M., and Howe, M. E. (1974). Anatomy of the descending auditory system in auditory system. *In* W. D. Keidel and W. D. Neff (Eds.), "Handbook of Sensory Physiology" (Vol V/1, pp. 363–388). Springer-Verlag, Berlin.

66. Harrison, R. V., and Evans, E. F. (1982). Reverse correlation study of cochlear filtering in normal and pathological guinea-pig ears. *Hear. Res.* **6**:303–314.

67. Hashimoto, I. (1982a). Auditory evoked potentials from the humans midbrain: Slow brain stem responses. *Electroencephalog. Clin. Neurophysiol.* **53**:652–657.

68. Hashimoto, I. (1982b). Auditory evoked potentials recorded directly from the human VIIIth nerve and brain stem: Origins of their fast. and slow components. *Kyoto Symp.* (*EEG Suppl.*) **36**:305–314.

69. Hashimoto, I., Ishiyama, Y., Yoshimoto, T., and Nemoto, S. (1981). Brainstem auditory evoked potentials recorded directly from humans brain-stem and thalamus. *Brain* **104**:841–859.

70. Hatayama, T., Sekiya, T., Suzuki S., and Iwabuchi, T. (1999). Effect of compression on the cochlear nerve: A short- and long-term electrophsysiological and histological study. *Neurological Research* **21**:559-610.

71. Heil, P., Rajan, R., Irvine, D. R. F. (1992a). Sensitivity of neurons in cat primary auditory cortex to tones and frequency-modulated stimuli. I. Effects of variation of stimulus parameters. *Hear. Res.* **63**:108–134.

72. Heil, P., Rajan, R., and Irvine, D. R. F. (1992b). Sensitivity of neurons in cat primary auditory cortex to tones and frequency-modulated stimuli. I. Organization of response properties along the 'isofrequency' dimension. *Hear. Res.* **63**:135–156.

73. Hong-Bo, Z., and Zhi-An, L. (1996). Processing of modulation frequency in the dorsal cochlear nucleus of the guinea pig: Sinusoidal frequency-modulated tones. *Hear. Res.* **95**:120–134.

74. Honrubia, V., and Ward, P. H. (1968). Longitudinal distribution of the cochlear microphonics inside the cochlear duct (guinea pig). *J. Acoust. Soc. Am.* **44**:951–958.

75. Huffman, R. F., Argeles, P. C., and Covey, E. (1998). Processing of sinusoidally frequency modulated signals in the nuclei of the lateral lemniscus of the big brown bat. *Eptesicus fuscus. Hear. Res.* **126**:161–180.

76. Irvine, D. R. F. (1987). Interaural intensity differences in the cat: Changes in sound pressure level at the two ears associated with azimuthal displacements in the frontal horizontal plane. *Hear. Res.* **26**:267–286.

77. Irvine D. R. F. (1992). Physiology of the auditory brainstem. *In* Popper AN, Fay RR. (Eds.), "The Mammalian Auditory Pathway: Neurophysiology," (pp 153–231), New York, Springer.

78. Jeffress, L. A. (1948). A place theory of sound localization. *J. Comp. Physiol.* **41**:35–39.

79. Jen, P. H., and Suga, N. (1976). Coordinated activities of middle-ear and laryngeal muscles in echolocating bats. *Science* **91**(4230):950–952.

80. Jewett, D. L. (1987). The 3-channel Lissajous' trajectory of the auditory brain-stem response. IX. Theoretical aspects. *Electroencephalogr. Clin. Neurophysiol.* **68**:386–408.

81. Jewett, D. L., and Williston, J. S. (1971). Auditory-evoked far fields averaged from scalp of humans. *Brain* **94**:681–696.

82. Johnstone B. M., Patuzzi R., and Yates. G. K. (1986). Basilar membrane measurements and the travelling wave. *Hear. Res.* **22**:147–153.

83. Joris, P. X., and Yin, T. C. T. (1992). Responses to amplitude-modulated tones in the auditory nerve of the cat. *J. Acoust. Soc. Am.* **91**:215–232.

84. Katsuki, Y., Sumi, T., Uchiyama, H., and Watanabe, T. (1958). Electric responses of auditory neurons in cat to sound stimulation. *J. Neurophysiol.* **21**:569–588.

85. Khanna, S. M., and Teich, M. C. (1989a). Spectral characteristics of the responses of primary auditory-nerve fibers to amplitude-modulated signals *Hear. Res.* **39**:143–158.

86. Khanna, S. M., and Teich, M. C. (1989b). Spectral characteristics of the responses of primary auditory-nerve fibers to frequency-modulated signals. *Hear. Res.* **39**:159–176.

87. Kiang, N. Y.-S. (1961). The use of computers in studies of auditory neurophysiology. *Transact. Am. Acad. Ophthalmol. Otolaryngol.* **65**:735–747.

88. Kiang, N. Y.-S. (1975). Stimulus representation in the discharge patterns of auditory neurons. *In* E. L. Eagles (Ed.), "The Nervous System" (pp. 81–96). Raven, New York.

89. Kiang, N. Y.-S., Christ, A. H., French, M. A., and Edwards, A. G. (1963). Postauricular electric response to acoustic stimuli in humans. *Research laboratory of electronics. MIT Q. Prog. Rep.* **68**: 218–225.

90. Kiang, N. Y.-S., Morest, D. K., Godfrey, D. A., Guinan, J. J., and Kane, E. C. (1973). Stimulus coding at caudal levels of the cat's auditory nervous system. I. Response characteristics of single units. *In* A. R. Møller (Ed.), "Basic Mechanisms in Hearing" (pp. 455–478). Academic Press, New York.

91. Kiang, N. Y.-S, and Peake, W. T. (1962). Cochlear responses to condensation and rarefaction clicks. *Biophys. J.* **2**:23–34.

92. Kiang, N. Y.-S., Watanabe, T., Thomas, E. C., and Clark, L. F., (1965). "Discharge Patterns of Single Fibers in the Cat's Auditory Nerve." MIT Press, Cambridge, MA.

93. Kidd R. C. Weiss T. F. (1990). Mechanisms that degrade timing information in the cochlea. *Hear. Res.* **9**:181–208.

94. Kim, D. O., Sirianni, J. G., and Chang, S. O. (1990). Responses of DCN-PVCN neurons and auditory nerve fibers in unanesthetized decrebrate cats to AM and pure tones: Analysis with autocorrelation/power-spectrum. *Hear. Res.* **45**:95–114.

95. Kraus, N., Odzamar, O., Heir, D., and Stein, L. (1982). Auditory middle latency responses (MLRs) in patients with cortical lesions. *Electroencephalogr. Clin. Neurophysiol* **54**:275–287.

96. Kuroki, A., and Møller, A. R. (1995). Microsurgical anatomy around the foramen of Luschka with reference to intraoperative recording of auditory evoked potentials from the cochlear nuclei. *J. Neurosurg.* **82**:933–939.

97. Lang, J. (1981). Facial and vestibulocochlear nerve, topographic anatomy and variations. *In* M. Samii and P. J. Jannetta (Eds.) "The Cranial Nerves" (pp. 363–377). Springer-Verlag, New York.

98. Lang, J. (1991). Clinical Anatomy of the Posterior Cranial Fossa and Its Foramina. Thieme-Verlag, Stuttgart.

99. Langner. G. (1992). A Review: Periodicity coding in the auditory system. *Hear. Res.* **60**:115–142.

100. Langner, G., Schreiner, C., and Merzenich, M. M. (1987). Covariation of latency and temporal resolution in the inferior colliculus of the cat. *Hear. Res.* **31**:197–201.

101. Langner, G., and Schreiner, C. E. (1988a). Periodicity coding in the inferior colliculus of the cat. I. Neuronal mechanisms. *J. Neurophysiol.* (1987). **60**:1799–1822.
102. Langner, G., and Schreiner, C. E. (1988b). Periodicity coding in the inferior colliculus of the cat. II. Topographical organization. *J. Neurophysiol.* **60**:1823–1840.
103. Lazorthes, G., Lacomme, Y., Ganbert, J., and Planel, H. (1961). La constitution du nerf auditif. *Presse Med.* **69**:1067–1068.
104. Lee, Y. S., Lueders, H. Dinner, D. S. Lesser, R. P. Han, J. and Klemm, G. (1984). Recording of auditory evoked potentials in man using chronic subdermal electrodes. *Brain,* **107**:102.
105. Lepore *et al.* (1997). Cortical and callosal contribution to sound localization. *In* J. Syka (Ed.), "Acoustical Signal Processing in the Central Auditory System" (Chapter 35, pp. 389–398). Plenum, New York.
106. Liberman, M. C. (1978). Auditory-nerve response from cats raised in low-noise chamber. *J. Acoust. Soc. Am.* **63**: 442–455.
107. Licklider, J. C. R. (1959). Three auditory theories. *In* S. Koch (Ed.), "Psychology: A Study of a Science" (Vol. 1) NYC: McGraw Hill.
108. Lorente de No, R. (1933). Anatomy of the eighth nerve III: General plan of structure of the primary cochlear nuclei. *Laryngoscope* **43**: 327–350.
109. Lorente de No, R. (1947). Analysis of the distribution of action currents of nerve in volume conductors. *Studies Rockefeller Inst. Med. Res.* **132**:384–482.
110. Markand, O. N, Farlow, M. R., Stevens, J. C., and Edwards, M. K. (1989). Brain-stem auditory evoked potential abnormalities with unilateral brain-stem lesions demonstrated by magnetic resonance imaging. *Arch. Neurol.* **46**:295–299.
111. Martin, W. H., Pratt, H., and Schwegler, J. W. (1995). The origin of the human auditory brainstem response wave. II. *Electroencephalogr. Clin. Neurophysiol.* **96**:357–370.
112. Mast, T. E. (1965). Short latency human evoked responses to clicks. *J. Appl. Physiol.* **20**:725–730.
113. Mast, T. E. (1973). Binaural interaction and contralateral inhibition in dorsal cochlear nucleus of chinchilla. *J. Neurophysiol.* **62**:61–70.
114. McAnally, K. I., and Calford, M. B. (1989). Spectral hyperacuity in the cat: Neural response to frequency modulated tone pairs. *Hear. Res.* **41**:237–248.
115. McGee, T., Kraus, N., Littman, T., and Nicol, T. (1992). Contributions of medial geniculate body subdivisions to the middle latency response. *Hear. Res.* **61**: 147–154.
116. Middlebrooks, J. C. (1987). Binaural mechanisms of spatial tuning in the cat's superior colliculus distinguished using monaural occlusion. *J. Neurophysiol.* **57**:688–701.
117. Miller, R. L., Calhoun, B. M., and Young, E. D. (1999). Discriminability of vowel representations in cat auditory-nerve fibers after acoustic trauma. *J. Acoust. Soc. Am.* **105**:311–325.
118. Møller, A. R. (1969a). Unit responses in the cochlear nucleus of the rat to pure tones. *Acta Physiol. Scand.* **75**:530–541.
119. Møller, A. R. (1969b). Unit responses in the rat cochlear nucleus to repetitive transient sounds. *Acta Physiol. Scand.* **75**:542–551.
120. Møller, A. R. (1970). The use of correlation analysis in processing neuro-electric data. *In* "Progress in Brain Research," (Vol. 33, pp. 87–99). J. P. Schade and J. Smith (Eds.). Elsevier Biomedical Press, Amsterdam, The Netherlands, .
121. Møller, A. R. (1971). Unit responses in the rat cochlear nucleus to tones of rapidly varying frequency and amplitude. *Acta Physiol. Scand.* **81**:540–556.
122. Møller, A. R. (1972). Coding of amplitude and frequency modulated sounds in the cochlear nucleus of the rat. *Acta Physiol. Scand.* **86**:223–238.
123. Møller, A. R. (1973). Statistical evaluation of the dynamic properties of cochlear nucleus units using stimuli modulated with pseudorandom noise. *Brain Res.* **57**:443–456.

124. Møller, A. R. (1974a). Responses of units in the cochlear nucleus to sinusoidally amplitude modulated tones. *Exp. Neurol.* **45**:104–117.

125. Møller, A. R. (1974b). Coding of sounds with rapidly varying spectrum in the cochlear nucleus. *J. Acoust. Soc. Am.* **55**:631–640.

126. Møller, A. R. (1975a). Latency of unit responses in the cochlear nucleus determined in two different ways. *J. Neurophysiol.* **38**:812–821.

127. Møller, A. R. (1975b). Dynamic properties of excitation and inhibition in the cochlear nucleus. *Acta Physiol. Scand.* **93**:442–454.

128. Møller, A. R. (1976). Dynamic properties of primary auditory fibers compared with cells in the cochlear nucleus. *Acta Physiol. Scand.* **98**:157–167.

129. Møller, A. R. (1977). Frequency selectivity of single auditory-nerve fibers in response to broadband noise stimuli. *J. Acoust. Soc. Am.* **62**:135–142.

130. Møller, A. R. (1979). Coding of increments and decrements in stimuli intensity in single units in the cochlear nucleus of the rat. *J. Neurosci. Res.* **4**:1–8.

131. Møller, A. R. (1983a). "Auditory Physiology." Academic Press, New York.

132. Møller, A. R. (1983b). Frequency selectivity of phase looking of complex sounds in the auditory nerve of the rat. *Hear. Res.* **11**:267–284.

133. Møller, A. R. (1985). Origin of latency shift of cochlear nerve potentials with sound intensity. *Hear. Res.* **17**:177–189.

134. Møller, A. R. (1988). Use of zero-phase digital filters to enhance brainstem auditory evoked potentials (BAEPs). *Electroencephalogr. Clin. Neurophysiol.* **71**:226–232.

135. Møller, A. R. (1995). "Intraoperative Neurophysiologic Monitoring." Harwood, Luxembourg.

136. Møller, A. R. (1998). Neural Generators of the Brainstem Auditory Evoked Potentials. *In* Kibbe (Ed.), Twenty-Five Years of ABR: Historical Perspective, Curent Issues, and Future Directions. Seminars in Hearing (Vol 19, pp. 11–27).

137. Møller, A. R. (1999). Review of the roles of temporal and place coding of frequency in speech discrimination. *Acta Otolaryngol.* **119**:424–430.

138. Møller, A. R., and Burgess, J. E. (1986). Neural generators of the brain-stem auditory evoked potentials (BAEPs) in the rhesus monkey. *Electroencephalogr. Clin. Neurophysiol.* **65**:361–372.

139. Møller, A. R., Colletti, V., and Fiorino, F. G. (1994). Neural conduction velocity of the human auditory nerve: Bipolar recordings from the exposed intracranial portion of the eighth nerve during vestibular nerve section. *Electroencephalogr. Clin. Neurophysiol.* **92**:316–320.

140. Møller, A. R., and Jho, H. D. (1989). Response from the exposed intracranial human auditory nerve to low-frequency tones: Basic characteristics. *Hear. Res.* **38**:163–175.

141. Møller, A. R., and Jho, H. D. (1990). Late components in the compound action potentials (CAP) recorded from the intracranial portion of the human eighth nerve. *Hear. Res.* **45**:75–86.

142. Møller, A. R., and Jho. H. D. (1991). Compound action potentials recorded from the intracranial portion of the auditory nerve in man: Effects of stimulus intensity and polarity. *Audiology* **30**:142–163.

143. Møller, A. R., Jho, H. D., and Jannetta, P. J. (1994). Preservation of hearing in operations on acoustic tumors: An alternative to recording BAEP. *Neurosurgery* **34**:688–693.

144. Møller, A. R., Jho, H. D., Yokota, M., and Jannetta, P. J. (1995). Contribution from crossed and uncrossed brainstem structures to the brainstem auditory evoked potentials (BAEP): A study in human. *Laryngoscope* **105**:596–605.

145. Møller, A. R., and Jannetta, P. J. (1981). Compound action potentials recorded intracranially from the auditory nerve in man. *Exp. Neurol.* **74**:862–874.

146. Møller, A. R., and Jannetta, P. J. (1982a). Evoked potentials from the inferior colliculus in man. *Electroencephalogr. Clin. Neurophysiol.* **53**:612–620.

147. Møller, A. R., and Jannetta, P. J. (1982b). Auditory evoked potentials recorded intracranially from the brainstem in man. *Exp. Neurol.* **78**:144–157.

148. Møller, A. R., and Jannetta, P. J. (1983a). Auditory evoked potentials recorded from the cochlear nucleus and its vicinity in man. *J. Neurosurg.* **59**:1013–1018.

149. Møller, A. R., and Jannetta, P. J. (1983b). Interpretation of brainstem auditory evoked potentials: Results from intracranial recordings in humans. *Scand. Audiol.* **12**:125–133.

150. Møller, A. R., Jannetta, P. J., and Jho, H. D. (1994). Click-evoked responses from the cochlear nucleus: A study in human. *Electroencephalogr. Clin. Neurophysiol.* **92**:215–224.

151. Møller, A. R., Jannetta, P. J., and Møller, M. B. (1981). Neural generators of brainstem evoked potentials: Results from human intracranial recordings. *Ann. Otol. Rhinol. Laryngol.* **90**(6):591–596.

152. Møller, A. R., Jannetta, P. J., and Sekhar, L. N. (1988). Contributions from the auditory nerve to the brainstem auditory evoked potentials (BAEPs): Results of intracranial recording in man. *Electroencephalogr. Clin. Neurophysiol.* **71**:198–211.

153. Moore, J. K. (1987a). The human auditory brain stem: A comparative view. *Hear. Res.* **29**:1–32.

154. Moore, J. K. (1987b). The human auditory brain stem as a generator for auditory evoked potentials. *Hear. Res.* **29**:33–44.

155. Morest, D. K. (1964). The neuronal architecture of the medial geniculate body of the cat. *J. Anat. (Lond.)* **98**:611–630.

156. Morest, D. K. (1965). The laminar structure of the medial geniculate body of the cat. *J. Anat. (Lond.)* **99**:143–160.

157. Morest, D. K., and Oliver, D. L. (1984). The neuronal architecture of the inferior colliculus in the cat. *J. Comp. Neurol.* **222**:209–236.

158. Mountain, D. C., and Cody, A. R. (1999). Multiple modes of inner hair cell stimulation. *Hear. Res.* **132**:1–14.

159. Moushegian, G., Rupert, A. L., and Gidda, J. S. (1975). Functional characteristics of superior olivary neurons to binaural stimuli. *J. Neurophysiol.* **38**:1037–1048.

160. Moushegian, G., Rupert, A. L., and Stillman, R. D. (1973). Laboratory note: Scalp-recorded early responses in man to frequencies in the speech range. *Electroencephalogr. Clin. Neurophysiol.* **35**(6):665–667.

161. Moushegian, G., Rupert, A. L., and Stillman, R. D. (1978). Evaluation of frequency following potentials in man: Masking and clinical studies. *Electroencephalogr. Clin. Neurophysiol.* **45**:711–718.

162. Müller, M., Roberston, D., and Yates, G. K. (1991). Rate-versus-level functions of primary auditory nerve fibres: Evidence of square law behavior of all fibre categories in the guinea pig. *Hear. Res.* **55**:50–56.

163. Musiek, F. E., and Geurkink, N. A. (1982). Auditory brainstem response and central auditory test findings for patients with brain stem lesions: A preliminary report. *Laryngoscope* **92**:891–900.

164. Neff, W. D., Diamond, J. T., and Casseday, J. H. (1975). Behavioral studies of auditory discrimination: Central nervous system. *In* W. D. Keidel, and W. D. Neff (Eds.), "Handbook of Sensory Physiology" (Vol. 5/2, pp. 307–400). Springer-Verlag, Berlin.

165. Oliver, D. L., and Morest D. K. (1984). The central nucleus of the inferior colliculus in the cat. *J. Comp. Neurol.* **222**:237–264.

166. O'Neill, W. E. and Suga, N. (1991a). Target range-sensitive neurons in the auditory cortex of the mustached bat. *Science* **203**:69–73.

167. O'Neill, W. E. and Suga, N. (1991b). Encoding of target range information and its representation in the auditory cortex of the mustached bat. *J. Neurosci* **2**:17–31.

168. Osman, E., and Galambos, R. (1967). Activation of auditory cortex by clicks after bilateral lesions of the brachium of the inferior colliculus. *J. Acoust. Soc. Am.* **42**:512–514.

169. Ota, C. Y., and Kimura, R. S. (1980). Ultrastructural study of the human spiral ganglion. *Acta Otolaryngol.* 53–62.

170. Palmer, A. R. (1982). Encoding of rapid amplitude fluctuations by cochlear-nerve fibers in the guinea-pig. *Arch. Otorhinolaryngol.* 236:197–202.

171. Peake, W. T., Goldstein, M. H., and Kiang, N. Y.-S. (1962). Responses of the auditory nerve to repetitive stimuli. *J. Acoust. Soc. Am.* 34:562–570.

172. Pfeiffer, R. R. (1966). Classification of response patterns of spike discharges for units in the cochlear nucleus: Tone-burst stimulation. *Exp. Brain Res.* 1:220–235.

173. Pickles, J. O. (1988). "An Introduction to the Physiology of Hearing" (2nd ed.). Academic Press, London.

174. Picton, T. W., Hillyard, S. A., Krausz, H. I., and Galambos, R. (1974). Human auditory evoked potentials. I. Evaluation of components. *Electroencephalogr. Clin. Neurophysiol.* 36:176–190.

175. Portmann, M., Cazals, Y., Negrevergne, M., and Aran, J. M. (1980). Transtympanic and surface recordings in the diagnosis of retrocochlear lesions. *Acta Otolaryngol.* 89:362–369.

176. Pratt, H., Bleich, N., and Martin, W. H. (1985). Three-channel Lissajous' trajectory of humans auditory brain-stem evoked potentials. I. Normative measures. *Electroencephalogr. Clin. Neurophysiol.* 61:530–538.

177. Pratt, H., Martin, W. H., Schwegler, J. W., Rosenwasser, R. H., and Rosenberg, S. J. (1992). Temporal correspondence of intracranial, cochlear and scalp-recorded humans auditory nerve action potentials. *Electroencephalogr. Clin. Neurophysiol.* 84:447–455.

178. Rees, A., and Møller, A. R. (1983). Responses of neurons in the inferior colliculus of the rat to AM and FM tones. *Hear. Res.* 10:301–330.

179. Rees, A., and Møller, A. R. (1987). Stimulus properties influencing the responses of inferior colliculus neurons to amplitude-modulated sounds. *Hear. Res.* 27:129–144.

180. Rhode, W. S. (1994). Temporal coding of 200% amplitude modulated signals in the ventral cochlear nucleus of cat. *Hear. Res.* 77:43–68.

181. Rhode, W. S., and Greenberg, S. (1994a). Lateral suppression and inhibition in the cochlear nucleus of the cat. *J. Neurophysiol.* 71(2):493–514.

182. Rhode, W. S., and Greenberg, S. (1994b). Encoding of amplitude modulation in the cochlear nucleus of the cat. *J. Neurophysiol.* 71(5):1797–1825.

183. Rhode, W. S., and Kettner, R. E. (1987). Physiological study of neurons in the dorsal and posteroventral cochlear nucleus of the unanesthetized cat. *J. Neurophysiol.* 57(2):414–442.

184. Rodrigues-Dagaeff, C., Simm, G., de Ribaupierre, Y., Villa, A., de Ribaupierre F., Rouiller, E. M. (1989). Functional organization of the ventral division of the medial geniculate body of the cat: Evidence for a rostro-caudal gradient of response properties and cortical projections. *Hear. Res.* 39:103–126.

185. Rose, J. E., Galambos, R., and Hughes, J. R. (1959). Microelectrode studies of the cochlear nuclei in the cat. *Bull. Johns Hopkins Hosp.* 104:211–251.

186. Rose, J. E., Gross, N. B., Geisler, C. D., and Hind, J. E. (1966). Some neural mechanisms in the inferior colliculus of the cat which may be relevant to localization of a sound source. *J. Neurophysiol.* 29:288–314.

187. Rose, J. E., Hind, J. E., Anderson, D. J., and Brugge, J. F. (1971). Some effects of stimulus intensity on response of auditory fibers in the squirrel monkey. *J. Neurophysiol.* 34:685–699.

188. Rouiller, E. M., Ribaupierre de Y, Toros-Morel A, Ribaupierre de F. (1981). Neural coding of repetitive clicks in the medial geniculate body of cat. *Hear. Res.* 3:81–100.

189. Rouiller, E. M., Rodrigues-Dagaeff, C., Simm, G., de Ribaupierre, Y., Villa, A., de Ribaupierre, F. (1989). Functional organization of the medial division of the medial geniculate body of the cat: Tonotopic organization, spatial distribution of response properties and cortical connections. *Hear. Res.* 39:127–142.

190. Ruben R. J., and Walker A. E. (1963). The VIIIth nerve action potential in Ménière's disease. *Laryngoscope* **11**:1456–1464.

191. Ruggero, M. A., and Rich, N. C. (1983). Chinchilla auditory-nerve responses to low-frequency tones. *J. Acoust. Am.* **73**:2096–2108.

192. Ruggero, M. A., and Rich, N. C. (1989). Peak splitting: Intensity effects in cochlear afferent responses to low frequency tones. In "Cochlear Mechanisms—Structure, Function, and Models." Plenum, New York.

193. Ryugo, D. K. (1992). The auditory nerve: Peripheral innervation, cell body morphology, and central projections. In D. B. Webster, A. N. Popper, and R. R. Fay (Eds.), "The Mammalian Auditory Pathway: Neuroanatomy." Springer-Verlag, New York.

194. Sachs, M. B., and Abbas, P. J. (1974). Rate versus level functions for auditory-nerve fibers in cats: Tone burst stimuli. *J. Acoust. Soc. Am.* **56**:1835–1847.

195. Sachs, M. B., and Kiang, N. Y.-S. (1968). Two-tone inhibition in auditory nerve fibers. *J. Acoust. Soc. Am.* **43**:1120–1128.

196. Sachs, M. B., and Young, E. D. (1979). Encoding of steady-state vowels in the auditory nerve: Representation in terms of discharge rate. *J. Acoust. Soc. Am.* **66**:470–479.

197. Scharf, B., Magnan, J., Collet, L., Ulmer, E., and Chays, A. (1974). On the role of the olivocochlear bundle in hearing: A case study. *Hear. Res,* **75**:11–26.

198. Scherg, M., and von Cramon, D. (1985). A new interpretation of the generators of BAEP waves I–V: Results of a spatio-temporal dipole. *Electroencephalogr. Clin. Neurophysiol.* **62**:290–299.

199. Schreiner, C. E., and Urbas, J. V. (1986). Representation of amplitude modulation in the auditory cortex of the cat. I. The anterior auditory field (AAF). *Hear. Res.* **21**:227–242.

200. Sellick, P. M., Patuzzi, R., and Johnstone, B. M. (1982a). Modulation of responses of spiral ganglion cells in the guinea pig cochlea to low frequency sound *Hear. Res.* **7**:199–221.

201. Sellick, P. M., Patuzzi, R., and Johnstone, B. M. (1982b). Measurement of basilar membrane motion in the guinea pig using the Mossbauer technique. *J. Acoust. Soc. Am.* **72**:131–141.

202. Silverstein, H., Norrell, H., Haberkamp, T., and McDaniel, A. B. (1986). The unrecognized rotation of the vestibular and cochlear nerves from the labyrinth to the brain stem: Its implications to surgery of the eighth cranial nerve. *Otolaryngol. Head Neck Surg.* **95**:543–549.

203. Sinex, D. G., and Geisler, C. D. (1981). Auditory-nerve fiber responses to frequency-modulated tones. *Hear. Res.* **4**:127–148.

204. Smith, R. L., and Brachman, M. L. (1980). Response modulation of auditory-nerve fibers by AM stimuli: Effects of average intensity. *Hear. Res.* **2**:123–134.

205. Snyder, R.-L., and Schreiner, C. E. (1984). The auditory neurophonic: Basic properties. *Hear. Res.* **15**:261–280.

206. Snyder, R. L., and Schreiner, C. E. (1985). Forward masking of the auditory nerve neurophonic (Ann) and the frequency following response (FFR). *Hear. Res.* **20**:45–62.

207. Snyder, R. L., and Schreiner, C. E. (1987). Auditory neurophonic responses to amplitude-modulated tones: Transfer functions and forward masking. *Hear. Res.* **31**:79–92.

208. Sokolich, W. G., Hamernick, R. P., Zwisklocki, J. J., and Schmiedt, R. A. (1976). Inferred response polarities of cochlear hair cells. *J. Acoust. Soc. Am,* **59**:963–974.

209. Spire, J. P., Dohrmann, G. J., and Prieto, P. S. (1982). Correlation of brainstem evoked response with direct acoustic nerve potential. In J. Courjon, F. Manguiere, and M. Reval (Eds.), "Advances in Neurology: Clinical Applications of Evoked Potentials in Neurology" (Vol. 32), (pp. 159–167). Raven, New York.

210. Spoendlin, H., and Schrott, A. (1989). Analysis of the human auditory nerve. *Hear. Res.* **43**:25–38.

211. Stapells, D. R., Picton, T. W., and Smith, A. D. (1982). Normal hearing thresholds for clicks. *J. Acoust. Soc. Am.* **72**:74–79.

212. Starr, A., and Zaaroor, M. (1990). Eighth nerve contributions to cat auditory brainstem responses (ABR). *Hear. Res.* **48**:151–160.

213. Stevens, S. S. (1935). The relation of pitch to intensity. *J. Acoust. Soc. Am.* **6**:150–154.

214. Suga, N. (1988). Auditory neuroetology and speech processing: complex-sound processing by combination-sensitive neurons. *In* Edelman, G. M., Gall, W. E., Cowan, W. M. (Eds.), "Auditory function." (Chapter 23, pp 679–720). John Wiley & Sons, New York.

215. Suga, N. (1989). Principles of auditory information processing derived from neuroethology. *J. Exp. Biol.* **146**:277–286.

216. Suga, N. (1994). Multi-function theory for cortical processing of auditory information: implications of single-unit and lesion data for future research. *J. Comp. Physiol. A* **175**:135–144.

217. Suga, N. (1995). Sharpening of frequency tuning by inhibition in the central auditory system—Tribute to Yasuji Katsuki. *Neurosci. Res.* **21**:287–299.

218. Suga, N. (1997). Parallel–hierarchical processing of complex sounds for specialized auditory function. *In* M. J. Crocker (Ed.), "Encyclopedia of Acoustics" (pp. 1409–1418). Wiley, New York.

219. Suga, N., Zhang, Y., and Yan, J. (1997a). Sharpening of frequency tuning by inhibition in the thalamic auditory nucleus of the mustached bat. *J. Neurophysiol.* **77**:2098–2114.

220. Suga, N., Yan, J., and Zhang, Y. (1997b). Cortical maps for hearing and egocentric selection for self-organization. *Trends Cog. Sci.* **1**:13–19.

221. Sullivan, W. E. (1982). Possible neural mechanisms of target distance coding in auditory system of the echolcating bat Myotis lucifugus. *J. Neurophysiology.* **48**:1033–1047.

222. Sullivan, W. E. (1986). Processing of acoustic temporal patterns in barn owls and echolocating bats: Similar mechanisms for generation of neural place representations of auditory space. *Brain Behav. Evol.* **28**:109–121.

223. Tobias, J. V., and Zerlin, S. (1959). Lateralization threshold as a function of stimulus duration. *J. Acoust. Soc. Am.* **31**:1591–1594.

224. Volman, S. F., and Konishi, M. (1989). Spatial selectivity and binaural responses in the inferior colliculus of the great horned owl. *J. Neurosci.* **9**:3083–3096.

225. Wallach, H., Newman, E. B., and Rosenzweig, M. R. (1949). The precedence effect in sound localization. *Am. J. Psychol.* **62**:315–336.

226. Walton, J. P., Frisina, R. D., and Meierhans, L. R. (1995). Sensorineural hearing loss alters recovery from short-term adaptation in the C57BL/6 mouse. *Hear. Res.* **88**:19–26.

227. Weinberg, R. J., and Rustioni, A. (1987). A cuneocochlear pathway in the rat. *Neuroscience* **20**:209–219.

228. Wever, E. G. (1949). "Theory of Hearing." Wiley, New York.

229. Whitfield, I. C., and Evans, E. F. (1965). Responses of auditory cortical neurons to stimuli of changing frequency. *J. Neurophysiol.* **28**:655–672.

230. Williston, J. S., Jewett, D. L., and Martin, W. H. (1981). Planar curve analysis of three-channel auditory brain stem response: A preliminary report. *Brain Res.* **223**:181–184.

231. Wise, L. Z., and Irvine, D. R. F. (1985). Topographic organization of interaural intensity difference sensitivity in deep layers of cat superior colliculus: Implications for auditory spatial representation. *J. Neurophysiol.* **54**:185–211.

232. Yan, J., and Suga, N. (1996a). Corticofugal modulation of time-domain processing of biosonar information in bats. *Science* **273**:1100–1103.

233. Yan, J., and Suga, N. (1996b). The midbrain creates and the thalamus sharpens echo-delay tuning for cortical representation of target-distance information in the mustached bat. *Hear. Res.* **93**:102–110.

234. Yin, T. C. T., Chan, J. C. K., and Carney, L. H. (1987). Effects of interaural time delays of noise stimuli on low frequency cells in the cat's inferior colliculus. III. Evidence for cross-correlation. *J. Neurophysiol.* **58**:562–583.

235. Yin, T. C. T., and Chan, J. C. K. (1990). Interaural time sensitivity in medial superior olive of cat. *J. Neurophysiol.* **64**:465–488.

236. Yin, T. C. T., Chan, J. C. K. and Irvine, D. R. F. (1986). Effects of interaural time delays of noise stimuli on low-frequency cells in the cat's inferior colliculus. I. Responses to wideband noise. *J. Neurophysiol.* **55**:280–300.

237. Yin, T. C. T., and Kuwada, S. (1984). Neuronal mechanisms of binaural interaction. *In* G. M. Edelman, W. E. Gall, and W. M. Crowan (Eds.), "Dynamic Aspects of Neocortical Function" (pp. 263–313). Wiley, New York.

238. York, D. H. (1986). Correlation between a unilateral midbrain-pontine lesion and abnormalities of the brain-stem auditory evoked potential. *Electroencephalogr. Clin. Neurophysiol.* **65**:282–288.

239. Young, E. D., and Sachs, M. B. (1979). Representation of steady-state vowels in the temporal aspects of the discharge patterns of populations of auditory nerve fibers. *J. Acoust. Soc. Am.* **66**:1381–1403.

240. Zanette G., Carteri, A., and Cusumano, S. (1990). Reappearance of brainstem auditory evoked potentials after surgical treatment of brain-stem hemorrhage: contributions to the question of wave generation. *Electroencephalogr. Clin. Neurophysiol.* **77**:140–144.

241. Zappia, M., Cheek, J. C., and Luders, H. (1996). Brain-stem auditory evoked potentials (BAEP) from basal surface of temporal lobe recorded from chronic subdural electrodes. *Electroencephalogr. Clin. Neurophysiol.* **100**:141–151.

242. Zhang, Y., Suga. N., and Yan, J. (1997). Corticofugal modulation of frequency processing in bat auditory system. *Nature* **387**:900–903.

243. Zhao, H., and Liang, Z. (1995). Processing of modulation frequency in the dorsal cochlear nucleus of the guinea pig: Amplitude modulated tones. *Hear. Res.* **82**:244–256.

244. Zwislocki, J. J. (1986). Are nonlinearities observed in firing rates of auditory-nerve afferents reflections of a nonlinear coupling between the tectorial membrane and the organ of Corti? *Hear. Res.* **22**:217–222.

245. Zwislocki, J. J. (1992). What is the cochlear place code for pitch? *Acta Otolaryngol.* **111**:256–262.

246. Zwislocki, J. J., and Sokolich, W. G. (1973). Velocity and displacement responses in auditory nerve fibers. *Science* **182**:64–66.

SECTION III

Acoustic Reflexes

Involuntary muscle contractions that are elicited by sound are known as acoustic reflexes. The best known is the acoustic middle ear reflex, which involves one or both of the two middle ear muscles. This reflex has been studied extensively, as it plays an important role in modern audiological testing. The basic characteristics of this reflex are described in this chapter, while its use in diagnosis of disorders of the middle ear, the cochlea, and the auditory nervous system is discussed in Chapters 13, 14, and 15. There are many types of acoustic reflexes besides the acoustic middle ear reflex. Contractions of face and neck muscles commonly occur in response to loud sounds. Small invisible contractions can be detected by recording EMG potentials in response to sounds in a wide range of intensities. The startle reflex is another acoustic reflex where a loud and unexpected sound causes contraction of many different skeletal muscles. The movement of the eyes toward the source of strong impulsive sounds may also be regarded as an acoustic reflex. The raise in heart rate in response to a strong sound is not commonly regarded as an acoustic reflex but it has most of the characteristics of an acoustic reflex. Whether seizures that, under rare circumstances, can be evoked by strong sounds (audiogenic seizures) can be called acoustic reflexes is a matter of definition. These other acoustic reflexes are not discussed in this book.

CHAPTER 12

Acoustic Middle Ear Reflex

ABSTRACT

1. The threshold of the acoustic middle ear reflex is approximately 85 dB HL in individuals with normal hearing. The strength of the muscle contraction increases gradually with increasing stimulus intensity.

2. The acoustic middle ear reflex involves only the stapedius muscle in humans, but both the tensor tympani and the stapedius muscles contract in animals, an acoustic reflex commonly used in auditory research.

3. When elicited by sounds presented to one ear, the stapedius muscle in humans contracts in both ears but the response is slightly stronger in the ipsilateral ear and its threshold is lower.

4. The contraction of the stapedius muscle occurs with a latency that decreases from approximately 25 ms for high-intensity stimuli to 100 ms or more near the threshold of the reflex.

5. Contractions of the stapedius muscle decrease sound transmission to the cochlea, more for low frequencies than for high frequencies.

6. The acoustic middle ear reflex acts as a control system that makes the sound that reaches the cochlea vary less than the sound that reaches the tympanic membrane. This is a form of amplitude compression.
7. Contraction of the stapedius muscle changes the ear's acoustic impedance, which can be recorded using noninvasive methods.
8. Recording of changes in the ear's acoustic impedance is the most common method used to study the acoustic middle ear reflex and is used in clinical diagnosis.
9. The acoustic middle ear reflex response is reduced after intake of sedative (hypnotic) drugs such as barbiturates and alcohol.
10. The middle ear muscles contract in other ways than in response to sounds: (1) some few individuals can voluntarily contract their middle ear muscles and (2) the stapedius muscle contracts before vocalization.

INTRODUCTION

The acoustic middle ear reflex has been studied extensively in both humans and animals. The effector organ of the acoustic middle ear reflex in humans is the stapedius muscle and in some animal species, the tensor tympani muscle also contracts in response to loud sounds. Thus, despite the fact that the stapedius and the tensor tympani muscles are innervated by two different cranial nerves (the facial and the trigeminal nerves, respectively) these two muscles contract together as an acoustic reflex, at least in the animals often used in auditory research such as the cat, guinea pig, and rat.

NEURAL PATHWAY OF THE ACOUSTIC MIDDLE EAR REFLEX

The stapedius muscle is innervated by the stapedius nerve, which is a branch of the facial nerve that takes off from the main trunk of this nerve in the facial canal peripheral to the petrosal nerve and the geniculate ganglion, but centrally to the chorda tympani. In humans, the nerve to the stapedius muscle branches off the facial nerve approximately 1 cm from the stylomastoid foramen (Fig. 12.1). The tensor tympani muscle is innervated by the mandibular branch (V_3) of the Vth cranial nerve (the trigeminal nerve).

Studies in the rabbit [2] have shown that the auditory nerve and the ventral cochlear nucleus are common to both the contralateral and the ipsilateral pathways of the acoustic stapedius reflex. The stapedius reflex and the tensor tympani reflex share the first synapse in the ventral cochlear nucleus. The dorsal cochlear nucleus is not involved in the acoustic middle ear reflex. Two

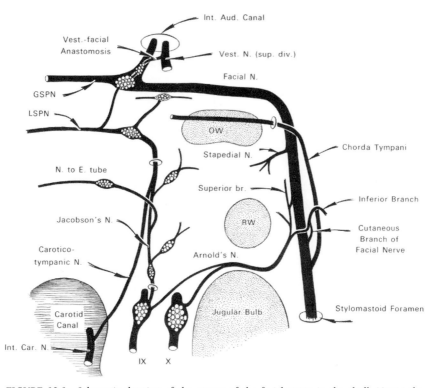

FIGURE 12.1 Schematic drawing of the course of the facial nerve in the skull. Notice the stapedius nerve. GSPN, greater superficial petrosal nerve; LSPN, lesser superficial petrosal nerve; OW, oval window of the cochlea; RW, round window of the cochlea (from Schuknecht (1974) with permission).

main parallel ipsilateral pathways of the stapedius reflex exist (Figs. 12.2 and 12.3). One pathway continues from the ventral cochlear nucleus directly to the facial motonucleus on the same side without any further synapses. The

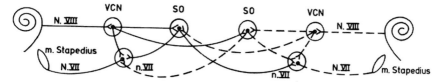

FIGURE 12.2 Schematic drawing of the reflex arc of the acoustic middle ear reflex (stapedius reflex). Abbreviations: N.VIII, auditory nerve; N.VII, facial nerve; n.VII, facial motonucleus; VCN, ventral cochlear nucleus; SO, superior olivary complex.

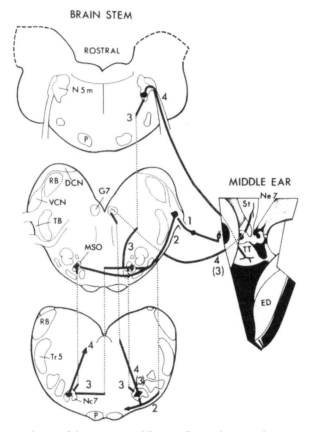

FIGURE 12.3 Pathways of the acoustic middle ear reflex as shown in three transverse sections through the brainstem of a rabbit. Solid lines represent nerve tracts and dotted lines show the connections between the sections. Abbreviations: CR, restiform body; DCN, dorsal cochlear nucleus; ED, ear drum; G7, internal geniculum of the seventh cranial nerve; MSO, medial superior olive; Nc 7, nucleus of the VIIth cranial nerve; N. VII, VIIth cranial nerve; P, pyramidal tract; St., stapedius; TB, trapezoid body; Tr 5, spinal trigeminal tract; TT, tensor tympani; VCN, ventral cochlear nucleus. Numbers refer to the order of neurons in the stapedius reflex and the tensor tympani reflex (from Borg, E. (1973). On the neuronal organization of the acoustic middle ear reflex: A physiological and anatomical study. *Brain Res.* 49:101–123).

other pathway makes connections from the ventral cochlear nucleus through the trapezoid body to the facial motonucleus on the same side and on the opposite side. The second synapse of this pathway of the stapedius reflex is in the lateral superior olivary complex (LSO) and medial superior olivary complex (MSO). From here both the crossed and the uncrossed pathways lead to the facial moto-nucleus. The neurons in the part of the facial motonucleus that are close to the superior olivary complex innervate the stapedius muscle [16]. The crossed pathway of the stapedius reflex (mediator of the contralateral response) has a synapse in the MSO from where the pathway leads to the facial motonucleus.

A more diffuse multisynaptic pathway of the acoustic middle ear reflex exists [2]. While the direct pathway of the ipsilateral acoustic middle ear reflex has two and three synapses and the contralateral reflex has at least three synapses, this indirect pathway has a larger number of synapses. The reflex response mediated by this indirect pathway is slower and more sensitive to anesthesia and is probably also affected by the degree of wakefulness. It is not known in detail which neural structures comprise the indirect pathway but ablation of the inferior colliculus (IC) did not have any noticeable effect on the reflex response [2], nor did lesions in the pyramidal tract noticeably affect the response. Therefore, the acoustic middle ear reflex probably does not involve the cerebellum, nor the midbrain or the forebrain.

The reflex pathways for the tensor tympani are slightly different from those of the stapedius reflex. Fibers from the MSO connect to the Vth cranial nerve motonucleus on both sides and these connections are probably the most important ones for the tensor tympani reflex [2]. No connections from the trapezoidal body to the Vth cranial nerve motonucleus have been found.

PHYSIOLOGY

The threshold of the human acoustic middle ear reflex is approximately 85 dB above the normal hearing threshold (hearing level, HL). A sound above the threshold of the reflex presented to one ear elicits a contraction of the stapedius muscle in both ears but the threshold is slightly lower for the ipsilateral ear. Because contraction of the middle ear muscles reduces sound transmission through the middle ear, the reflex comes to act as a control system that tends to keep the sound input to the cochlea constant. When this amplitude compression becomes absent through paralysis of the stapedius muscle, speech discrimination at high sound levels becomes impaired. Some individuals are able to contract their middle ear muscles voluntarily.

RESPONSES TO STIMULATION WITH TONES

When the acoustic middle ear reflex is studied by recording changes in the ear's acoustic impedance the response amplitude increases gradually after a brief latency to attain a plateau after approximately 500 ms at stimulus levels just above threshold. The increase in the response amplitude occurs more rapidly when elicited by sounds of higher intensity (Fig. 12.4). The amplitude

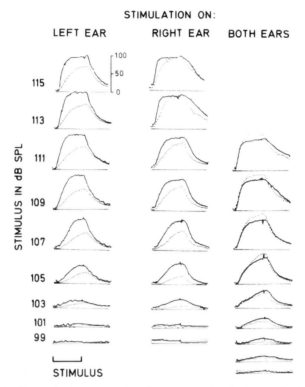

FIGURE 12.4 Change in the acoustic impedance recorded in both ears simultaneously as a result of contraction of the middle ear muscles elicited by tone bursts of different intensities. In the two left-hand columns, one ear was stimulated. The solid lines are the impedance change in the ipsilateral ear and the dashed lines are the impedance change in the contralateral ear. The right-hand columns show responses of both ears when both ears were stimulated simultaneously. The solid lines show contractions of the middle ear muscles in the right ear and the dashed lines are the responses in the left ear. The stimulus sound was 1450-Hz pure tones presented in bursts of 500-ms duration. The intensity of the sound is given in dB SPL. The results were obtained in an individual with normal hearing (from Møller, A. R. (1962a). The acoustic reflex in man. J. Acoust. Soc. Am. **34**(II):1524–1534).

of the reflex response elicited by high-frequency sounds decreases over time (adaptation) but normally the reflex response elicited by tones below 1500 Hz shows little adaptation.

Techniques for Recording the Contractions of the Middle Ear Muscles

Recording changes in the ear's acoustic impedance is a convenient and noninvasive method to record the contractions of the middle ear muscles and it is the method commonly used to study the acoustic middle ear reflex. Another noninvasive method makes use of recordings of the displacement of the tympanic membrane as an indicator of contractions of the middle ear muscles. Yet, recording of changes in the ear's acoustic impedance is the most used method for studies of the function of the acoustic middle ear reflex both in humans and in animals. Its use is based on the fact that contractions of the middle ear muscles changes the ear's acoustic impedance (Chapter 2).

Geffcken [11] was probably the first to report that the ear's acoustic impedance changed when the middle ear muscles were brought to contraction by a loud sound. Metz, who was one of the first to use measurements of the ear's acoustic impedance for clinical purposes [20], also pioneered the use of measurement of changes in the ear's acoustic impedance to record the contractions of the middle ear muscles [21]. Since then it has been used by numerous investigators for clinical studies of the acoustic middle ear reflex [9, 17, 22, 23] and for research purposes [8, 14, 22, 24, 25]. While Metz [20] and Jepsen [15] used the Schuster bridge, the investigators who followed mainly used an electroacoustic method [22, 24, 25, 30] and that is also the basis for the equipment used clinically. The equipment that is in clinical use at present for recording the response of the acoustic middle ear reflex and for tympanometry use test tones of approximately 220 Hz, but some investigators of the function of the acoustic middle ear reflex have used a 800-Hz probe tone [23, 24, 25].

Recording electromyographic (EMG) potentials from the middle ear muscles or recording the change in the cochlear microphonic (CM) potentials have been used to study the function of the acoustic middle ear reflex in animals. Recording of EMG potentials makes it possible to discriminate between the contraction of the two muscles, which is not possible by recording of the acoustic impedance. Recording CM makes it possible to measure the change in sound transmission through the middle ear that contractions of the middle ear muscles cause. Both the EMG and the CM methods are invasive and are not practical for use in humans except in special situations where the middle ear cavity becomes exposed in an operation.

One of the strengths of the impedance method for recording the response of the acoustic middle ear reflex is that it can be used in humans as well as in animals. Changes in the impedance seen in animal models can thus be compared to human conditions. The study of the neural pathways of the acoustic middle ear reflex in the rabbit [2] described above was done by recording changes in the acoustic impedance simultaneously in both ears while one ear is being stimulated [5] in a similar way as shown in Fig. 12.4. The changes were measured before and after making controlled experimental lesions in fiber tracts and nuclei of the brainstem. These studies provided information about the neural pathway of the acoustic middle ear reflex for both the ipsilateral and the contralateral reflex as shown in Figs. 12.2 and 12.3.

The amplitude of the reflex response to stimulation with pure tones, measured 500 ms after the onset of the stimulus tone, increases with the stimulus intensity at approximately the same rate (Fig. 12.5), when elicited from the ipsilateral ear compared with the contralateral ear. The response to ipsilateral stimulation is slightly larger and the stimulus–response curves to ipsilateral and contralateral stimulation are shifted 2–10 dB, indicating that the sensitivity of the reflex is 2–10 dB higher when elicited from the ipsilateral ear. The amplitude of the responses to bilateral stimulation is greater and the stimulus response curve is shifted approximately 3 dB relative to that of ipsilateral

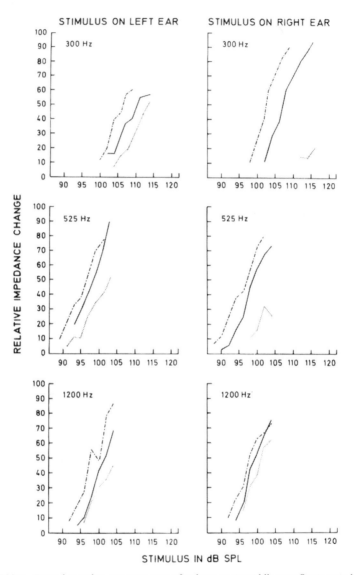

FIGURE 12.5 Typical stimulus–response curves for the acoustic middle ear reflex in an individual with normal hearing. Dashes and dots show the amplitude of the response to bilateral stimulation, solid lines are the response to ipsilateral stimulation, and the dots are the contralateral response. Results from both ears are shown. The stimuli were 500-ms tone bursts. In these experiments the stimulus intensity was first raised (in by 2-dB steps) from below threshold to the maximal intensity used and then lowered again (in 2-dB steps) to below threshold. The change in the ear's impedance given is the mean of two determinations, one when the stimulus was increased

stimulation (Fig. 12.5), indicating a slightly higher sensitivity to bilateral stimulation [23, 24]. The difference between the response to ipsilateral and contralateral stimulation is greater when the reflex response is elicited by low-frequency tones than by tones above 500 Hz. There is also a considerable individual difference in that respect. The amplitude of the reflex responses to ipsilateral stimulation reaches a plateau approximately 20 dB above the threshold. The maximal response amplitude that can be obtained is higher when recorded from the ear from which the reflex is elicited than when recorded from the contralateral side (Fig. 12.5).

The slope of the stimulus responses curves is affected by the duration of the stimulus tones. The stimulus response curves are steeper for long tones than for short tones (Fig. 12.6). The difference between the response to bilateral, ipsilateral, and contralateral stimulation is greater when the reflex is elicited by short tones than by long tones (Fig. 12.6). The response to short tones also reaches a plateau at a lower response amplitude than that to long tones and that is much more pronounced for contralateral stimulation than for ipsilateral and bilateral stimulation (Fig. 12.6). That the response to ipsilateral stimulation produces a stronger contraction of the stapedius muscle than contralateral stimulation does was also shown in animal experiments (cat) [12]. It is noteworthy that most studies of the acoustic middle ear reflex, including its use in clinical diagnosis, have been restricted to studies of the contralateral responses.

THRESHOLD OF THE REFLEX

The threshold of the acoustic middle ear reflex is poorly defined because small irregular responses are obtained in a large range of stimulus intensities near threshold (Fig. 12.7). The variability of these responses makes it difficult to determine the absolute threshold of the acoustic middle ear reflex accurately. The "threshold" of the acoustic middle ear reflex, defined as the sound intensity necessary to elicit a response with an amplitude of 10% of the maximal response, is a more reproducible measure of the sensitivity of the reflex [23]. While the sound intensity needed to elicit a response with a small amplitude

---

from below threshold and the other when the stimulus intensity was decreased from the maximal used intensity to the threshold. The change in the ear's acoustic impedance is given in percentage of the maximally obtained response at any stimulus frequency and situation (usually bilateral stimulation) (from Møller, A. R. (1962). The acoustic reflex in man. *J. Acoust. Soc. Am.* 34(II):1524–1534).

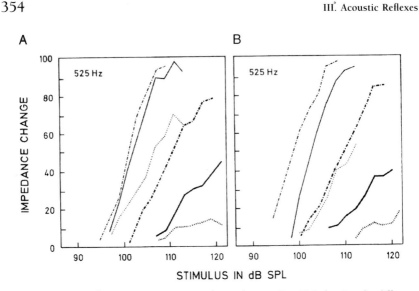

FIGURE 12.6 Stimulus–responses curves similar to those in Fig. 12.5 showing the difference between the response to tones of 500-ms duration (thin lines) and the responses to short tones (25-ms duration, thick lines). Dots and dashes indicate bilateral stimulation; solid lines indicate ipsilateral stimulation; dotted lines indicate contralateral stimulation. The stimulus frequency was 525 Hz. (A) Stimulation of the left ear; (B) stimulation of the right ear (from Møller, A. R. (1962). The acoustic reflex in man. *J. Acoust. Soc. Am.* **34**(II):1524–1534).

(e.g., 10% of the maximal response) shows a high degree of reproducibility in the same individual when recorded at different times (Fig. 12.8), the reflex threshold determined in that way shows considerable individual variations. The reflex threshold, as defined here is approximately 85 dB above hearing threshold in young individuals with normal hearing for stimulation of the contralateral ear [25] (Fig. 12.9). These measures have large individual variations, even in young individuals with normal hearing and without a history of middle ear disorders (Fig. 12.9). The large individual variation should be considered when the threshold of the acoustic middle ear reflex is used for diagnostic purposes.

It is not known how the threshold of the acoustic middle ear reflex is set but it is interesting to note that individuals whose auditory nerve is injured have an elevated acoustic middle ear reflex threshold and a poor growth of the reflex response amplitude with increasing stimulus intensity (see Chapter 15). Such injuries mainly affect the synchronization of neural activity in the auditory nerve, thus indicating that the function of the middle ear reflex may depend on synchronization of neural activity in many nerve fibers (temporal coherence).

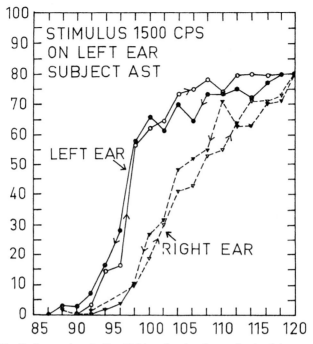

FIGURE 12.7   Similar graph as in Fig. 12.5 but showing the amplitude of the response to each stimulus. The stimulus was increased from below threshold to 115 dB SPL (in 2-dB steps and then reduced by 2-dB steps to below threshold) (from Møller, A. R. (1961). "Bilateral Contraction of the Tympanic Muscles in Man" (pp. 1–15). Royal Institute of Technology (KTH), Div. of Telegraphy-Telephony, Report No. 18, Speech Transmission Laboratory, Stockholm).

External Factors Affecting the Acoustic Middle Ear Reflex

The response of the acoustic middle ear reflex is affected by drugs, such as alcohol, and sedative drugs, such as barbiturates [4] (Figs. 12.10 and 12.11). The threshold of the response increases as a function of the concentration of alcohol in the blood or the amount of barbiturate ingested. A blood alcohol concentration of 1/10 of 1% results in an average 5-dB elevation of the reflex threshold.

LATENCY OF THE REFLEX RESPONSE

The latency of the earliest detectable response of the acoustic middle ear reflex (recorded as a change in the ear's acoustic impedance) decreases with increasing stimulus intensity. The latency of the response to 1500-Hz tones is shorter

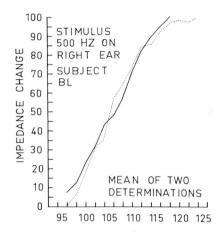

FIGURE 12.8 Illustration of the reproducibility of the responses of the acoustic middle ear reflex. The changes in the ear's impedance are expressed in percentages of the maximally obtainable response. Amplitude is shown at two occasions, 2 months apart (from Møller, A. R. (1961). "Bilateral Contraction of the Tympanic Muscles in Man" (pp. 1–15). Royal Institute of Technology (KTH), Div. of Telegraphy-Telephony, Report No. 18, Speech Transmission Laboratory, Stockholm).

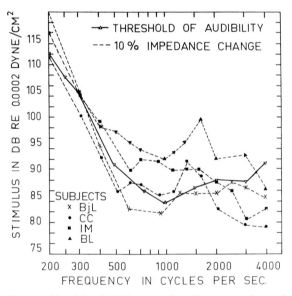

FIGURE 12.9 The sound level (in dB SPL) required to elicit an impedance change of 10% of the maximal obtainable response amplitude at different frequencies of stimulation. The results were obtained in young individuals with normal hearing. The thick line shows the sound levels (in dB SPL) that are 80 dB above the normal threshold of hearing (80 dB HL) (from Campbell, K. C. M., and Abbas, P. J. (1994). Electrocochleography and postural changes in perilymphatic fistula. *Annals Otol. Rhin. Laryngol.* 103:474–482).

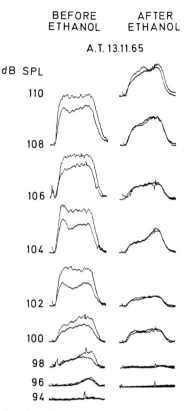

BEFORE   AFTER
ETHANOL   ETHANOL

A.T. 13.11.65

dB SPL

110

108

106

104

102

100

98

96

94

FIGURE 12.10   Effect of ethanol on the ipsilateral and contralateral response of the acoustic middle ear reflex in a person with normal hearing before (left column) and 30 min after ingestion of 105 ml alcohol, when the blood alcohol concentration was 0.12% (from [4] with permission).

than it is to 500-Hz tones [22]. The shortest latency is approximately 25 ms and the longest is over 100 ms. The individual variation as well as the variation in the results obtained in the same individual are large. The latency of the ipsilateral and the contralateral responses are similar.

The latency of the change in the acoustic impedance is the sum of the neural conduction time and the time it takes for the stapedius muscle to develop sufficient tension to cause a measurable change in the ear's acoustic impedance. The neural conduction time has been measured by recording electromyographic (EMG) potentials from the stapedius muscle in animals and in humans. The latency of the EMG response is shorter than the change in the acoustic impedance, which involves the time it takes to build up contraction strength of the stapedius muscle. Some of the first investigators who reported

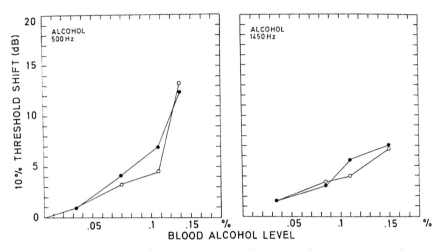

FIGURE 12.11   Mean value of the increase in stimulus intensity that is necessary to obtain a reflex response that is 10% of the maximally obtainable response as a function of blood alcohol concentration for two different frequencies of the stimulus tones. Open circles are the ipsilateral response and solid circles the contralateral response (from [4] with permission).

on recordings of EMG potentials from the stapedius muscle in human were Perlman and Case [26], who recorded the response to "loud" tones. They reported a mean latency of 10.5 ms based on recordings from several patients. Similar methods were used by Zakrisson [31] who also studied the response from the stapedius muscle during vocalization.

FUNCTIONAL IMPORTANCE OF THE ACOUSTIC MIDDLE EAR REFLEX

Many hypotheses about the functional importance of the acoustic middle ear reflex have been presented. Maybe the most plausible hypothesis is that at least one of the important functions of the reflex is to keep the input to the cochlea nearly constant for sounds above sound levels of a certain value, while allowing rapid changes in the sound level to be preserved. The middle ear reflex thus acts as a relatively slow automatic volume control that keeps the mean level of sound that reaches the cochlea within narrow limits (amplitude compression). Studies of the function of the cochlea with the acoustic middle ear reflex inactivated indicate that the cochlea does not function properly at sound levels above threshold for the acoustic reflex (approximately 85 dB above hearing threshold).

Acoustic Reflex as a Control System

Contractions of the stapedius muscle reduce sound transmission through the middle ear (Chapter 2). The acoustic middle ear reflex therefore functions as a control system that makes the input to the cochlea vary less than the sound that reaches the tympanic membrane, thus amplitude compression. The compression of the input to the cochlea is most effective for low-frequency sounds. The compression occurs with a latency that is equal to the time it takes the stapedius muscle to contract after sound stimulation. That means that the latency of the reduction in sound transmission through the middle ear is at least 25 ms for sounds 20 dB or more above the threshold of the reflex. The stapedius muscle is relatively slow to attain its full strength (in the order of 100 ms, Fig. 12.4) and it therefore affects fast changes in sound intensity relatively little. Therefore, the amplitude compression is most effective for steady-state sounds or sounds with slowly varying amplitude.

The initial damped oscillation seen in the reflex response to low-frequency tone bursts (Fig. 12.12) is a sign that the reflex regulates the input to the cochlea [24]. These oscillations are a result of the fact that the contractions of the middle ear muscles reduce the input to the cochlea slowly, and with a small delay. The decrease in the input to the cochlea decreases the strength of the contraction of the middle ear muscles and thus their attenuation of the sound that reaches the cochlea decreases and the input to the cochlea again increases. This process repeats but the oscillations decay after a short period and the reflex response becomes constant. The reflex response to tones above approximately 800 Hz do not show such oscillations, which is a sign the that contraction of the stapedius muscle does not change the sound transmission through the middle ear noticeably above that frequency and thus the acoustic middle ear reflex is not an efficient control system for sounds above 800 Hz.

Another sign that the acoustic middle ear reflex attenuates low-frequency sounds was shown in studies of individuals with Bells palsy in whom the stapedius muscle was paralyzed on one side. When the reflex responses was elicited by stimulating the ear on the paralyzed side with a low-frequency tone and the response recorded on the nonparalyzed side, the impedance change increased at a steeper rate, as a function of the stimulus intensity, than it did when the reflex was activated from the nonparalyzed side [1] Fig. 12.13). This difference in the slope of the stimulus response curves was less when the reflex was elicited by a tone of a higher frequency (1450 Hz) (Fig. 12.13).

Another indication that the acoustic middle ear reflex reduces the input to the cochlea comes from a study of the temporary threshold shift in response to exposure to loud noise. It was shown that the resulting temporary threshold shift (TTS) was much greater in an ear where the stapedius muscle is paralyzed

300 Hz   **500 Hz**

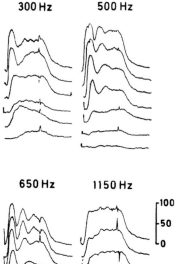

650 Hz   **1150 Hz**

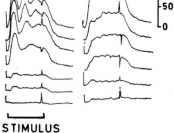

S TIMULUS

FIGURE 12.12   Graphs showing the change in the ear's acoustic impedance in response to stimulation of the ipsilateral ear with tones of different frequencies. The duration or the stimulus tones was 500 ms (from Møller, A. R. (1962a). The acoustic reflex in man. *J. Acoust. Soc. Am.* 34(II):1524–1534).

than it is in an ear with a normally functioning stapedius muscle [31] (Fig. 12.14).[1] These studies were performed in individuals with Bell's palsy, who had paralyzed stapedius muscle on one side. The two curves show the TTS in the ear with the paralyzed stapedius muscle compared with the individual's other ear, where the acoustic middle ear reflex was normal (Fig. 12.14A). While the TTS increases as an almost linear function of the level of the noise

[1] In this study the noise exposure consisted of a band of noise, centered at 500 Hz, and a width of 0.3 kHz. The exposure time was 5 or 7 min. Hearing threshold was measured at 750 Hz before exposure and 20 s after the end of the exposure using continuous pure tone audiometry (Békésy audiometry). The ability to regulate the input to the cochlea depends on the spectrum of a sound because contraction of the stapedius muscle affects the sound transmission to the cochlea, mostly for low-frequency sounds (Chapter 2, Fig. 2.27). Quantitative studies of the acoustic reflex as a control system [1, 8] have shown that this reflex can keep the input to the cochlea nearly constant for low-frequency sounds with slowly varying intensity above the reflex threshold.

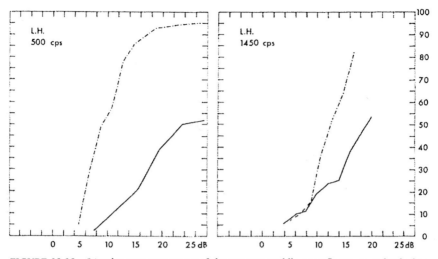

FIGURE 12.13   Stimulus response curves of the acoustic middle ear reflex in an individual in whom the stapedius muscle was paralyzed elicited from the side of the paralysis (dots and dashes) and from the nonparalyzed side (solid lines) (from [1] with permission).

to which the ear with the paralyzed stapedius muscle is exposed, the noise levels used caused little TTS in the ear with the normally functioning acoustic reflex. However, the individual variations were considerable (Fig. 12.14B). Similar studies of the TTS after exposure to noise centered at 2000 Hz showed that the TTS was not noticeably affected by the paralysis of the stapedius muscle [31]. This is in agreement with the findings that the sound attenuation from contraction of the stapedius muscle is small at frequencies higher than 1000 Hz [1] (Chapter 2). The results shown in Figs. 12.13 and 12.14 were obtained in individuals in whom the acoustic middle ear reflex response was absent when recorded in the affected ear, thus showing that the stapedius muscle was paralyzed.

It is important to understand that contractions of the middle ear muscles affect (attenuates) low-frequency sounds of any intensity. Thus if the reflex is elicited by the high-frequency components of a sound that also contains energy at low frequencies, the low-frequency components of the sound will be attenuated even when these components are not sufficiently intense to activate the reflex. When the acoustic middle ear reflex is elicited by complex sounds such as speech sounds the contraction of the stapedius muscle will thus affect all components of the sound, independent of whether the spectral components contribute to activating the reflex. The ability of the reflex to compress the intensity of sounds of high intensity has been referred to as the

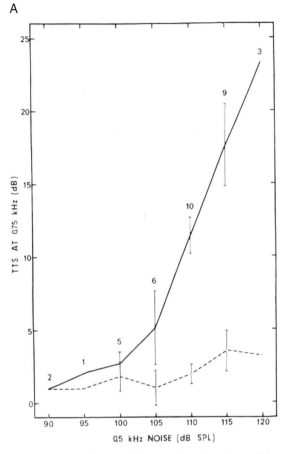

FIGURE 12.14 (A) TTS in the affected ear during unilateral paralysis of the stapedius muscle (solid line) compared with the TTS in the other ear (dashed line) as a result of exposure to band-pass-filtered noise (centered at 500 Hz, 300 Hz wide) for 5 min. Mean values from 18 subjects and standard error of the mean are shown as a function of the intensity of the noise. The TTS was measured 20 s after the end of the exposure (from [31]). (B) Individual TTS values in an ear with a paralyzed stapedius muscle as a result of 5-min exposure to the same noise as in A (from Zakrisson, J. E. (1974). The role of the stapedius reflex in poststimulatory auditory fatigue. *Acta Otolaryngol. (Stockh.)* **3**:1–10).

perceptual theory of the action of the acoustic middle ear reflex [4], which relates to the proposal by Simmons [29], who recognized that the stapedius muscle can modulate the amplitude of sounds that reach the cochlea and by that help to separate specific sounds from a noise background. Attenuation of low-frequency components of a broadband sound by contraction of the staped-

B

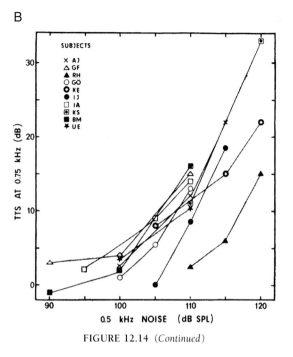

FIGURE 12.14 (*Continued*)

ius muscle may reduce masking of high-frequency components from strong low-frequency components such as sounds from one's own vocalizing and chewing. These features may have exerted evolutionary pressure to develop the acoustic middle ear reflex. The threshold of the reflex is generally lower in animals in which it has been studied and its role in reducing masking may have been important for its phylogenetic development.

The functional importance of the acoustic middle ear reflex has been studied by comparing speech discrimination in individuals who have paresis of the stapedius muscle in one ear [6]. It was found that discrimination of speech at high sound levels is impaired when the acoustic middle ear reflex is not active (Fig. 12.15) in individuals with facial palsy where the stapedius muscle is affected. Speech discrimination tests in individuals that lack the reflex show that their discrimination of loud speech is impaired (Fig. 12.15). Normally speech discrimination is nearly 100% in the range of speech intensities from 60 dB to 120 dB SPL, but when the stapedius muscle is paralyzed, speech discrimination deteriorates above 90 dB SPL. This illustrates one important function of the acoustic middle ear reflex that may be related to reducing masking from low-frequency components of speech sounds that may impair

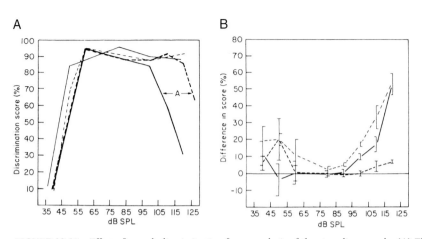

FIGURE 12.15 Effect of speech discrimination from paralysis of the stapedius muscle. (A) The dependence of speech discrimination on the function of the stapedius muscle (the average of results obtained in 13 patients). Speech discrimination scores (articulation scores) are shown as a function of the intensity for monosyllables (maximal levels, in dB SPL) during paralysis of the stapedius muscle (from Bell's palsy) (thick solid line) and after recovery of the paralysis (thin line). The thick dashed line shows the discrimination scores in the opposite (unaffected) ear during the paralysis. (B) Average difference in discrimination scores during and after paralysis of the stapedius muscle. The thick solid line shows the difference between the discrimination scores when the sound was led to the unaffected ear and obtained when the sounds were led to the affected ear at the time of paralysis. The thin dashed line shows the difference between the discrimination scores in the affected ear at the time of paralysis and after recovery for 6 of the subjects who participated in this study (from Borg, E., and Zakrisson, J. E. (1973). Stapedius reflex and speech features. *J. Acoust. Soc. Am.* **54**:525–527).

discrimination of speech of high intensity. The fact that speech discrimination does not become impaired unless the sound intensity is above 90 dB, a high sound intensity that does not normally exist, may mean that the acoustic middle ear reflex has little importance under normal listening conditions.

Several studies have shown that the acoustic middle ear reflex gives protection against noise-induced hearing loss. It is, however, questionable if reduced noise-induced hearing loss could have played any role in the development of the acoustic middle ear reflex. The type of noise it would protect against, i.e., long-duration, high-intensity sounds, are not common in nature.

NONACOUSTIC WAYS TO
ELICIT CONTRACTION OF THE
MIDDLE EAR MUSCLES

The tensor tympani muscle can be brought to contract by stimulating the skin around the eye, for instance by air puffs [18]. (These investigators believed that it was the stapedius muscle that contracted, while it in fact most likely was the tensor tympani muscle). The response they observed was elicited by stimulation of the trigeminal nerve, which innervates the skin around the eye and the cornea. Stimulation of that area of the face also elicits the blink reflex that is a natural protective reflex. (The blink reflex is frequently used in neurologic diagnosis.) The tensor tympani muscle normally contracts while swallowing.

Voluntary Control over Middle Ear Muscles

In most individuals contraction of the middle ear muscles can only be induced involuntarily as an acoustic reflex, but a few people can voluntarily contract their middle ear muscles. The changes in the ear's acoustic impedance in individuals who can voluntary contract their middle ear muscles may be different (Fig. 12.16). The impedance change recorded in one of the individuals

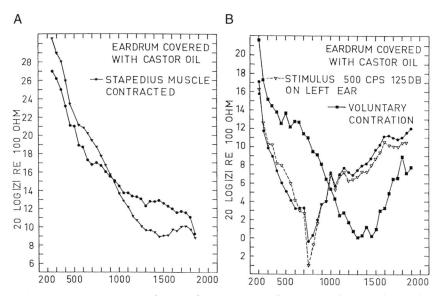

FIGURE 12.16 Acoustic impedance at the tympanic membrane in two human subjects who could voluntarily contract their middle ear muscles. The tympanic membrane had been covered with a thin layer of castor oil before the recordings were done. (A) Solid circles show acoustic impedance without contraction of the middle ear muscles. Triangles show the impedance during voluntary contraction of the middle ear muscles. (B) Solid circles show the acoustic impedance without middle ear muscle contraction. Open triangles show sound elicited middle ear muscle contraction. Filled squares show acoustic impedance during voluntary contraction of the middle ear muscles (from Møller, A. R. (1961). Bilateral contraction of the tympanic muscles in man. Royal Institute of Technology (KTH), Div. of Telegraphy–Telephony, Report No. 18, Speech Transmission Laboratory, 1–51, with permission.)

shown in Fig. 12.16A was similar to that seen in response to stimulation with a strong sound and thus assumed to be the result of a contraction of the stapedius muscle while in the another individual (Fig. 12.16B), the impedance change was much larger and that change was assumed to have been caused by contraction of the tensor tympani muscle. In this individual, sound-evoked contraction gave a much smaller impedance change [23].

STAPEDIUS CONTRACTION MAY BE ELICITED BEFORE VOCALIZATION

In humans the stapedius muscle contracts a brief period before vocalization [7], as evidenced by EMG recordings from the stapedius muscle in a patient in whom the tympanic membrane had been deflected as part of a middle ear operation so that recording electrodes could be placed on the stapedius muscles (Fig. 12.17). The EMG potentials are present before the start of vocalization (recorded by a microphone close to the patient's mouth). This means that the contractions of the stapedius muscle are not a result of an acoustic reflex but that the muscle must have contracted in response to activation of the facial motonucleus from the brain center, which is involved in controlling vocalization. It may be assumed that the stapedius muscle is activated by the same neural circuitry that activates vocalization. Recordings of EMG potentials from the laryngeal muscles and the middle ear muscles in the flying bat show that contractions of the middle ear muscles in the flying bat are coordinated with the laryngeal muscles, so that the middle ear muscles contract before the echolocating sound is emitted [13].

CLINICAL USE OF THE ACOUSTIC MIDDLE EAR REFLEX

Recording the acoustic middle ear reflex response provides information about the function of the middle ear and it can help differentiate between hearing loss caused by cochlear injury and that caused by injury of the auditory nerve. The use of the acoustic middle ear reflex in diagnosis of middle ear disorders is based on the fact that contraction of the stapedius muscle does not cause any noticeable change in the ear's impedance if the stapes is immobilized or if the ossicular chain is interrupted (see Chapter 13). The threshold of the acoustic middle ear reflex is elevated in patients with injuries to the auditory nerve but (see Chapter 15) it is nearly normal in patients with hearing loss of cochlear origin (see Chapter 14). The acoustic middle ear reflex is a valuable

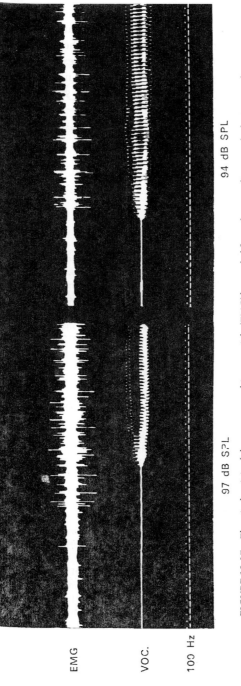

EMG

VOC.

100 Hz

97 dB SPL

94 dB SPL

FIGURE 12.17 Electrical activity [electromyographic potentials (EMG)] recorded from the stapedius muscle during vocalization (upper trace). The sound of the vocalization (lower trace) was recorded near the patient's mouth. The timing impulses shown below have intervals of 10 ms. (from [7] with permission).

17. Klockhoff, I. (1961). Middle-ear muscle reflexes in man. *Acta Otolaryngol. (Stockh.)* Suppl. 164.

18. Klockhoff, I., and Anderson, H. (1959). Recording of the stapedius reflex elicited by cutaneous stimulation. *Acta Otolaryngol. (Stockh.)* 50:451–454.

19. Liberman, M. C., and Guinan, J. J., Jr. (1998). Feedback control of the auditory periphery: Anti-masking effects of middle ear muscles vs. olivocochlear efferents. *J. Commun. Disorders* 31(6):471–482.

20. Metz, O. (1946). The acoustic impedance measured on normal and pathological ear. *Acta Otolaryngol. (Stockh.)* Suppl. 63.

21. Metz, O. (1951). Studies of the contraction of the tympanic muscles as indicated by changes in the impedance of the ear. *Acta Otolaryngol. (Stockh.)* 39:397–405.

22. Møller, A. R. (1958). Intra-aural muscle contraction in man, examined by measuring acoustic impedance of the ear. *The Laryngoscope* LXVIII(1):48–62.

23. Møller, A. R. (1961). "Bilateral Contraction of the Tympanic Muscles in Man" (pp. 1–15). Royal Institute of Technology (KTH), Div. of Telegraphy-Telephony, Report No. 18, Speech Transmission Laboratory, Stockholm.

24. Møller, A. R. (1962a). The acoustic reflex in man. *J. Acoust. Soc. Am.* 34(II):1524–1534.

25. Møller, A. R. (1962b). The sensitivity of contraction of the tympanic muscles in man. *Ann. Otol. Rhinol. Laryngol.* 71:86–95.

26. Perlman, H. B., and Case, T. J. (1939). Latent period of the crossed stapedius reflex in man. *Ann. Otol. Rhinol. Laryngol.* 48:663–675.

27. Pilz, P. K., Ostwald, J., Kreiter, A., and Schnitzler, H. U. (1997). Effect of the middle ear reflex on sound transmission to the inner ear of the rat. *Hear. Res.* 105(1–2):171–182.

28. Schuknecht, H. F. (1974). "Pathology of the Ear". Harvard Univ. Press, Cambridge, MA.

29. Simmons, F. B. (1964). Perceptual theories of middle ear muscle function. *Ann. Otol. Rhinol. Laryngol.* 73:724–739.

30. Terkildsen, K. (1960). Acoustic reflexes of the human musculus tensor tympani. *Acta Otolaryngol. (Stockh.)* Suppl. 158.

31. Zakrisson, J. E. (1974). The role of the stapedius reflex in poststimulatory auditory fatigue. *Acta Otolaryngol. (Stockh.)* 3:1–10.

Disorders of the Auditory System and Their Pathophysiology

Understanding the functional changes of the diseased auditory system can facilitate treatment and provide better patient information. Disorders of the auditory system are commonly divided into two broad groups, conductive hearing loss and sensorineural hearing. Conductive hearing loss consists of impairment of the sound conduction to the cochlea and it can be completely described by the pure-tone audiogram. It is the simplest form of hearing loss and it acts in the same way as an attenuation of sounds, though the attenuation is different for different frequencies. Sensorineural hearing loss is the name used for different kinds of hearing loss that are caused by disorders of the cochlea and the auditory nervous system. Sensorineural hearing loss has different attributes than conductive hearing loss, with many different expressions. Sensorineural hearing loss cannot be fully described by the pure tone audiogram. While speech discrimination can be accurately predicted on the basis of the pure-tone audiogram in patients with conductive hearing loss, speech discrimination may be more impaired in patients with sensorineural hearing loss than in patients that have a conductive hearing loss with the same threshold elevation. The degree to which speech discrimination is altered from hearing loss caused by disease or trauma to the auditory nervous system is more difficult to predict than that which occurs due to hearing loss from cochlear causes, at least when the latter is moderate.

While conductive hearing loss can be successfully treated, little can be done medically for patients with sensorineural hearing loss. However, such patients can be helped in other ways, such as with hearing aids. This section discusses the pathophysiology of disorders of the auditory system and how to correct or compensate for different disorders of the ear and the auditory nervous system.

When hearing disorders are considered, usually disorders of the ear are discussed and little attention has been devoted to disorders of the auditory nervous system. Most attention has been devoted to disorders that are associated with morphologic changes. It has become evident only relatively recently that the function of the auditory nervous system can change without any detectable changes in morphology. Such changes occur as a result of neural plasticity, which means that the function of specific parts of the nervous system change more or less permanently as a result of how they are activated. Such changes are now considered important factors in many disorders of the auditory system. It is now recognized that deprivation from stimulation such as from hearing deficits, conductive or sensorineural, can severely affect the normal development of the auditory nervous system during childhood and even change the function of the mature nervous system. This means that development of the auditory nervous system depends on appropriate stimulation early in life. Since deficits from insufficient stimulation during development cannot be reversed by sound exposure later in life, it is imperative to test hearing in neonates and any hearing loss detected should be compensated for in an early stage of life so that appropriate stimulation of the auditory system can be established. This also has made it important to understand the pathophysiology of hearing deficits in general and how to test those individuals who cannot cooperate in traditional behavioral audiometry.

Plasticity of the auditory nervous system comes in many forms and can be compensatory or detrimental. Thus, severe tinnitus may be caused by neural plasticity, as discussed in Chapter 16. Hearing loss from exposure to loud noise is affected by prior sound stimulation. Studies of mice indicate that the progression of age-related hearing loss (presbycusis) can be slowed by appropriate sound stimulation. Hearing impairments from disorders of the central nervous system are more difficult to assess than disorders affecting the conductive apparatus and the cochlea. Our limited knowledge about the normal function of the auditory system, and particularly about the diseased auditory system, has hampered identification of disorders that affect the auditory nervous system. Deficits in neural processing of sound therefore affect more people than earlier estimated. Hearing deficits from neural plasticity occur without any detectable morphologic abnormalities and that is important to recognize for those who treat hearing disorders as well as for researchers of pathologies of the auditory system.

The terms "psychogenic dysacusis," "functional deafness," or "nonorganic deafness" have been used for disorders of the nervous system, the (organic) cause of which could not be demonstrated morphologically. That, however, does not mean that these disorders do not have an organic pathology. They are different from malingering, where the patient knows that (s)he can hear but pretends not to hear. If no pathology can be found it does not mean that patients' complaints are false, it simply means that we are unable to find it with present knowledge and technology.

This section discusses the underlying pathologies for hearing deficits. In the first chapter (Chapter 13) we discuss conductive hearing loss and in the following two chapters (Chapter 14 and 15) we discuss sensorineural hearing loss. Tinnitus is the topic of Chapter 16.

CONDUCTIVE HEARING LOSS

Conductive hearing loss has similar effects on hearing as reducing the intensity of a sound and is similar to turning down the volume of a loudspeaker. The reduction in volume affects different frequencies differently, however. Various audiological tests make it possible to determine the nature and the anatomical location of this pathology. Many forms of conductive hearing loss will resolve on their own or can be successfully treated by surgery. Surgical treatment of conductive hearing loss requires knowledge about the normal function of the conductive apparatus the mechano-acoustical properties of the sound conducting apparatus, and how these are altered by disease processes, trauma, and birth defects. Such knowledge is also important when conductive hearing loss is to be managed by amplification or by leading sounds to the cochlea though paths other than the normal route.

SENSORINEURAL HEARING LOSS

The term "sensorineural hearing loss" covers hearing loss caused by changes in function of the cochlea, the auditory nerve, and the entire auditory nervous system. Hearing loss from cochlear pathologies is sometimes called "cochlear hearing loss" and hearing loss caused by disorders of the auditory nervous system is called "retrocochlear hearing loss." In cochlear hearing loss, impairment of the function of outer hair cells is the most common problem, followed by injuries to inner hair cells. Hearing loss from disorders of the auditory nerve differs from hearing loss caused by cochlear impairments in the way that they affect the patient and in how such disorders alter the outcome of audiometric tests. Hearing loss associated with detectable morphologic changes in the central auditory pathways is extremely rare. However, studies are now demonstrating that cochlear hearing loss may cause plastic changes in the central auditory nervous system that impair hearing. Hearing disorders from changes in the function of more central portions of the auditory nervous system manifest themselves with even more complex symptoms and signs than those caused by injury to the auditory nerve.

The pathophysiology of sensorineural hearing loss is thus far more complex than conductive hearing loss and it can often affect other aspects of hearing beyond the sensitivity of the ear. The degree of hearing impairment associated with sensorineural hearing can therefore not be completely described by the pure-tone audiogram, and other tests, such as speech discrimination tests, are necessary to fully describe sensorineural hearing loss, but even that may not fully describe disorders that affect functions of the auditory nervous system. Identifying the anatomical location of the physiologic abnormality of disorders of the auditory nervous system requires more complex tests, thorough assessment of the patient's symptoms and history is necessary for proper diagnosis.

Sensorineural hearing loss is often associated with tinnitus and hyperacusis. Recently it has become evident that hyperacusis, tinnitus, and, to some degree, impaired speech discrimination may be caused by changes in the function of the central auditory nervous system that are the result of neural plasticity and they may develop as a result of decreased input from the ear or other yet-unknown factors. Since these changes are assumed to occur without any morphologic changes it is likely that such changes can be reversed by proper sound treatment. One of the best known examples of intervention that affect hearing loss is the modification of noise-induced hearing loss that can be achieved by prior sound stimulation. Animal studies also indicate that the progression of age related hearing loss may be slowed by sound stimulation.

The possibilities of treating sensorineural hearing loss have been few and treatment is often limited to amplification of the sound (hearing aids), but cochlear implants now offer the possibility to restore some forms of hearing in people with profound hearing loss if the auditory nerve is intact. More recently, cochlear nucleus implants have been introduced to aid people in whom both auditory nerves are severely injured or surgically removed, such as is often the case after operations for bilateral acoustic tumors. Treatment of other aspects of hearing disorders, such as tinnitus and hyperacusis, has also increased through better understanding of the underlying mechanisms for these disorders. Treatments using exposure to specific sounds are already in use and progress in the treatment of disorders that involve neural plasticity can be expected to improve rapidly.

Hearing loss that is associated with cochlear pathologies are discussed in Chapter 14, and hearing disorders from diseases that affect the auditory nervous system are covered in Chapter 15. The coverage of disorders of the auditory nervous system in Chapter 15 concerns the auditory nerve, the brainstem, and the primary auditory cortex. The fact that injuries to the cochlea are often associated with changes in the auditory nervous system has blurred the border between cochlear hearing loss and hearing loss caused by disorders of the auditory nervous system.

OTHER DISORDERS OF THE AUDITORY SYSTEM

The disorders of the auditory system discussed in Chapters 13 and 14 all have morphologic signs. There is a group of hearing disorders that do not produce detectable morphologic changes, nor do they show any clear signs of functional abnormalities. The most prevalent of such disorders are subjective tinnitus and abnormal perception of sounds, such as hyperacusis and phonophobia. These disorders are some of the most diverse and complex disorders of the auditory system and their causes are often obscure. Often it is not even possible to identify the anatomical location of the pathologies that cause these sensations. The causes of the different kinds of abnormal perception of sounds that may be described as hyperacusis or phonophobia, which often accompany tinnitus, are even more obscure. Tinnitus and hyperacusis may be associated with identifiable diseases but usually they are the only symptoms. Since tinnitus is perceived as a sound, the ear has often been assumed to be the location of the pathology but recently it has become increasingly evident that the pathology often is located in the central nervous system.

While patients with severe tinnitus and hyperacusis or phonophobia are obviously miserable, it is not obvious which medical specialty is best suited for treating them. It is, however, certain that whomever is to take care of such patients must have the best possible knowledge and understanding of the changes in the function of the auditory system that can lead to tinnitus and hyperacusis in order to be able to help these individuals. Chapter 16 discusses current hypotheses about the generation of tinnitus and hyperacusis and describes the mechanisms that are believed to be involved in generating tinnitus and causing hyperacusis and phonophobia.

CHAPTER 13

Sound-Conducting Apparatus

Abstract

1. Disorders of the conductive apparatus cause elevation of the hearing threshold and affect speech discrimination in the same way as reducing the sound intensity.
2. Hearing impairment can be caused by an impairment in sound conduction in the outer ear or the middle ear.
3. Common causes of conductive hearing loss are obstruction of the ear canal by cerumen.
4. Hereditary malformations of the ear canal (ear canal atresia) can block sound conduction to the middle ear.
5. Accumulation of fluid in the middle ear or the air pressure in the middle ear cavity being different from the ambient pressure cause conductive hearing loss.
6. Otosclerosis is a disease that impairs sound conduction through the middle ear by bone growth around the stapes footplate, which ultimately becomes immobilized.

7. Various disorders and trauma can cause interruption of the ossicular chain or perforation of the tympanic membrane, which results in conductive hearing loss.
8. Hearing without the middle ear involves loss of the transformer action of the middle ear and additional hearing loss because the difference between the sound that reaches the two windows of the cochlea becomes smaller.
9. Some rare instances of conductive hearing loss are accompanied by tinnitus.
10. Diagnosis of conductive hearing loss is made from pure-tone audiometry, tympanometry, and recording the acoustic middle ear reflex response.
11. Some forms of conductive hearing loss reverse without treatment. Other forms are treatable with medicine or surgery.

INTRODUCTION

The anatomical location of impairment of sound transmission to the cochlea can be the ear canal, the tympanic membrane, or the ossicular chain. Various audiologic tests can determine the anatomical location of the impairment. Correct interpretation of such tests requires knowledge and understanding about the normal function of the sound-conducting apparatus as well as about how various disease processes and trauma can alter the function of the sound-conducting apparatus.

EAR CANAL

The simplest form of hearing loss that is most easily treated is caused by a build-up of cerumen, which blocks the ear canal (impacted wax) and thereby obstructs sound conduction to the tympanic membrane (Fig. 13.1). This results in a nearly flat hearing loss that varies between 20 and 30 dB but can reach 40 dB at higher frequencies. Hearing is restored to normal by removing the cerumen. In frequent swimmers the ear canal often narrows because of formation of new bone (exostosis). This makes it easier for cerumen to accumulate and obstruct the ear canal and it makes it difficult to clean the ear canal.

EAR CANAL COLLAPSE

With age, the outer (cartilaginous) portion of the ear canal in many individuals changes from a nearly circular cross section to an oval shape and consequently it may become totally occluded. If the earphone used for audiometry has a supraaural cushion (AR/MX41) it may cause a nearly collapsed ear canal to

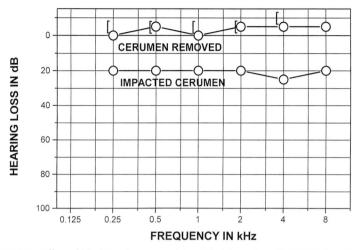

FIGURE 13.1 Effect of blocking the ear canal (modified from Sass, K. (1998). Sensitivity and specificity of transtympanic electrocochleography in Ménière's disease. *Acta Otolaryngol.* (*Stockh.*) **118**:150–156).

become totally occluded due to assertion of pressure on the ear canal by the earphone, thus causing the audiometry results to be erroneous. The hearing loss is similar to that caused by impacted cerumen (approximately 25 dB). Placing a short plastic tube in the ear canal during hearing tests can solve this problem. The use of insert earphones is the best solution.

EAR CANAL ATRESIA

Ear canal atresia is a condition where one or both ear canals have not opened during prenatal life. The mild form of this congenital malformation is character- ized by a small ear canal and a nearly normal middle ear. In a more severe form the ear canal is totally occluded (or actually missing) and the ossicular chain is malformed. In the most severe form the middle ear space is small or absent in addition to the ear canal being occluded. Ear canal atresia impairs transmission of airborne sound to the tympanic membrane and the function of the middle ear may be impaired. If it occurs on both sides, the hearing loss will be 55–70 dB (Fig. 13.2). Such hearing loss implies a listening distance of less than 10 cm (4 inches) (hearing loss of 60 dB, which is in the speech frequency range, results in a required listening distance of 10 cm from the ear). A person with such a condition will require a hearing aid. The bone conduction is little affected (Fig. 13.2) and therefore bone-conduction hearing aids have been used to help such patients as an alternative to surgical interven- tion. Ear canal atresia on one side will allow a person to hear with one ear,

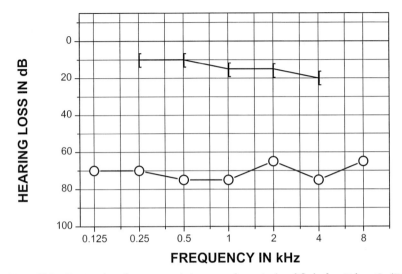

FIGURE 13.2 Hearing loss from congenital ear canal atresia (modified after Liden, G. (Ed.) (1985). "Audiologi." Almquist and Wiksell, Stockholm).

but such a person will have difficulties in determining the direction to a sound source and have difficulties in the discrimination of speech in noisy environments.

MIDDLE EAR

The middle ear is the site of most disorders that affect sound transmission to the cochlea. The air pressure in the middle ear cavity being different from the ambient pressure is probably the most common cause of impairment of sound transmission to the cochlea. Accumulation of fluid in the middle ear cavity and pressure in the middle ear cavity that is different from the ambient pressure are the effects of some of the most common disorders that can impair sound transmission to the cochlea. Less common and more serious pathologies of the middle ear include perforation of the tympanic membrane or interruption or fixation of the ossicular chain. Each one of these conditions, affect sound transmission through the middle ear in specific ways.

AIR PRESSURE

Sound transmission to the cochlea is impaired when the air pressure in the middle ear cavity is different from that in the ear canal (the ambient pressure), as was discussed in Chapter 2. The effect is a decrease in transmission that is greatest for low frequencies. Negative pressure in the middle ear cavity causes a larger hearing loss than the same value of positive pressure. Negative pressure in the middle ear cavity is caused by malfunction of the eustachian tube and often occurs in connection with middle ear infections. Positive pressure in the middle ear cavity may occur in the ascending phase of flying because of a decrease in the ambient pressure, but it usually equalizes spontaneously even with a partly functioning eustachian tube. Negative pressure that occurs during descent is more difficult to equalize because the higher ambient pressure exerts a closing pressure on the opening of the eustachian tube in the pharynx. Therefore, a person is more likely to have problems equalizing pressure in the middle ear on landing than on take-off. The air pressure in the middle ear cavity can be determined by tympanometry. The pressure in the middle ear cavity is equal to the pressure in the ear canal at which the peak in the tympanogram occurs.

FLUID IN THE MIDDLE EAR (OTITIS MEDIA WITH EFFUSION)

The mucosa of the middle ear cavity has attracted attention because it is involved in a common disorder known as otitis media with effusion (OME), which is an inflammation of the lining of middle ear cavity, the mastoid cell system, and the eustachian tube. A large number of children acquire this condition. It has been estimated that approximately 90% of children acquire OME within the first 3 years of life [12, 175–177]. At an early age OME may also disturb the pneumatization of mastoid air cells [176, 177]. Small mastoid cell systems promote middle ear infections later in life. The incidence of OME decreases rapidly with age and OME occurs infrequently in adults.

The cause of OME is inflammation of the middle ear mucosa that prevents the eustachian tube from opening normally, which creates a negative air pressure in the middle ear cavity and an accumulation of fluid in the middle ear cavity. The development of negative pressure in the closed middle ear cavity is a result of absorption of oxygen by the mucosa. If the eustachian tube does not open properly, clear fluid may effuse from the mucosa of the middle ear and fluid may accumulate in the middle ear cavity if not drained through the eustachian tube. Fluid that is more viscous may accumulate as a result of the inflammatory processes. The major reasons that OME is more frequent in children than adults are that the eustachian tube is shorter in children up to age 5–6 years than in adults and that the direction of the eustachian tube is nearly horizontal rather than pointing 45° downward, as it does in adults (Fig. 1.6B).

Fluid in the middle ear cavity affects sound conduction only when it covers the tympanic membrane. Fluid that covers the backside of the tympanic membrane prevents it from moving; when the entire tympanic membrane is covered by fluid sound must be transferred directly to a fluid in order to reach the cochlea, which is very inefficient, as was discussed in Chapter 2. Hearing, however, is likely to be essentially unaffected by fluid that fills the middle ear cavity incompletely as long as there is air behind the tympanic membrane. The air behind the tympanic membrane acts as a cushion that adds stiffness to the middle ear and impedes the motion of the tympanic membrane for low frequencies only. The stiffness of that air cushion increases when the volume decreases but it causes only slight hearing loss at low frequencies and therefore often escapes detection from normal audiometric testing. Clear (low-viscosity) fluid that covers the ossicles has minimal effect on their movement and fluid covering the round window of the cochlea will not affect the motion of the cochlear fluid noticeably.

Since hearing loss from an incompletely fluid-filled middle ear depends on how large a portion of the tympanic membrane is covered with fluid, the resulting hearing loss will depend on the head position, provided that the fluid has a low viscosity so that it can move freely in the middle ear cavity. Hearing loss may only be evident when such a patient is lying down with the head turned toward the side of the fluid-filled ear. In that body position hearing loss may become evident even when the amount of fluid in the middle ear cavity is small. Hearing loss is independent of head position in patients whose middle ear is totally fluid filled or in patients whose ears are filled with highly viscous fluid. Fluid with a gel-like consistency, as is often the case in middle ear infections, may impair hearing noticeably even without covering the tympanic membrane because it impedes the motion of the middle ear ossicles. Such "glue ears" thus typically present with hearing loss that is independent of the position of the head.

When fluid fills the middle ear cavity sound must be transferred to the fluid and it exerts approximately the same force on both the round and the oval window, thus a situation similar to hearing without a middle ear. Fluid in the middle ear can therefore cause large hearing losses. However, statistics show that the average hearing loss is approximately 30 dB with nearly normal bone-conduction thresholds [85] (Fig. 13.3). Only few patients have hearing losses of 50 dB.

Fluid in the middle ear can be diagnosed by tympanometry. If the fluid covers the entire backside of the tympanic membrane, the tympanogram will be flat because the acoustic impedance of the ear does not change with changing air pressure in the ear canal. It is not possible to record the response of the acoustic middle ear reflex in such an ear even when elicited from a normal opposite ear. The reflex response can be elicited from an ear with fluid in the middle ear cavity and recorded from the opposite ear, provided that the middle

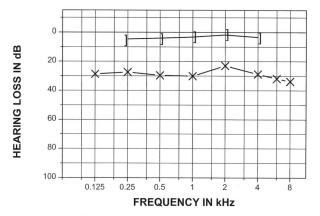

FIGURE 13.3   Hearing loss from chronic otitis media (modified after [85] with permission).

ear is normal in that ear. The reflex threshold is elevated by the amount of hearing loss in the affected ear. The tympanogram will have a small peak in an ear where fluid does not completely cover the backside of the tympanic membrane but it is unlikely that a response of the acoustic middle ear reflex can be recorded.

In children, when hearing loss caused by fluid in the middle ear cavity impairs speech discrimination, it is important that hearing is restored sufficiently because exposure to speech and other natural sounds is important for the normal development of the auditory nervous system (see Chapter 15). This is one of the reasons that OME is routinely treated by placing a small tube in the tympanic membrane so that the fluid can drain and the air pressure in the middle ear cavity can equalize to that of the ambient pressure. The main purpose is to restore hearing. Although such a tube acts as a hole in the tympanic membrane, it will not interfere noticeably with hearing because of its small opening (see discussion about perforated tympanic membrane, Fig. 13.4).

The fluid in the middle ear cavity may contain toxic substances produced by bacteria and these substances may enter the cochlea by diffusing through the membranes of the round and the oval windows. Studies using evoked potentials in rats with middle ear effusion showed signs of cochlear involvement in addition to conductive hearing loss [161]. Whether these substances cause permanent injury is unknown.

CHOLESTEATOMAS

Cholesteatomas are examples of other growths in the middle ear cavity that may affect hearing. Cholesteatomas are benign growths that may develop in the middle ear after long-term recurrent or chronic middle ear infections or

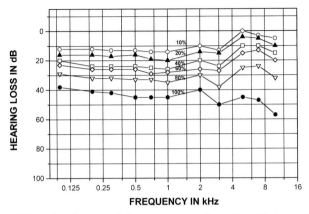

FIGURE 13.4   Effect of perforation of the tympanic membrane (modified after Payne, M. C., and Githler, F. J. (1951). Effects of perforations of the tympanic membrane on cochlear potentials. *Arch. Otolaryngol.* 54:666–674).

may occur with no apparent (known) cause. When a cholesteatoma grows in the middle ear cavity, the extent of the hearing loss it causes varies with the size of the growth and whether it is in contact with the ossicular chain or to what extent it may cause erosion of the ossicular chain and interruption of the ossicular chain.

Perforation of the Tympanic Membrane

It is the difference between the sound that reaches the front side and the backside of the tympanic membrane that causes it to move (vibrate). A hole in the tympanic membrane will allow some sound to reach the backside of the tympanic membrane, which reduces the difference between the sound on the two sides of the tympanic membrane. The result is that the force that causes the tympanic membrane to vibrate becomes reduced. The reduction in vibration of the tympanic membrane caused by a perforation depends on the size of the hole in the tympanic membrane and the size of the middle ear cavity. A small hole in the tympanic membrane acts as a low-pass filter and therefore causes hearing loss at low frequencies because only low frequencies will reach the middle ear cavity and decrease the force that acts on the tympanic membrane.[1] A larger hole permits sounds within a larger frequency range to

[1] Acoustically, a hole in the tympanic membrane acts in the same way as an electrical inductance and the middle ear cavity acts as a capacitance (the mechanical analogy is a mass and a spring respectively). This explains why a hole acts as a low-pass filter, attenuating high frequencies. This is why the effect of a hole in the tympanic membrane is greatest for low frequencies. The cut-off frequency of this low-pass filter is lower for a small hole than for a large one and it is lower for larger middle ear cavities.

reach the middle ear cavity and thus impair hearing over a larger range of frequencies (Fig. 13.4).

Experimental studies in the cat of the effect of a perforation of the tympanic membrane [128] have confirmed that the effect is largest at low frequencies. A small hole in the tympanic membrane causes hearing loss mainly at frequencies below 4 kHz, whereas a larger hole also affects higher frequencies (Fig. 13.4). The effect of a large perforation is not limited to the effect of sound reaching the backside of the tympanic membrane. When large parts of the tympanic membrane are lost the perforation also affects the way the manubrium of malleus vibrates because some of the suspension of the malleus is lost. The effect on the hearing threshold depends not only on the size of the perforation but also on its location on the tympanic membrane. Animal experiments have shown that the largest effect of a hole of a certain size occurs when it is placed in the posterior or superior part of the tympanic membrane and the least effect occurres when it is located in the anterior, inferior portion.

Clinical experience is generally in good agreement with the results of animal experiments. Thus, in humans, a large hole in the tympanic membrane commonly results in a 40- to 50-dB, mostly flat, hearing loss (a 45-dB hearing loss, which is in the speech range, corresponds to a maximal listening distance of approximately 1.5 m, approximately 5 feet, for normal speech). When the tympanic membrane is totally missing the hearing loss can reach 60 dB (corresponding to a listening distance of approximately 10 cm) [129, 149].

The acoustic middle ear reflex cannot be recorded in an ear with a perforated tympanic membrane because stiffening of the middle ear from contraction of the stapedius muscle will not affect the acoustic impedance of the ear at the low frequency it is usually measured (220 Hz), unless the perforation is very small. The acoustic reflex may be elicited from the affected ear and recorded in the opposite ear, provided that the hearing loss in the affected ear is not excessive at the frequency of the stimulus tone and that the middle ear is normal in the opposite ear. A tympanogram is flat in an ear with a perforated tympanic membrane because pressure in the ear canal is transferred to the middle ear cavity and thus does not change the ear's acoustic impedance.

INTERRUPTION OF THE OSSICULAR CHAIN

Interruption of the ossicular chain may occur as a result of trauma or because of disease processes that erode the middle ear ossicles, such as cholesteatomas. The conductive hearing loss may exceed 60 dB when the ossicular chain is interrupted with an intact tympanic membrane (Fig. 13.5). This hearing loss is thus greater than when sound reaches the two windows of the cochlea

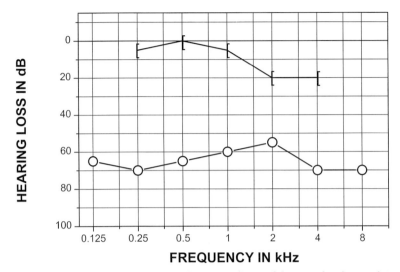

FIGURE 13.5 Effect on hearing threshold from interruption of the ossicular chain with intact tympanic membrane (modified from Liden, G. (Ed.) (1985). "Audiologi." Almquist and Wiksell, Stockholm).

directly such as occurs when the entire middle ear, including the tympanic membrane, is missing (cf. p. 390 and Fig. 13.7). The greater hearing loss is caused by the attenuation of the sound that reaches the middle ear cavity by the intact tympanic membrane. Individuals with an interrupted ossicular chain thus hear better when the tympanic membrane is perforated. The hearing loss from an interrupted ossicular chain observed in humans (Fig. 13.5) is of the same order of magnitude as that obtained in animal experiments (cats) where interruption of the incudo-stapedial joint results in a hearing loss of 50–70 dB [129, 189].

Interruption of the ossicular chain can be diagnosed by tympanometry because the acoustic impedance of the ear is abnormally low and changes more than normal when the pressure in the ear canal is altered. (cf. Fig. 2.23 in Chapter 2). Tympanometry thus shows a larger than normal peak. The acoustic middle ear reflex response is absent in an ear with an interrupted ossicular chain because contraction of the stapedius muscle does not change the ear's acoustic impedance. Considering the large conductive hearing loss that results from interruption of the ossicular chain, it is also unlikely that the reflex can be elicited from the affected ear and recorded in the opposite ear, even if that is normal.

In patients with an interrupted ossicular chain, a connection between the incus and the stapes may be reestablished by soft tissue. In such a situation, the patient will have improved hearing for low frequencies because an elastic connection between the middle ear bones, transmits high frequencies poorly. Such reestablishment of the connections in the middle ear may occur as a result of the growth of a cholesteatoma or in otospongiosis, which involves the ossicles. Further growth of a cholesteatoma may thus result in a paradoxical improvement of hearing despite a progression of the disorder. If left untreated, further growth of a cholesteatoma may cause sensorineural hearing loss due to erosion into the cochlea.

FIXATION OF THE OSSICULAR CHAIN

The normal motion of the stapes is impaired in a disorder known as otosclerosis in which the stapes footplate becomes fixated in the round window because new bone is constantly forming around the stapes footplate. Hearing loss in patients with otosclerosis is largest for low frequencies and increases with the progression of the disease, usually over many years. Typically, hearing loss in patients who have had otosclerosis for many years is 50 dB at low frequencies and less at high frequencies (Fig. 13.6). The bone-conduction threshold is nearly normal but often has a dip (up to 30 dB) around 2 kHz. This dip, known as Carhart's notch, is not a result of cochlear (sensorineural) involvement because it disappears after a successful operation. It is an indication that

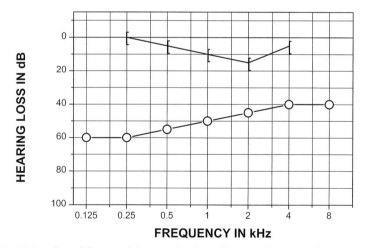

FIGURE 13.6   Effect of fixation of the ossicular chain (from otosclerosis) on hearing threshold (modified from Liden, G. (Ed.) (1985). "Audiologi." Almquist and Wiksell, Stockholm).

fixation of the stapes footplate affects sound conduction to the cochlea through bone conduction.

The response of the acoustic middle ear reflex is absent in an ear in which the stapes is immobilized (e.g., in otosclerosis). A tympanogram of an otosclerotic ear has a small peak because the mobility of the middle ear is reduced by the fixation of the stapes.

Patients with otosclerosis may, over time, develop cochlear hearing loss because of formation of new bone inside the cochlea. Otosclerosis may also result in sensorineural hearing loss that adds to the approximately 60-dB conductive hearing loss. The total hearing loss may reach 80–85 dB after 20–25 years of untreated otosclerosis. New bone formation in the cochlea may be slowed by treatment with fluor compounds.

The earliest treatments of otosclerosis involved making an artificial route for sound to the cochlea by making an opening in the bone of the semicircular canal. This operation, known as fenestration, was replaced by a procedure where the new bone around the stapes was removed, but that method has also been abandoned because the relief was short due to rapid formation of new bone.

At present, the common treatment of otosclerosis is to replace the middle ear bones with a prosthesis. This procedure was introduced by Shea (1958) [157]. The stapes is removed and replaced with a prosthesis that connects the incus to an artificial stapes placed in the oval window. This solves the problem of recurring hearing loss from bone growth. It is now the common treatment for otosclerosis. Techniques that are now being developed include making a small hole in the stapes footplate to insert a prosthesis. It is less traumatic than replacing the entire stapes footplate and it produces the same results.

Other treatments of middle ear pathologies also use prostheses. Such prostheses must be properly sized and must be placed so that maximal sound transfer to the cochlea is accomplished. With modern microsurgical techniques and modern designs of middle ear prostheses, hearing can be restored to nearly normal levels. However, middle ear prostheses do not replace the function of the stapedius muscle in regulating sound transmission through the middle ear and patients with middle ear prostheses may have a higher susceptibility to noise-induced hearing loss (see Chapter 14).

HEARING WITHOUT THE MIDDLE EAR

An understanding of the hearing deficits that result when the tympanic membrane or the entire middle ear is missing so that sound reaches both cochlear windows directly requires an understanding of how the cochlear fluid is set into motion with and without the middle ear. The improvement of sound conduction to the cochlea by the transformer action of the middle ear was discussed in Chapter 2. However, that does not explain all ramifications on hearing from total or partial loss of the middle ear.

When the middle ear transformer action is absent, the hearing loss exceeds that of the transformer gain (approximately 30 dB) because the sound reaches the two windows of the cochlea at approximately the same intensity. Since it is the difference in pressure between the two cochlear windows that causes the cochlear fluid to move, hearing depends on the size of the differences between the sound that reaches the two cochlear windows. Normally, the force that acts on the oval window is much larger than that acting on the round window because of the gain of the middle ear transformer. If the resulting sound pressure at the two windows is exactly equal there would theoretically be total deafness, because there would be no motion of the cochlear fluid at all. However, in practice there will always be some difference in the amplitude and the phase of the sound that reaches the two windows and it is that difference that sets cochlear fluid in motion and makes it possible for individuals without a middle ear to hear, although at a much elevated threshold.

Without surgical restoration of hearing, individuals with an open middle ear cavity and no middle ear ossicles have hearing loss for air-conducted sounds of 50–60 dB (Fig. 13.7), thus approximately 30-dB in addition to the approximately 30 dB loss from the transformer action of the middle ear. The hearing loss is often less for high frequencies because the phase difference between the sound that reaches the two windows of the cochlea is larger for high frequencies than low

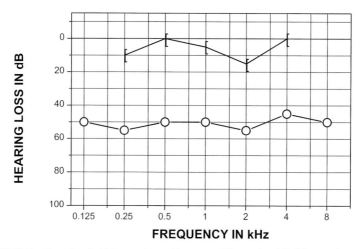

FIGURE 13.7 Tone threshold in a patient without a middle ear (modified from Liden, G. (Ed.) (1985). "Audiologi." Almquist and Wiksell, Stockholm).

frequencies, which creates a larger difference between the force that acts on the two windows for high frequencies than for low frequencies.

It has been confirmed in animal experiments that it is the difference between the amplitude and phase of the pressure that acts on the two windows that drives the cochlear fluid [182]. In these experiments, pure tones were applied independently to the round window and the oval window of the cochlea in cats. The amplitude and the phase angle between the sound at the two windows were varied independently, while the cochlear microphonic (CM) potentials were recorded. Recall that the CM is a valid measure of the volume velocity of the cochlear fluid (Chapters 2 and 4). The force that sets the cochlear fluid into motion is the vector difference between the forces at the two cochlear windows (Fig. 13.8). This means that the largest

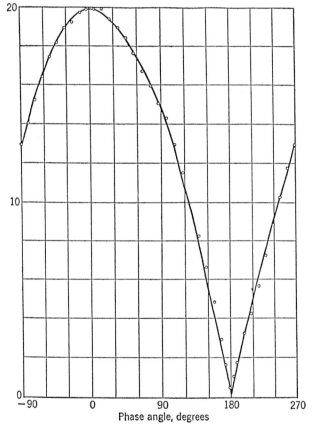

FIGURE 13.8   Illustration of the vectorial summation of sound that reaches both the oval and the round windows. The results were obtained in a cat using the cochlear microphonic potentials to measure the motion of the cochlear fluid (from Wever, E. G., and Lawrence, M. (1950). The acoustic pathways to the cochlea. *J. Acoust. Soc. Am.* **22**:460–467).

motion of the cochlear fluid is induced when the two sounds are precisely out of phase. Thus, a sinusoidal force (pure tone) produces the largest motion of the cochlear fluid when applied with a phase difference of 180° (opposite phase).

Individuals in whom restoring the function of the middle ear using a prosthesis is not possible can be helped by operations that aim to make the difference between the sound that reaches the two windows of the cochlea as large as possible. Since the distances in the middle ear in relation to the wavelength of the sound are small, the sound level will be nearly uniform in the middle ear cavity in an ear without a tympanic membrane. This means that the difference between the sound that reaches the two windows of the cochlea will be small. The phase angle is 180° for 1/2 wavelength and at 1000 Hz, 1/2 wavelength is 17 cm; at 10 kHz it is 1.7 cm. However, a smaller phase shift can cause the cochlear fluid to move considerably (Fig. 13.8). Thus, a difference of only 10° produces a CM that is only 20 dB below its maximal value (i.e., a hearing loss of only 20 dB). A difference between the forces at the two cochlear windows of 10% (1 dB) will produce motion of the cochlear fluid that is only 20 dB less than its maximal value.

Many different surgical methods have been tried in order to shield the two windows from each other. In the type IV tympanoplasty operation, a tissue graft (e.g., fascia from a muscle) is placed to shield the round window from sound while the stapes footplate resting in the oval window is exposed to sound. (If the stapes footplate is removed and replaced by a graft, the same operation is called tympanoplasty type V). It is important that the cavity created over the round window, known as the cavum minor, is kept, ventilated, and the shield should be as rigid as possible to reduce the sound that reaches the round window as much as possible. If a successful shielding between the two windows is accomplished, the expected hearing loss would be equal to that of the loss of the transformer action of the middle ear, thus 25–30 dB. That ideal situation cannot be achieved surgically and in practice the results of such operations are moderate hearing loss of 35–40 dB for low frequencies and 20–25 dB at 2000 and 4000 Hz. (The results of such operations are usually expressed as the air–bone gap, i.e., the difference between the air-conduction threshold and the bone-conduction threshold, because the bone conduction threshold represents the threshold of the cochlea.)

IMPAIRMENT OF SOUND CONDUCTION IN THE COCHLEA

Sound conduction in the cochlea can be impaired, for instance, by bone growth, similar to that in otosclerosis (known as cochlear otosclerosis). Some of the attributes to this kind of hearing loss are similar as impairment of sound conduction in the middle ear and some are different. Since this condition occurs together with immobilization of the stapes, tests such as the tympanogram and the acoustic middle ear reflex will be similar to that seen in common oto-sclerosis. Patients with cochlear otosclerosis will, in addition, have signs of sensorineural hearing loss because bone growth in the cochlea affects the nerve supply to the hair cells.

ACCURACY OF MEASUREMENTS OF CONDUCTIVE HEARING LOSS

It is usually assumed that pure-tone audiograms are accurate measures of conductive hearing loss. However, the commonly used earphones such as the TDH39 with MX41/AR also conduct sound by bone conduction. It was shown in Chapter 2 that the average crosstalk for these earphones is approximately 60 dB between 500 and 4000 Hz (Fig. 2.7A in Chapter 2). This is the reason why it is necessary to mask the better hearing ear when testing the hearing of a person with much larger hearing loss in one ear than in the other. If an earphone can stimulate the opposite ear by bone conduction, bone conduction must be equally effective in stimulating the cochlea on the side where the earphone is applied. In a patient with a conductive hearing loss that exceeds 60 dB, the sound that reaches the cochlea by bone conduction will be stronger than the sound that reaches the cochlea by air conduction. Bone-conducted sound generated by earphones such as the TDH39 may therefore set a limit for the maximal conductive hearing loss that can be measured at approximately 60 dB. That means that conductive hearing loss that is greater than approxi-mately 60 dB cannot be assessed using this type of earphones and conductive hearing losses exceeding 60 dB may therefore be underestimated. Insert-earphones generate less bone-conducted sound (Fig. 2.7B) and, for frequencies below 1000 Hz, bone-conducted sounds are, on average, more than 85 dB below air-conducted sounds. The level of bone-conducted sound that these two types of earphones deliver varies from individual to individual. The difference between the air-conducted and the bone-conducted sounds delivered by the TDH39 earphone thus can be as small as 50 dB (Fig. 2.7A). In some individuals the bone-conducted sounds delivered by insert-earphones may be only 60 dB below the air-conducted sound (Fig. 2.7B).

These matters should be taken into account when interpreting hearing loss in individuals with large conductive components of hearing loss. The actual hearing loss associated with hearing without the middle ear, and that due to ear canal atresia may be higher than the measured air conduction threshold when that exceeds 60 dB, using conventional audiometry because of the contri-bution of bone-conducted sound.

CHAPTER 14

Disorders of the Cochlea

ABSTRACT

1. The most common cause of cochlear hearing loss is injury to outer hair cells, which impairs the cochlear amplifier causing elevation of the hearing threshold and often loudness recruitment and tinnitus.
2. Hearing loss caused by injury to outer hair cells usually develops gradually, beginning at high frequencies and extending to lower frequencies as the disorder progresses, but it rarely exceeds 50 dB.
3. Besides an elevation of threshold of hearing, the cochlear filters become broader following hair cell injury, which may increase masking and impair temporal coding of broadband sounds such as vowels.
4. Amplitude compression is impaired from injuries to outer hair cells and this may be the cause of loudness recruitment.
5. When tested under ideal conditions, speech discrimination remains normal for moderate threshold shifts if adequate amplification is provided but speech discrimination in noisy environments may deteriorate even for a moderate degree of hearing loss. The average speech discrimination is related to the threshold elevation for pure tones, but the individual variation is large.

6. Brainstem auditory evoked potentials (BAEP) and the acoustic middle ear reflex are little affected by injuries to outer hair cells.

7. Age-related changes are the most common cause of cochlear hearing loss.

8. Exposure to loud sounds can cause injuries to cochlear hair cells, as can drugs, such as diuretics and aminoglycoside antibiotics, trauma, and diseases.

9. Some forms of cochlear hearing loss are hereditary, and some forms of hearing loss worsen within the first year of life and may become severe.

10. Normal functioning of the cochlea requires that the pressure (or rather, volume) in the different compartments of the cochlea remain within close limits.

11. The cochlear hearing loss that is one of a triad of symptoms that defines Ménière's disease is assumed to be caused by an imbalance of pressure (or volume) in the compartments of the cochlea. In early stages of the disease, hearing loss mostly affects low frequencies and fluctuates.

12. Cochlear hearing loss does not reverse by itself, except in the case of the temporary threshold shift (TTS) that occur after noise exposure and in the early stages of Ménière's disease, where the hearing loss fluctuates.

13. Cochlear hearing loss cannot be restored medically or surgically but can often be compensated for by wearing a hearing aid.

14. Cochlear implants can possibly restore useable hearing in people with severe cochlear damages as long as the auditory nerve is intact.

INTRODUCTION

The decline in hearing with age, known as presbycusis, is the most common form of sensorineural hearing loss. It is assumed to be caused by impaired function of cochlear hair cells as a part of the normal aging process. It is mainly outer hair cells that are affected, which impairs the cochlear amplifier, with little effect on sensory transduction in the cochlea (see Chapter 3). The decline in hearing sensitivity with age progresses at different rates in different individuals, and other types of hearing loss, such as impaired function of the auditory nerve, may contribute to age-related hearing impairments. There is increasing evidence that changes may occur in the central auditory nervous system in connection with cochlear impairments. These changes are thought to occur as a result of neural plasticity and it may be the deprivation of input from the cochlea in certain frequency ranges of hearing that elicit such changes (see Chapter 15). Causes of reduced function of outer hair cells other than presbycusis include exposure to noise and administration of drugs such as certain antibiotics, certain diuretics, aspirin, quinine, and many other substances. Certain diseases such as Ménière's disease are associated with cochlear hearing loss. The hearing loss in Ménière's disease is probably not associated with injury to hair cells, at least not the fluctuation in hearing loss that is typical in the beginning of the disease. It is more likely that the low-frequency fluctuating hearing loss is caused by a distension of the basilar membrane that occurs as a result of cochlear hydrops.

Injury of cochlear hair cells affects hearing in several ways. The most obvious is reduced sensitivity, i.e., an elevated pure-tone threshold. Speech discrimination is not profoundly affected by moderate cochlear hearing loss and it is closely related to pure-tone hearing loss. Loudness recruitment is common in people with cochlear hearing loss. This means that the loudness of a sound increases more rapidly than normal. A person with cochlear hearing loss hears a sound that is slightly above the elevated threshold almost as loud as a person with normal hearing and changes in a sound's intensity are perceived to be larger than what a person with normal hearing perceives. Increased masking may also be a result of cochlear hearing loss.

In this chapter, we discuss the pathophysiologic basis for these changes. Specifically, we discuss how the impairment of speech perception, increased masking, and recruitment of loudness are related to the impairments of the function of cochlear hair cells that commonly occur as a result of aging (presbycusis), exposure to loud noise, administration of ototoxic substances, and various disease processes.

GENERAL AUDIOMETRIC SIGNS OF HAIR CELL INJURIES

Some general features about cochlear hearing loss can be identified. Thus, most forms of hearing loss caused by injury to cochlear hair cells affect the hearing threshold at high frequencies more than low frequencies. Typically, the hearing loss caused by injuries to cochlear hair cells begins at the highest frequencies that are tested by clinical audiometry (8 kHz) and progresses toward lower frequencies as it becomes more severe. It is generally assumed that many forms of cochlear hearing loss, such as presbycusis, begin at the highest frequencies of the audible range. The normal hearing range in humans extends to about 20 kHz, but the hearing threshold is normally not tested at frequencies above 8 kHz. Therefore data about hearing loss in the frequency range above 8 kHz is sparse and the beginning of the progression of hearing loss usually escapes detection. Hearing loss caused by exposure to noise is an exception because it normally affects the hearing threshold the most at about 4 kHz. Hearing loss from Ménière's disease mostly affects low frequencies. Certain forms of hereditary hearing loss mainly affect the mid-frequency range that is commonly tested ("cookie-bite" audiogram).

Speech discrimination is usually little affected in patients with cochlear hearing loss when the hearing loss is not excessive. Speech discrimination only becomes affected when the threshold elevation at frequencies below 2000 Hz becomes noticeable (Fig. 14.1). The data shown in Fig. 14.1 can only serve as a guide in estimating speech discrimination in an individual because they represent the average decrease in speech discrimination obtained in a large number of individuals and the relationship between pure-tone thresholds and speech discrimination varies considerably between different individuals. Only hearing loss of the high-frequency type was included in the data shown in Fig. 14.1 because that is the commonly occurring type of hearing loss associated with injuries to cochlear hair cells. The relationship between pure-tone threshold and speech discrimination may be different in other forms of cochlear hearing loss.

The reason that speech discrimination is only affected to a small degree when the hearing loss is moderate is that only outer hair cells are affected, which only impairs the cochlear amplifier, leaving sensory transduction unaffected. Recall from Chapter 3 that the cochlear amplifier is mainly effective at low sound levels and its effect is small at physiological sound levels. Since speech tests are performed at physiologic sound levels, impairment of the cochlear amplifier has little effect on speech discrimination. Also, most speech tests are done in quiet. If they were done in noise, deficits due to cochlear

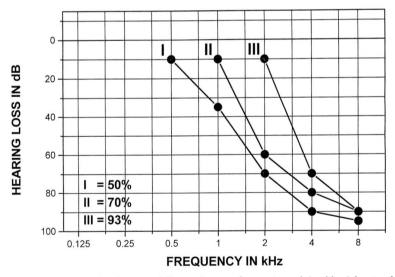

FIGURE 14.1   Relationship between different degrees of sensorineural (cochlear) hearing loss and speech discrimination (modified from Liden, G. (Ed.) (1985). "Audiologi." Almquist and Wiksell, Stockholm).

hearing loss would often appear more severe and perhaps more similar to what the individual experiences in normal everyday listening conditions. Individuals who have greater impairments of speech discrimination than the average person with the same hearing loss may have impairments of other structures in addition to cochlear impairments. Thus, injury to the auditory nerve, which is present in some patients with presbycusis, may be one such factor that is responsible for poor speech discrimination in some patients.

Brainstem auditory evoked potentials are unaffected by moderate degrees of cochlear hearing loss and the threshold of the acoustic middle ear reflex is within normal limits. As is discussed in Chapter 15, some forms of cochlear hearing loss have changes in the auditory nervous system in addition to hair cell injuries. In such individuals, both the BAEP and the acoustic middle ear reflex may be abnormal and speech discrimination is likely to be lower than what it is in individuals with pure cochlear hearing loss.

AGE-RELATED HEARING LOSS (PRESBYCUSIS)

AUDIOMETRIC SIGNS

Presbycusis appears as a gradually sloping hearing loss toward higher frequencies. It is often regarded as part of the normal aging process. Several studies [70, 115, 164] have shown that high-frequency hearing loss increases with age (Fig. 14.2). The data in Fig. 14.2 are the averages of eight published studies comprising data from more than 7600 men (Fig. 14.2A) and almost 6000 women (Fig. 14.2B) [164]. Such studies rarely define which criteria were used for inclusion in the studies and it is therefore possible that the results may reflect hearing loss that is caused by factors other than age. In large population studies such as those compiled by Spoor [164], many individuals have been exposed to noise, which results in greater hearing loss at 4 kHz than at other frequencies. A cross-sectional and longitudinal population study of hearing loss and speech discrimination in an unselected population of individual ages 70 and 75 years (Fig. 14.3) [115] showed that both these groups of individuals had high speech discrimination scores (Fig. 14.4), somewhat lower in men than women. Exposure to noise affected hearing in men more than in women and hearing loss is slightly greater for high frequencies (Fig. 14.3). The reason for this gender difference may thus be that many men had noise-induced hearing loss but there may be other reasons related to hormonal influences on the progression of age-related changes in the cochlea and possibly differences in the age-related change in neural processing of sounds.

Møller [115] provided the distributions of hearing loss among the individuals of the study (Fig. 14.3) and these show that hearing loss between the men and

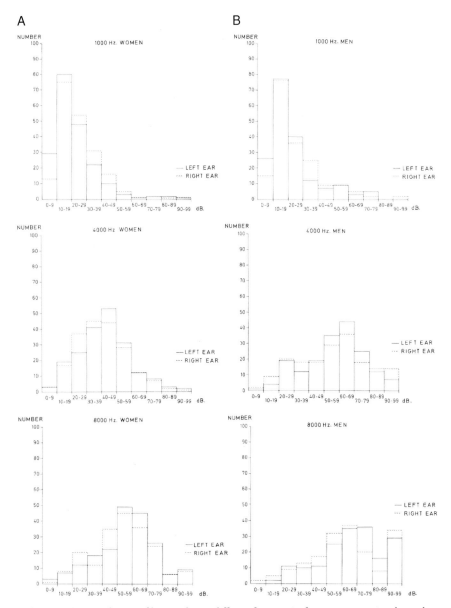

FIGURE 14.3 Distribution of hearing loss at different frequencies from a cross-sectional population study of hearing in people of age 70; (A) for women and (B) men. Solid lines represent left ear and dashed lines represents right ears (from Møller, M. B. (1981). Hearing in 70- and 75-year-old people: Results from a cross-sectional and longitudinal population study. *Am. J. Otolaryngol.* 2:22–29).

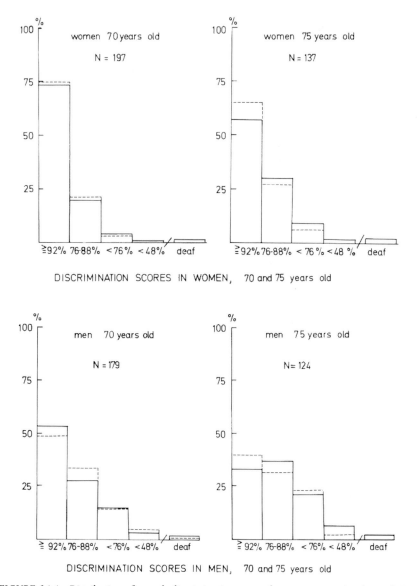

FIGURE 14.4 Distribution of speech discrimination scores from a cross-sectional population study of hearing in people of ages 70 and 75. The speech discrimination scores were obtained using phonetically balanced word lists presented at 30 dB SL or at the most comfortable level. Solid lines represent left ears and dashed lines represent right ears (from Møller, M. B. (1981). Hearing in 70- and 75-year-old people: Results from a cross-sectional and longitudinal population study. *Am. J. Otolaryngol.* **2**:22–29).

loss that is affected (slowed), but the degenerative process does not seem to be reversed by such sound exposure.

NOISE-INDUCED HEARING LOSS

Noise-induced hearing loss (NIHL) is normally associated with noise exposure in industry and thus is thought of as a product of modern civilization. It was earlier believed to be caused solely by injury to cochlear hair cells, but as our knowledge about disorders of the auditory system increases, the effect of noise exposure becomes increasingly complex. Mainly the loss of hearing sensitivity has been studied, but noise exposure has many other effects on hearing, such as affecting the perception of sounds. Tinnitus may accompany any of the different forms of cochlear hearing deficits but it is more common in noise-induced hearing loss and, in fact, most incidences of tinnitus are associated with NIHL (see Chapter 16).

AUDIOMETRIC SIGNS

Exposure to a moderately loud noise causes hearing loss that decreases gradually after the end of the noise exposure. The hearing threshold may return to its normal value after minutes, hours, or days, depending on the intensity and duration of the noise exposure and the individual person's susceptibility to noise exposure. Exposure to noise above a certain intensity and duration results in hearing loss that does not fully recover to its preexposure level. This remaining hearing loss is known as a permanent threshold shift (PTS). Hearing loss that resolves is known as a temporary threshold shift (TTS).

Hearing loss caused by noise exposure affects high frequencies more than low frequencies. While the hearing loss increases as a function of frequency, starting above a certain frequency, it also usually has a dip at or near 4 kHz (Fig. 14.5) and the hearing threshold at 8 kHz is better than it is at 4 kHz, at least for moderate degrees of noise-induced hearing loss. This distinguishes noise-induced hearing loss from age-related hearing loss (presbycusis), which results in threshold elevation that increases with frequency above a certain frequency (Fig. 14.2). The 4-kHz dip in NIHL is more or less pronounced, depending on the noise exposure, and it is most pronounced in individuals who have been exposed to impulsive noise, thus noise with broad spectra. Exposure to pure tones or to sounds with a narrow spectrum causes the greatest hearing loss at about one-half-octave above the frequency of the highest energy of the sound.

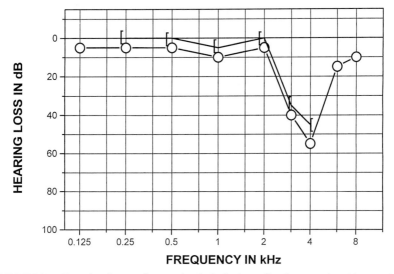

FIGURE 14.5 Typical audiogram for an individual who has suffered noise-induced hearing loss (modified from Liden, G. (Ed.) (1985). "Audiologi." Almquist and Wiksell, Stockholm).

The amount of acquired hearing loss depends not only on the intensity of the noise and the duration of exposure but also on the character of the noise (frequency spectrum and time pattern). The hearing loss from noise exposure is thus distinctly related to the physical characteristics of the noise exposure, but great individual variations exist. The combination of noise level and duration of exposure is known as the immision level and is used as a measure of the effectiveness of noise in causing hearing loss (PTS). However, the PTS caused by exposure to noise with the same immision level shows large individual variations (Fig. 14.6).

What Is the Cause of the Half-Octave Shift of Hearing Loss?

The reason for the half-octave shift in NIHL is most likely the shift of the maximal vibration of the basilar membrane toward the base of the cochlea with increasing sound intensity (see Chapter 3). Exposure to loud noise is expected to cause the most damage to hair cells on the basilar membrane at the location where the noise gives rise to the largest amplitude of vibration. This means that the most damage is done at the location that is tuned to the frequency of the maximal energy of the noise. This location is not the same as that where a weak sound used to measure the hearing loss gives rise to maximal amplitude. The frequency to which a certain location along the basilar membrane is tuned decreases with increasing stimulus intensity and the point on the basilar membrane that was tuned to 3 kHz at a high sound intensity (e.g., 90 dB) will be tuned to a higher frequency when tested near the threshold. This is why the largest threshold shift from exposure to a 3-kHz sound occurs at a higher frequency (approximately 4 kHz).

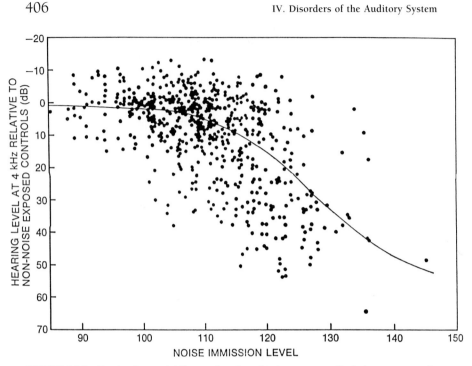

FIGURE 14.6 Hearing loss at 4 kHz as a function of noise exposure. Each dot represents the elevation in hearing threshold at 4 kHz for one ear. The solid line is the mean value. The horizontal axis represents both the sound level and the time of exposure [known as the noise immision level, which is equal to the noise level (in dB) + 10 times the logarithm of the duration of exposure] (from [22]).

Why Does the Greatest Noise-Induced Hearing Loss Usually Occur at 4 kHz?

The audiograms obtained in individuals who have been exposed to many different kinds of noise are similarly shaped, but the 4-kHz dip is probably most pronounced for exposure to impulsive noise. It is assumed that the resonance of the ear canal, amplifies sounds in the region of 3 kHz (cf. Chapter 2). That the greatest hearing loss from exposure to sound with their highest energy around 3 kHz occurs near 4 kHz can be explained by the half-octave shift discussed above.

Studies of the correlation between the resonance frequency of the ear canal and the frequency of the greatest hearing loss in people with noise-induced hearing loss [132] have shown that the mean resonance frequency of the ear canal in the group of people studied was 2814 Hz and the maximal hearing loss occurred at 4481 Hz. Assuming that the maximal energy of broadband noise occurred at the resonance frequency of the ear canal (2814 Hz) and that the greatest hearing loss occurs at a frequency that is 1.5 times the frequency of the maximal energy of the noise exposure, the maximal hearing loss would be expected to occur at 4221 Hz. The mean frequency of maximal hearing loss (4481 Hz) is thus very close to the expected value. This study also showed a high correlation between ear canal resonance frequency and the frequency of the maximal hearing loss in individuals. Earlier, Caiazzo and Tonndorf [23] found that extending the ear canal by a tube that caused the resonance frequency to decrease caused a similar decrease in the frequency of the maximal TTS in volunteers who were exposed to broadband noise. The greatest hearing loss (TTS) occurred at frequencies about one-half-octave higher than the frequency of maximal sound energy. These studies thus support the hypothesis that the typical 4-kHz dip in the audiograms of individuals who have suffered noise-induced hearing loss is a result of the resonance of the ear canal. (It has been pointed out by Rosowski [142] that the maximal transfer of sound power to the cochlea does not necessarily occur at the frequency of the ear canal resonance but depends on other factors that are frequency dependent, such as the transformer ratio of the middle ear.)

FACTORS THAT AFFECT NOISE-INDUCED HEARING LOSS

The considerable individual variation in susceptibility to noise-induced hearing loss has many sources. Genetic variations are one, but age and health status are also important factors that affect injury to hair cells from noise exposure. Drugs of various kinds most likely also increase susceptibility to noise-induced hearing loss. Animal experiments have shown that essential hypertension may increase the risk for NIHL [17–19]. Hearing loss of the conductive type also affects the risk of NIHL [126]. Absence or impairment of the acoustic middle ear reflex will result in increased hearing loss from noise exposure [195]. The protective effect of stapedius muscle contraction, as studied in rabbits, is considerable (cf. Chapter 12). Ingestion of alcohol and other drugs that impair the function of the acoustic middle ear reflex (cf. Chapter 12) may thus also affect susceptibility to noise-induced hearing loss.

> Experiments in rats [17–19] have shown that spontaneous hypertensive rats acquire more PTS from noise exposure than normotensive rats. However, hypertension caused by impaired blood supply to the kidney does not show such increased PTS [16]. Thus, hypertension in itself is probably not the cause of the higher susceptibility to noise-induced hearing loss. The increase in susceptibility to noise-induced hearing loss seen in the spontaneous hypertensive rats is probably related to factors that occur together with the predisposition for hypertension. The fact that the presence of any one of these factors in an individual is unknown adds to the uncertainty in prediction of hearing loss from noise exposure in an individual person.

Conductive hearing loss in a person who is exposed to noise will act as an ear protector and actually decrease the person's hearing loss from exposure to noise (Fig. 14.7) [126]. Hearing can be considerably better in the ear with conductive hearing loss than in the ear without conductive loss. The conductive hearing loss has not affected hearing to any great extent at high frequencies

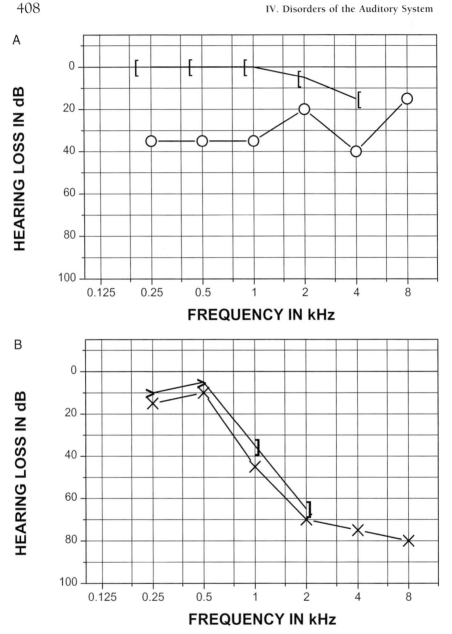

FIGURE 14.7   (A) Audiogram of a welder exposed to shipyard noise for 30 years and who had conductive hearing loss in one ear (top audiogram). (B) Audiogram from the ear without conductive hearing loss (from Nilsson, R., and Borg, E. (1983). Noise-induced hearing loss in shipyard workers with unilateral conduction hearing loss. *Scand. Audiol.* **12**:135).

but the protective effect from low-frequency conductive hearing loss against noise-induced hearing loss is substantial. Many individuals have small conductive hearing loss that may go unnoticed but affect a person's susceptibility to NIHS noticeably and thus contribute to individual variations in NIHS.

MORPHOLOGICAL AND FUNCTIONAL SIGNS OF INJURY FROM NOISE EXPOSURE

Noise-induced hearing loss has many similarities with presbycusis. It mainly affects outer hair cells and speech discrimination is little affected when hearing loss is moderate. Morphological studies of the cochlea reveal characteristic signs of injury to cochlear hair cells from noise exposure. It is mainly outer hair cells in the basal portion of the cochlea that are injured or totally destroyed, thus causing impairment of the cochlear amplifier. While NIHL has usually been assumed to be caused only by the loss or injuries of outer hair cells it has been shown that NIHL is also associated with specific morphologic changes in the central nervous system [81, 123]. In addition to that, neural plasticity may result in functional changes in the nervous system because of the deprivation of input to specific groups of neurons that is caused by the injury to the cochlea [57] (see Chapter 16).

MORPHOLOGIC CHANGES IN THE COCHLEA

Light-microscopic studies of cochlear hair cells in animals that have been exposed to a moderately loud noise show loss of some hair cells, mainly outer hair cells (Fig. 14.8). Exposure to more intense sounds for longer periods causes more extensive damage and inner hair cells may be affected. An increment of only 5 dB in the intensity of the sound to which the animals were exposed caused a considerable increase in the injury of hair cells and in the PTS (Fig. 14.8). Pure tones or noise that has a narrow spectrum causes lesions within a restricted region of the basilar membrane. Thus, loud sounds with large amounts of energy around 3 kHz will injure the hair cells at a location along the basilar membrane that has its highest vibration amplitude at 3 kHz at the exposure level. Cell counts using surface preparation of the cochlea (cytocochleograms) reveal damage mainly to outer hair cells in the first row in an animal where the loss of sensitivity was moderate (30–40 dB), while high-resolution light microscopy reveal abnormalities in stereocilia in both outer and inner hair cells (Fig. 14.8). An animal exposed to the same noise but studied at different times after noise exposure (right hand graphs in Fig. 14.9) showed much greater hearing loss and more extensive hair cell damage,

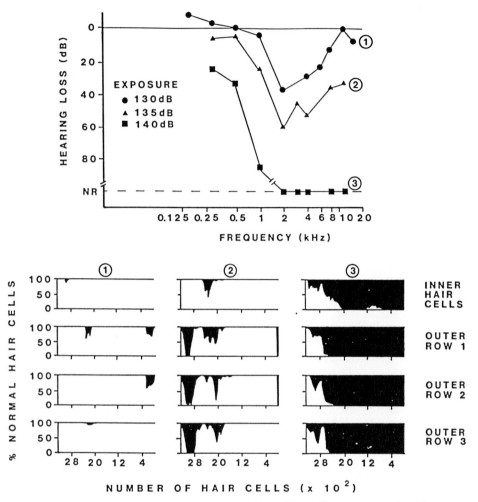

FIGURE 14.8 Relationship between hearing loss and loss of hair cells in cats exposed to 2-kHz tones for 1 h and at three different intensities (from Dolan, T. R., Ades, H. W., Bredberg, G., and Neff, W. D. (1975). Inner ear damage and hearing loss after exposure to tones of high intensity. *Acta Otolaryngol. (Stockh.)* 80:343–352).

including missing inner hair cells. There is a clear correlation between loss of hair cells and threshold shift at the CF but there is considerable individual variation in the extent of the damage, even in animals that are genetically similar and treated in similar ways.

Only exposure to extremely loud noise causes other structural damage. Thus, exposure to sounds with levels in excess of 125 dB SPL seems to be

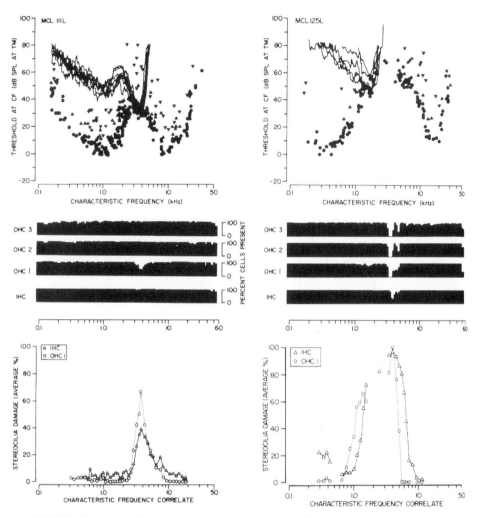

FIGURE 14.9 Results of recordings from single auditory nerve fibers and morphologic examination of the cochleae of two cats after exposure to 2 h of noise, 2 octaves wide, centered at 3 kHz, and with an intensity of 115 dB SPL. The cats were examined 620 (left panel) and 63 days (right panel) after noise exposure. (Upper graphs) Sample tuning curves, centered at approximately 3.6 kHz, of single auditory nerve fibers and threshold at CF. (Middle graphs) Cytocochleograms of the cochleae showing loss of hair cells. (Bottom graphs) Stereocilia damage in the first row of outer hair cells and inner hair cells as revealed by high resolution (Nomarsky) light microscopy with 100X objectives (from Liberman, M. C. (1987). Chronic ultrastructural changes in acoustic trauma: Serial-section reconstruction of stereocilia and cuticular plates. *Hear. Res.* **26**:65–88).

necessary to cause mechanical damage to the cochlea of the guinea pig [162]. The level of noise exposure that causes structural damage varies between species and it may thus be different in humans.

> Little damage to the stereocilia can be detected by light-microscopic examination after noise exposure that produces 40- to 60-dB hearing loss [95]. In many forms of moderate degrees of cochlear hearing loss, inner hair cells are intact when examined using the light microscope. High-resolution light microscopy (using Nomansky optics) and scanning electron microscopy (SEM) (Fig. 14.10), however, have revealed more subtle morphologic changes, but it is not known what the physiologic implications of such changes are. Both SEM and special high-resolution light microscopy have shown that noise exposure causes a disarray of stereocilia on both inner and outer hair cells (Fig. 14.9). High-resolution light microscopy has revealed that the stereocilia of inner hair cells are altered to almost the same extent as the stereocilia of outer hair cells after exposure to moderate levels of noise. The abnormalities in the stereocilia occur at locations along the basilar membrane, where the loss of sensitivity is largest.

It has been shown that noise exposure causes disconnection between stereocilia of outer hair cells and the tectorial membrane. It should be noted that this is different from other types of insults to the cochlea, such as from ototoxic antibiotics, which affect the integrity of the cell bodies of hair cells.

PHYSIOLOGICAL CHANGES

Hearing loss caused by injury to outer hair cells does not affect sensory transduction but rather the mechanical properties of the basilar membrane. Recall

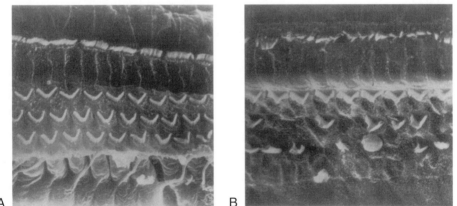

A   B

FIGURE 14.10   Scanning electronmicroscopic (SEM) picture of hair cells (A) in a monkey with an intact organ of Corti and (B) hair cells from a monkey that had been exposed to gunshot noise (courtesy of Dr. H. Engstrom).

from Chapter 3 that the outer hair cells function as "motors" that increase the sensitivity and the frequency selectivity of the ear and that it is the inner hair cells that transduce the motion of the basilar membrane and control the discharge pattern of auditory nerve fibers. Also, recall that the amplification caused by outer hair cells is most effective for sounds of low intensity and that it has little effect for sounds that are more than 50–60 dB above (normal) hearing threshold. This explains why hearing loss caused by functionally impaired outer hair cells rarely exceeds 50 dB. It is also the reason why tests that employ high-intensity sounds, such as BAEP and the acoustic middle ear reflex, are largely normal in patients with hearing loss caused by malfunction of outer hair cells. It is not known why hair cells located in the base of the cochlea are more susceptible to injuries from ototoxic agents, noise, and aging compared to hair cells in other portions of the cochlea.

The most prominent physiologic signs of noise-induced hearing loss, as revealed in animal studies, are deterioration of the tuning of single auditory nerve fibers, loss of sensitivity at the CF, and downward shift in frequency of the CF (Fig. 14.11). The widening of basilar membrane tuning after noise exposure is typical for loss of function of the active role of outer hair cells that normally act to reduce friction and thereby increase the sensitivity and frequency selectivity of the ear (cf. Chapter 3). The widening of the tuning of the basilar membrane broadens the "slices" of the spectrum of broadband sounds that the cochlea provides to the (temporal) analyzer in the central nervous system. This may cause interference between different spectral components (impair "synchrony capture") and may increase masking. The impairment of the cochlear amplifier from injury of the outer hair cells also impairs the amplitude compression that is prominent in the normal cochlea and may be the reason why loudness recruitment (hyperacusis) accompanies NIHL. The sensitivity of a single auditory nerve fiber may increase below a fiber's CF (in the tail region of the tuning curves) after noise exposure [96] and that may contribute to hyperactivity, such as tinnitus, that occurs in some individuals with cochlear hearing loss (Chapter 16).

CAUSE OF INJURY TO HAIR CELLS

While published reports of morphologic changes of the cochlea as a result of noise exposure are abundant, few studies that concern the cause of these changes have been published. It is thus poorly understood how noise exposure causes the observed damage to hair cells. It has been suggested that impairment of blood supply or simple metabolic exhaustion could be the cause of hair cell injury and destruction. These hypotheses have received little experimental support.

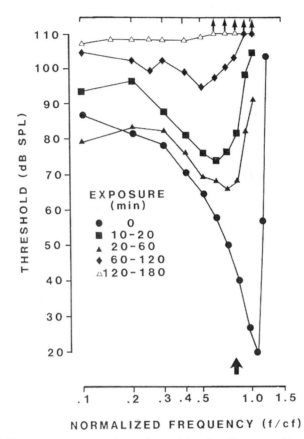

FIGURE 14.11 Deterioration in tuning and sensitivity of cochlear nerve fibers as a result of exposure to pure tones. The data were pooled from many nerve fibers and the frequency scale is normalized. The arrows show the frequency of the exposure tones and the different curves represent different exposure times (from Cody, A. R., and Johnstone, B. M. (1980). Single auditory neuron response during acute acoustic trauma. *Hear. Res.* 3:3–16).

Recently, oxygen free radicals have been implicated in causing injury to hair cells from noise exposure, aging, and ototoxic antibiotics [92, 194]. It has been shown that the level of glutathione, an enzyme that defends cells against the toxic effects of reactive oxygen species, decreases with age and the physiologic state and environmental challenges. It has been shown that oxygen free radical scavengers can reduce the effect of noise exposure on hearing. The best effect was obtained when the free radical scavenger was administered before noise exposure but some effect was also achieved when it was administered after noise exposure [194].

The finding that the cochlea can recover from severe noise-induced hearing loss shows that hair cells can cease to function, or show reduced function, without permanent injury occurring. This also explains the recovery of threshold shift after moderate-intensity noise exposure (TTS). Only when the insult has reached a certain level does the recovery become incomplete and the result is permanent injury.

A related question addresses individual variation in NIHL to the same exposure. The reason for this variation is also not well understood. Numerous hypotheses have been presented but published experimental evidence is rare. Besides variability in the exposure conditions, genetic differences, age, gender, pigmentation, differences in the sound-conducting apparatus (small conductive hearing loss), blood supply, and innervation of the cochlea have all been suggested as cause of the variability in NIHL to the same noise exposure. Spontaneous hypertensive rats acquired considerably more NIHL than normotensive rats when exposed to the same noise [17, 19] but it was not related to the elevation in blood pressure [16], thus probably related to genetic factors. The hypothesis that age is a factor has been supported by studies in mice [69].

REDUCING THE RISK OF NOISE-INDUCED HEARING LOSS

Establishing hearing conservation programs in industries where the noise level may cause hearing loss has been the most effective means of reducing the risk of NIHL. However, attempts to identify individuals who are more likely to acquire NIHL have not been successful. More recently it has been found that other means, such as preexposure to noise, can reduce the risk of NIHL, but that has not yet come into practical use. Attempts to find drugs that can reduce susceptibility to noise-induced hearing loss have so far also not yielded results that have gained practical use, although a few studies in animals have produced promising results [194].

HEARING CONSERVATION PROGRAMS

Hearing conservation programs reduce the risk of NIHL by reducing noise exposure and promoting personal protection devices in industry and other environments where individuals are exposed to noise at levels and duration that involve risks to hearing loss. Such programs depend on knowledge about the noise level, duration of exposure, and character of the noise, and how these factors affect the risk of noise-induced hearing loss. Reduction of noise exposure and promotion of personal protection have lowered the risk of NIHL considerably. Assessing other factors such as individual susceptibility to noise-induced hearing loss has been less successful.

Relation between Noise Exposure and PTS

While it is well established that the risk of hearing loss increases when the intensity of the noise is increased and when the duration of the exposure is prolonged, it is less certain whether the risk of PTS is directly proportional to the total energy of the noise exposure, i.e., the product of noise intensity and duration. If the risk of PTS is proportional to the total energy of the noise exposure, a doubling of the exposure time would be equivalent to increasing the noise level by 3 dB (i.e., a doubling of the intensity). Some investigators have interpreted the data on the effect of noise exposure on hearing threshold to mean that a doubling of exposure time increases the risk of hearing loss with the same amount as an increase of the noise level of 5 dB. This means that these investigators find evidence that the exposure time plays a lesser role than the intensity of the noise. One problem in relating noise levels and exposure time is related to the difference in the character of the noise that individuals are exposed to. Thus, the amount of PTS that a certain individual acquires not only depends on the intensity of the noise and the exposure time but also on the character of the noise (spectrum and time pattern). It may very well be that the relation between exposure time and noise level is different for different kinds of noise and it is most likely different for different intensities of the noise. These factors are only partly included in commonly used evaluations of risks of NIHL.

Measurements of Noise Level

Hearing conservation programs depend on accurate assessment of noise that a person is exposed to. Ideally, the units of measurement should be related to the risk of NIHL, but that is not possible because the risk of PTS is not a simple function of the energy of noise as it is measured physically. The spectrum of the noise and its temporal pattern are important factors and these factors are difficult to account for in practical measurements of noise. To somehow take the spectrum into consideration, measurements of noise are usually weighted with regard to spectrum before being measured.

The earliest noise-level meters had build-in filters that weighted the spectrum of the sound to provide measurements that were as close as possible to how the noise was perceived. The risk of noise-induced hearing loss is related to the spectrum of sounds in a rather different way than perception, and the properties of modern sound-level meters were therefore modified to better reflect the risk of hearing loss. The most commonly used weighting is the so-called A-weighting. The A-weighting places less weight on energy at low frequencies than at high frequencies. A-weighting follows approximately the 40-phon loudness curve. This weighting was originally used for measuring sounds about 40 dB above threshold to reflect the perception of such rather weak sounds. Experience later revealed that the same weighting is suitable for evaluation of the risk of NIHL, thus sound of much higher intensities. The unit of measurement of noise using that weighting is "dB (A)," to distinguish it from the measure that places equal weight on all spectral components known as "dB SPL."

Industrial noise that can cause hearing loss is not constant but typically varies over time. It can be slowly varying or impulsive in nature, such as gun shot noise.

Thus, noise level meters should integrate the measured noise over time in a way that is in accordance with the way the ear integrates noise because it is its effect on hair cells that is assumed to cause NIHL. The standard noise-level meter integrates sound energy over 100 or 125 ms, corresponding approximately to the temporal integration of perception of sounds that is related to the neural integration in the cerebral auditory cortex. This is a carryover from the time sound-level meters were used to estimate the perception (loudness) of sounds. However, sounds do their damage to hearing in the cochlea and it could therefore be expected that modern sound-level meters would have an integration time that corresponded to that of the cochlea, which is only a few milliseconds. This means that a noise-level meter with a 100-ms integration time will underestimate the risk from impulsive noise. Some noise-level meters have a setting with an integration time of 35 ms for measurements of impulsive noise. While that is shorter than 100 ms, it is still too long to adequately represent the integration of sound in the cochlea. The readings of currently used noise-level meters may therefore underestimate the risk posed by impulsive sounds.

Noise is commonly measured at the place where people are present and such measurements do not reflect the influence of the head on the sound that reaches the ear. The head increases the sound level at the entrance of the ear canal (cf. Chapter 2) and a person is therefore exposed to higher levels of noise than most measurements indicate. The fact that people often move between places where the noise levels are different introduces uncertainties in estimations of total noise exposure. This problem can be avoided by using noise dosimetry, which measures the noise level near the ear of workers with a device that is worn by individuals who are exposed to noise. Noise dosimeters integrate the noise level near the ear of a person over an entire day and provide information about the total noise exposure in a single measure, thus similarly to radiation dosimetry.

While noise dosimetry has advantages over the conventional way of measuring noise exposure, it relies on conversion between exposure time and sound level; this is not entirely correct and it naturally does not eliminate the uncertainty that is related to individual difference in susceptibility to NIHL.

Establishment of Noise Standards

Work on correlating exposure levels (sound intensity and duration of exposure) to the risk of NIHL has resulted in so-called "noise criteria" or "noise standards." Noise standards are rules about how noise affects hearing loss and such noise standards play important roles in legal matters [36]. Noise standards are supposed to provide estimates of the hearing loss a person will acquire from exposure to noises of different intensity and for different lengths of time. While the average hearing loss in a large population of individuals is clearly correlated with the exposure level and the exposure time, it is generally not possible to predict what degree of hearing loss any one individual will acquire when exposed to a certain level of noise for a certain period of time because of the considerable individual variation in the susceptibility to noise exposure (Fig. 14.6).

Therefore, when noise standards are used as a basis for setting limits of noise exposure that are deemed to be "safe," it in fact means that (only) a certain percentage of individuals will acquire hearing loss that is greater than a certain value. The large individual variation in susceptibility to NIHL prevents the development of noise criteria that provide accurate estimates of the hearing loss an individual person will acquire from exposure to certain noise for a certain time. One can only, at best, estimate the average hearing loss that a (large) group of people will acquire. Therefore, it is only possible to predict the probability of acquiring a certain amount of NIHL when exposed to noise of a certain level for a certain period of time.

Other sources of hearing loss contribute to errors in estimation of the effect of noise exposure on hearing. Noise standards are presumed to predict the hearing loss solely that an individual person will acquire from exposure to noise of a certain intensity for a certain period. It therefore seems natural that corrections should be made for hearing loss from other causes and subtracted from the total hearing loss acquired by a person who is exposed to noise. Hearing loss that can be attributed to presbycusis should thus be subtracted from the total hearing loss to arrive at the value of pure NIHL. However, such a "correction" of the hearing loss in an elderly person with noise-induced hearing loss may, under certain circumstances, result in a paradoxical result in that the calculated NIHL decreases with time of noise exposure. The reason for this paradoxical result is that these two contributions to hearing loss, presbycusis and noise exposure, do not add linearly. Thus, 2 + 2 is not 4 but, rather, less; perhaps 3. When one factor is subtracted from such a "sum" in order to obtain the other factor, an obvious error thus occurs.

MODIFICATION OF SUSCEPTIBILITY TO NOISE-INDUCED HEARING LOSS

It was shown by Miller *et al.* [105] that the temporary threshold shift caused by noise exposure decreased gradually during repeated exposures. This was taken to indicate that the ear's susceptibility to noise exposure is affected by previous exposure. This "toughening" of the ear with regard to TTS from noise exposure has been extensively studied in a variety of animals and in humans by several investigators [27, 160] and it has been confirmed that it is possible to reduce the effect of noise exposure on PTS by preexposure to noise. The exposure pattern of such "conditioning" is important for achieving this effect, but the underlying mechanisms are not completely understood. Several studies have suggested this toughening of the cochlea against noise-induced injury is related to induced changes in the hair cells by the "conditioning" noise exposure. It has also been suggested that the medial olivocochlear efferent system may be involved. However, none of these possibilities have been confirmed.

It has generally been assumed that exposure to loud sounds (noise) causes hearing loss only because it affects hair cells either by mechanical stress or by changing the chemical composition inside or outside the hair cells. The recent finding that prior noise exposure can affect the hearing loss from subsequent exposure to loud noise brought a new and unexpected angle to the relations between the physical noise exposure and acquired hearing loss and it became evident that the physiological mechanisms involved in noise-induced hearing loss are more complex than earlier believed. The finding that noise-induced hearing loss is affected by prior stimulation and by simultaneous stimulation of the opposite ear may explain some of the individual variation in susceptibility to noise-induced hearing loss.

HEARING LOSS CAUSED BY OTOTOXIC AGENTS

Many commonly used medications can cause hearing loss [54, 154, 190]. Antibiotics of the aminoglycoside type can cause permanent hearing loss. Streptomycin (dehydrostreptomycin) was the first of this family of antibiotics found to cause hearing loss, but now commonly used antibiotics of the same family, such as gentamycin, kanamycin, amikacin, and tobramycin, are also ototoxic but to a varying degree. Erythromycin and polypeptide antibiotics, such as vancomycin, can produce hearing loss, but it is mostly reversible once the drugs are terminated. Commonly used agents in cancer therapy (chemotherapy) such as cisplatin and carboplatin are also ototoxic.

> While most ototoxic substances mainly affect outer hair cells, carboplatin causes injury mainly to inner hair cells in one animal species, the chinchilla, leaving outer hair cells intact. In the guinea pig, carboplatin injures outer hair cells, mainly in the basal region of the cochlea, thus similarly to other ototoxic substances. This causes the nature of the resulting hearing loss caused by carboplatin to be different from that of other ototoxic substances because it affects neural transduction in the cochlea rather than the mechanical properties of the basilar membrane. Its effect in human is unknown. [71].

Aspirin (acetylsalicylic acid) can produce tinnitus and transient hearing loss but only at high dosages (2 to 5 g/day) and the hearing loss normally resolves when the drug is terminated or the dosage reduced [124]. Certain diuretic drugs, such as furosemide and ethacrynic acid, can produce transient hearing loss and tinnitus, but they rarely cause permanent hearing loss. The same is the case for quinine.

Causes of Hearing Loss from Ototoxic Antibiotics

> Ototoxic antibiotics cause hearing loss by changing important biochemical processes that lead to metabolic exhaustion of hair cells and that can eventually lead to cell death. It is generally assumed that oxygen free radicals are involved in causing injuries to the cochlea by ototoxic substances [54, 92, 154]. Attempts have been made to prevent the ototoxic effect of drugs by administration of substances developed to protect against the effect of radioactivity, but such drugs also reduce the ototoxic effect of these antibiotics [133] and, so far, practical application of this method has not been demonstrated. The effect of toxic substances such as salicylate is different from that of noise in that it affects the cell bodies of the outer hair cells, while noise also causes a decoupling between the outer hair cell stereocilia and the tectorial membrane.
>   The ototoxic effect of all these drugs depends on the dosage and a number other factors, most of which are unknown but which cause the considerable individual

differences in the risk of hearing loss from these drugs. There is evidence that older individuals have a higher susceptibility and certain diseases may render a person more susceptible. Whether aminoglycoside antibiotics such as gentamycin and kanamycin may cause hearing loss and vestibular disturbance depends on the way they are administered. Antibiotics that are ototoxic are often given in fixed dosages to treat life-threatening infections in individuals who are generally weakened and often have impaired kidney function. Impaired kidney function affects resorption and the rate of excretion of these drugs so that the plasma levels may become higher than they would in individuals with normal kidney function. Most studies of ototoxic substances have been done in healthy animals and the side effects in humans are usually assessed in healthy, often young individuals. These substances are used to treat diseases of different kinds and therefore they are almost always used in people with various kinds of illnesses that may increase their ototoxicity. The experimental results upon which recommendations on the safe limits of such drugs are based were obtained in healthy individuals and these recommendations may not be applicable to humans with diseases for which these substances are administrated. Many of these drugs are excreted through the kidneys and if kidney function is impaired, they are excreted more slowly than normal. Administration of these drugs is still mostly in fixed dosages. The blood levels of the drug will therefore increase and become higher than anticipated if the excretion is slower than normal. This means that the risk of hearing loss may increase in individuals with reduced kidney function. Since many of the ototoxic drugs are also nephrotoxic, a vicious circle may result from impaired kidney function that becomes aggravated by higher blood levels of an ototoxic drug. Monitoring of plasma levels of ototoxic antibiotics can reduce the risk of hearing loss considerably.

Effect of Ototoxic Drugs

Most ototoxic drugs induce hearing loss by injuring outer hair cells and thus impairing the feedback in the cochlea (the cochlear amplifier), which normally increases the sensitivity of the ear and its frequency selectivity at low sound intensities. Inner hair cells are usually unaffected. Hearing loss caused by ototoxic drugs therefore seldom exceeds 50–60 dB and it usually begins at high frequencies and extends gradually toward lower frequencies as it progresses. High-frequency audiometry (i.e., determination of the pure-tone threshold at frequencies above 8 kHz) may detect the beginning of hearing loss before it reaches frequencies that affect speech discrimination.

Studies in animals have shown that administration of ototoxic substances can affect cochlear frequency tuning. These changes of the tuning of auditory nerve fiber were later interpreted to be caused by impairment of the function of outer hair cells. Tuning of single auditory nerve fibers in animals treated with furosemide show similar changes to those caused by anoxia. Treatment with kanamycin also results in deterioration of the tuning of auditory nerve fibers similarly to that caused by metabolic insult to the cochlea (see Chapter 3, Fig. 3.6). Nerve fibers that did not respond at all to sound stimulation *did* respond to electrical stimulation of the cochlea, indicating that the nerve fibers were still excitable, thus showing the possibility of using electrical stimulation in cochlear prosthesis in individuals who are deaf due to loss of hair cells. Harrison and Evans [64] arrived at essentially similar results.

It is not possible to record from single auditory nerve fibers in humans but estimates of the cochlear tuning in humans can be obtained by recording the ECoG from the ear in connection with masking (two-tone masking [34]).[1] That method was used to study the effect of injuries to cochlear hair cells in humans and in animals [65]. The results confirmed that cochlear tuning becomes broader when hair cells are injured due to administration of kanamycin. Comparison between the results obtained using electrophysiologic methods and psychoacoustic methods shows good agreement, and the obtained tuning curves are similar to those obtained in recordings from single auditory nerve fibers. Interestingly, simultaneous masking and forward masking gave different results in individuals with hearing loss while in individuals with normal hearing the results of the two tests were similar.

Antibiotics usually enter the cochlear fluid space from systemic administration but these substances may also enter the cochlear fluid space when administered directly into in the middle ear, such as to treat infections. It is interesting that an ototoxic antibiotic, neomycin, which is not approved for systemic administration because of its ototoxicity, is approved for local administration, including in the ear. Evidence has been presented that inflamed mucosa of the middle ear act as a barrier for neomycin and prevent it from entering the cochlea [161]. It is also possible that toxic substances generated by bacterial activity in the middle ear fluid can enter the cochlear fluid space through the membranes of the round and oval windows and cause injuries to hair cells [63, 99, 161].

Some ototoxic drugs also affect the nervous system [42] (see Chapters 15 and 16). There may thus be neural components of hearing loss caused by administration of ototoxic substances as there are in connection with other causes of cochlear injuries. There is also a possibility that the drugs themselves act on the auditory nervous system (see Chapter 15). Most ototoxic drugs cause tinnitus and, in some patients, hyperacusis in addition to causing hearing

[1] The use of masking to determine the tuning of the cochlea is based on the assumption that a weak tone activates only a few auditory nerve fibers. To obtain a tuning curve, the electrophysiologic response (AP, CAP from the auditory nerve, or the BAEP) to a weak tone (a few decibels above threshold) is recorded while a masking tone is applied. The intensity of the masking tone is adjusted so that the test tone evokes a reduced response (e.g., 2/3 of the response without a test tone). The test tone and the masker are presented as short tone bursts, and the masker is usually applied immediately before the test tone (forward masking), but it can also be applied at the same time as the test tone (simultaneous masking). This procedure can be used in animals [34] as well as in humans [65]. A similar procedure can also be used to obtain psychoacoustic tuning curves in humans [196].

loss (see Chapter 16). Administration of salicylates has been used extensively in animal studies of tinnitus (see Chapter 16).

One aspect of the reversible effect of salicylates is that it abolishes otoacoustic emissions. Rather high dosages (5–10 g/day) are required to achieve that effect.

DISEASES THAT AFFECT THE FUNCTION OF THE COCHLEA

Several diseases may affect the function of the cochlea. The most common disease is Ménière's disease. Hearing loss may also result from hereditary causes. Infectious diseases such as meningitis and certain virus diseases can also cause hearing loss by destroying cochlear hair cells.

MÉNIÈRE'S DISEASE

Ménière's disease is a progressive disorder that is defined by a triad of symptoms, namely vertigo with nausea, fluctuating hearing loss, and tinnitus. In the beginning of the disease the patient experiences acute attacks of vertigo, hearing loss, and tinnitus that may last from several hours to days. Typically, hearing loss in the early stages of Ménière's disease affects only low frequencies, fluctuates, and increases during an acute attack (Fig. 14.12). It is thus one of a few kinds of

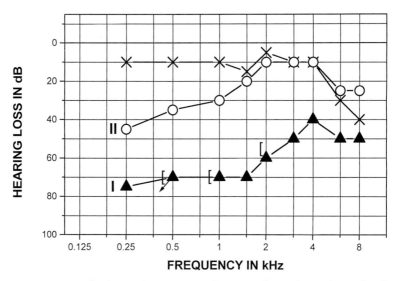

FIGURE 14.12   Typical audiogram from a person with Ménière's disease showing hearing loss during an attack (I) and between attacks (II). Crosses show hearing loss on the unaffected left ear (from Møller, M. B. (1994). Audiological evaluation. *J. Clin. Neurophysiol.* 11:309–318).

sensorineural hearing loss that affects hearing at low frequencies. Hearing returns to normal after each attack in the beginning of the disease but as the disease progresses, residual hearing loss from each attack accumulates and the hearing loss spreads to higher frequencies. After years of disease, some patients may experience "drop attacks," i.e., sudden severe vertigo that occurs without warning and which causes the patient to fall to the ground. The end stage of the disease, reached 10–15 years after its debut, is a flat hearing loss of about 50 dB.

The incidence of Ménière's disease is different in different geographic locations. A study made in Rochester, Minnesota showed an incidence of 15.3 per 100,000 people, with a small preponderance in women (16.3 vs 13.3 for men) [193]; a study in Italy [30] showed an incidence of 8.2; and a study in Sweden [165] arrived at an incidence of 46. Some of this variation is probably due to differences in the definition of the disease.

The hearing loss in Ménière's disease can be explained by a distension of the basilar membrane, which is largest where it is least stiff, i.e., in the apical portion (Chapter 3). Permanent damage to hair cells does not seem to occur, at least not in the early stage of the disease. The fluctuations in hearing are assumed to be caused by varying degrees of pressure imbalance in the fluid compartments of the cochlea (endolymphatic hydrops). This is supported by studies by Kimura [82], who showed that blocking the endolymphatic duct in guinea pigs mimicked the signs of Ménière's disease. It is possible that Reisners's membrane may rupture and cause the most violent symptoms because fluids of different ionic compositions mix. However, it is not known how the imbalance in pressure comes about and why it only occurs at certain times.

Over time, hearing loss progresses and extends to higher frequencies, but it rarely exceeds 50 dB. Speech discrimination is little affected in the early stages of the disease but may become affected in the advanced, late stage of the disease. Tinnitus, which is one of the triad of symptoms that defines Ménière's disease is discussed in detail in Chapter 16.

> Ménière's disease is probably not one disorder but rather a group of different disorders. Recent studies have indicated that some individuals with Ménière-like syndromes can be successfully treated with microvascular decompression (MVD) of the intracranial potion of the VIIIth cranial nerve. This indicates that irritation of the auditory vestibular nerve can cause similar symptoms as those of Ménière's disease. Other variations of Ménière's disease have predominantly cochlear signs and some investigators have called these diseases cochlear Ménière's disease. Thus, while the classic definition of Ménière's disease is a triad of symptoms (fluctuating hearing loss, tinnitus, and vertigo), over time physicians have accepted variations to that classic pattern and have labeled these disorders Ménière's disease as well.

Diagnosis of Ménière's Disease

Ménière's disease can be diagnosed by the patient's history and standard audiologic tests. Recently it has been hypothesized that the cochlear summating potential (SP) is abnormal in patients with Ménière's disease and recording the SP is commonly used to diagnose Ménière's disease and to monitor treatment. The SP is the sum of the cochlear distortion products (Chapter 4) and its amplitude depends on several factors, one of which is endolymphatic pressure; this is why it has been suggested as a way to detect endolymphatic hydrops. The SP has been reported to be high in patients with Ménière's disease, but other reports express doubt about the significance of such findings and refer to the large individual variation in the SP. Thus the value of recordings of SP as a diagnostic tool to identify Ménière's disease has been set in question by some studies [26, 40], whereas others [148] find that the SP anomaly and the ratio between the AP and the SP are important signs of the disease.

In response to clicks, the SP appears as a shoulder on the rising phase of the AP and it may be difficult to distinguish from the CM components of the ECoG. The ratio between the amplitude of the SP and the AP components is used as indication of cochlear hydrops (Fig. 14.13). Some investigators [148] have made the SP appear more clearly by using high-repetition-rate clicks as stimuli (Fig. 14.14), which reduces the amplitude of the AP component of the ECoG without affecting the SP. Subtracting the response to high stimulus rate from the response to low stimulus rate eliminates the SP component and thus shows a clean AP waveform. However, tone bursts would be more adequate stimuli than clicks for recording the SP. In the ECoG elicited by tone bursts, the SP appears as a plateau that occurs during and after the AP and it is thus easy to measure the amplitude of both AP and SP (Fig. 14.15). Note that the polarity of the SP is different in response to tones of different frequency. This is just one example of how the SP may vary. The SP also varies considerably between individuals without signs of Ménière's disease and it may also vary from time to time in the same individual. The large variability of the SP in individuals without cochlear hydrops hampers the use of SP in diagnosis of disorders with hydrops including Ménière's disease.

The ratio between the SP and the AP has been used as an indicator of endolymphatic hydrops and in 1998, Sass et al. [148] found a mean SP/AP ratio of 0.26 in individuals with normal hearing, with a standard deviation of 0.11. In patients with Ménière's disease, the mean SP/AP was 0.46, with a standard deviation of 0.15. The SP was significantly larger in Ménière's patients at 1 and 2 kHz but not at 4 and 8 kHz. Campbell et al. [26] used 6-kHz tones as stimuli and this may be the reason these investigators found little difference between the SP/AP ratio in patients with Ménière's disease compared with patients who did not have Ménière's disease. The ratio of SP/AP was not significantly correlated to the patient's hearing loss [148]. The sensitivity of transtympanic ECoG using the SP/AP ratio of the response to 1-kHz tone bursts [148] was 82% with specificity of 95%. Thus the choice of stimuli affects the sensitivity of the SP/AP ratio as an indicator of endolymphatic hydrops and may be one of the reasons why different investigators have arrived at different values of sensitivity of this test.

Recordings the SP during operations may be of value because it involves comparison of the SP over a short time period in the same individual and thus is not subject to the effect of interindividual variations. The SP is affected by vestibular nerve section [50], probably because also the olivocochlear bundle

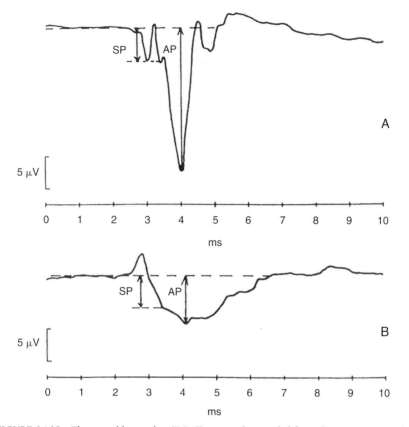

FIGURE 14.13 Electrocochleographic (ECoG) potentials recorded from the promontorium by an electrode that pierced the tympanic membrane in a person with normal hearing (A) and in a person with Ménière's disease (B). Abbreviations: SP, summating potential; AP, (nerve) action potential (from Sass, K., Densert, B., and Arlinger, S. (1998). Recording techniques for transtympanic electrocochleography in clinical practice. *Acta Otolaryngol.* (*Stockh.*) 118:17–25).

was severed in these operations. Neural activity in the olivocochlear bundle influences the function of outer hair cells, which contribute to the SP and this may explain the effect of a severed olivocochlear bundle on the SP.

Treatment of Ménière's Disease

One treatment aims at releasing the endolymphatic pressure and surgically establishing an artificial drainage of the endolymphatic sac (endolymphatic shunt [136, 155]). Vestibular nerve section was an early treatment to relieve the vertigo in patients with Ménière's disease [51] and it is still frequently

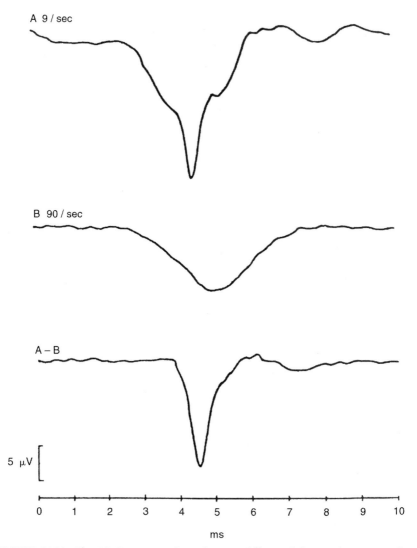

FIGURE 14.14 The ECoG response obtained at two different click rates (A: 9 pps and B: 90 pps) in a patient with Ménière's disease, and the difference between the two responses (A−B) (from Sass, K., Densert, B., and Arlinger, S. (1998). Recording techniques for transtympanic electrocochleography in clinical practice. *Acta Otolaryngol. (Stockh.)* **118**:17–25).

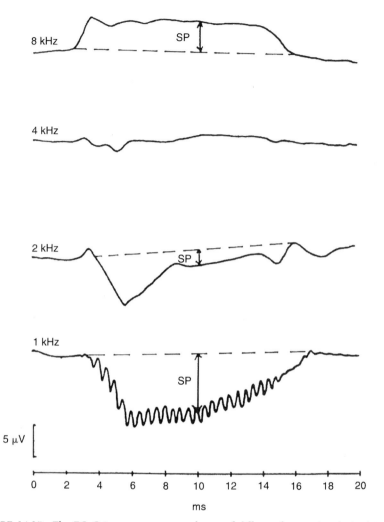

FIGURE 14.15 The ECoG in response to tone bursts of different frequencies obtained in a patient with Ménière's disease (from Sass, K., Densert, B., and Arlinger, S. (1998). Recording techniques for transtympanic electrocochleography in clinical practice. *Acta Otolaryngol.* (*Stockh.*) 118:17–25).

done [158]. Modern treatments of Ménière's disease now involve the use of infusion of streptomycin or gentamycin, ototoxic antibiotics, into the middle ear cavity [10, 91] to destroy parts of the sensory epithelium. More recently, a method was devised to infuse gentamycin into the cochlea through a catheter that is passed through the tympanic membrane and the end of which is placed over the round window.

Pressure Regulation in the Cochlear Fluid Compartments

The pressure in the different fluid compartments of the cochlea is normally kept within narrow limits by mechanisms that are poorly understood. It is known that imbalance of the pressure in the endolymphatic and perilymphatic spaces causes malfunction of the cochlea and results in symptoms from both the auditory and the vestibular systems. Ménière's disease is the best known example of a disease where imbalance of labyrinthine fluid pressure is the probable cause of symptoms. Thus, proper balance in the pressure, or, rather, the volume, of the fluid in these compartments is essential to achieve optimal functioning of the cochlea.

Little is known about the effects of elevated perilymphatic pressure on the function of the ear and it is a matter of diverse opinion whether moderately abnormal pressure in the perilymphatic space causes any signs of pathology. It is, however, generally recognized that an increase in the volume of the endolymphatic space is associated with disturbances of hearing and balance as seen in Ménière's disease, but it is not known whether the abnormal volume of inner ear fluid compartments is a cause of the disorder or a result of the pathology of Ménière's disease. Elevated endolymphatic volume causes Reissner's membrane to bulge and the basilar membrane to bow. That is assumed to give rise to the low-frequency hearing loss and perhaps tinnitus, which are two of the triad of symptoms that define Ménière's disease, at least in its early stage.

It is not known what mechanisms keep the endolymphatic pressure within its normal range but it seems reasonable to assume that pressure-sensitive areas of membranes that limit the endolymphatic space act as the sensors. Since the pressure in the perilymphatic space is closely coupled to that of the intracranial pressure (ICP), it seems unlikely that the pressure in the perilymphatic space can be regulated locally in the cochlea, at least in individuals in whom the cochlear aqueduct is patent. The role of the endolymphatic sac in pressure regulation in the inner ear is incompletely known but it is the target for some of the different treatments used in disorders that are believed to be caused by inner ear hydrops (excessive accumulation of fluid), for which Ménière's disease is one.

Measurement of the Pressure in the Cochlea

While measurements of the pressure in the cochlea have been done in animals for research purposes for many years, it is only recently that it has become possible to measure intralabyrinthine pressure noninvasively. A method for measuring intralabyrinthine pressure that makes use of the effect of contractions of the middle ear muscles on the displacement of the tympanic membrane was described recently [47]. Increased pressure in the perilymphatic space pushes the stapes footplate out of the oval window. This method is based on the assumption that the displacement of the incus by contraction of the stapedius muscle depends on how much the stapes is pushed out of the oval window. Displacement of the incus causes the tympanic membrane to displace and this can be measured as a small change in the air pressure in the sealed ear canal (Fig. 14.16). Normally, contraction of

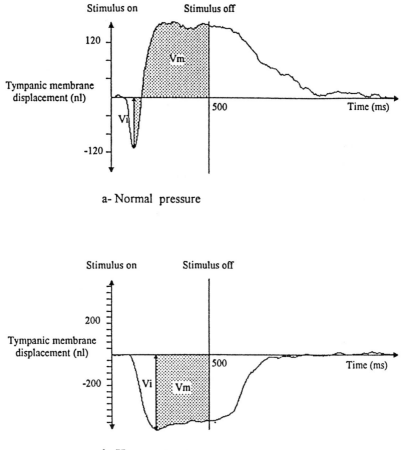

FIGURE 14.16 Illustration of how intracochlear (and intracranial) pressure can be measured by recording changes in the air pressure in the sealed ear canal (as a measure of the displacement of the tympanic membrane) during contraction of the stapedius muscle (from Wable, J., Collet, L., and Chery Croze, S. (1996). Age-related changes in perilymphatic pressure: preliminary results. *In* A. Ernst, R. Marchbanks, and M. Samii (Eds.), "Intracranial and Intralabyrinthine Fluids" (pp. 191–198). Springer-Verlag, Berlin).

the stapedius muscle causes only a very small shift of the position of the incus, as shown previously for animals such as the cat or the rabbit (Chapter 2, Fig. 2.26). This is because contraction of the stapedius muscle normally causes the stapes to move perpendicular to the surface of the flat portion of the incudo-stapedial joint. This prevents any noticeable movement of the incus and thus displacement of the tympanic membrane. If the pressure in the perilymphatic space is abnormally elevated, the stapes tilts because the elasticity in the two ligaments of the stapes footplate is different (Fig. 14.16). Contraction of the stapedius muscle then does not displace the stapes exactly perpendicular to the surface of the incudo-stapedial joint, causing the incus to displace, and the tympanic membrane moves. This means that increased intralabyrinthine pressure causes contraction of the stapedius muscle to displace the tympanic membrane and this can be measured as a small change in the air pressure in the sealed ear canal. This method is used in clinical tests that make it possible to measure intralabyrinthine pressure noninvasively. The outcome of the test thus depends on fine details of the anatomy of the stapes and its suspension in the oval window, the incudo-stapedial joint, and the orientation of its plane surface. This causes considerable individual variation in the displacement of the tympanic membrane from contraction of the stapedius muscle. This method is therefore best suited for measuring changes in the same individual that occur over time.

Measurement of the displacement of the tympanic membrane has also been proposed as a (noninvasive) method for measuring intracranial pressure or, rather, as an indicator of elevated ICP [47]. The validity of this method of measuring intracranial pressure assumes that the perilymphatic space communicates with the intracranial space, which depends on the patentcy of the cochlear aqueduct (see Chapter 1).

Measurements of the change in the air pressure in the sealed ear canal from contraction of the stapedius muscle are technically difficult and the air pressure in the ear canal may change due to other reasons, such as pulsation of the blood. These unrelated changes act as background noise that interferes with measurements of the displacement of the tympanic membrane from contraction of the stapedius muscle. These difficulties may be overcome by using laser interferometry to measure the displacement of the tympanic membrane (mentioned in Chapter 2).

Since that method of detecting elevated intracochlear pressure or ICP relies on contraction of the stapedius muscle, the method is limited to individuals who have an acoustic middle ear reflex. Conductive hearing loss; lesions to the auditory nerve; and the presence of drugs such as barbiturates, alcohol, and anesthetics are all factors that can affect or abolish the acoustic middle ear reflex.

INFECTIOUS DISEASES

Bacterial meningitis was one of the most common causes of childhood hearing loss before immunization came into common use. Several bacteria can cause meningitis and hearing loss is a result of inflammation of the labyrinth that destroys hair cells and replaces the membranous labyrinth with fibrous tissue. Hearing loss is usually bilateral and permanent. Sometimes the cochlea fills with bone after meningitis, which makes it difficult to use cochlear implants to improve hearing.

HEREDITARY HEARING LOSS

Hereditary hearing loss most often affects cochlear hair cells and results in sensorineural hearing loss. Hereditary hearing loss is usually bilateral and high frequencies are affected more often than low frequencies, but the audiograms may have widely different shapes. One common shape shows the largest hearing loss in the mid-frequency range ("cookie bite" audiograms) (Fig. 14.17). Hereditary hearing loss may progress after birth and it may reach various degrees of severity. Hearing loss may accompany other genetically related disorders.

OTHER CAUSES OF IMPAIRMENT OF COCHLEAR FUNCTION

PERILYMPHATIC FISTULAS

Perilymphatic fistulas are small perforations that develop around the cochlear windows. They are most likely a result of slight weakening of the membranes

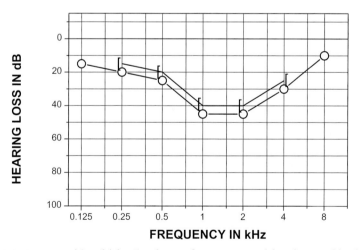

FIGURE 14.17   Typical ("cookie bite") audiogram from a person with hereditary cochlear hearing loss (modified from Liden, G. (Ed.) (1985). "Audiologi." Almquist and Wiksell, Stockholm).

that seal the cochlea fluid (perilymph). Such fistulas cause the perilymph of the cochlea to leak and the result is hearing loss and vestibular symptoms (vertigo). Perilymphatic fistula can appear spontaneously but they more often occur as a result of increased vein pressure from accidents, scuba diving, rapid descent by airplanes, or extreme strain. Hearing loss is purely cochlear, with normal or near-normal acoustic middle ear reflexes and normal BAEP. Many children with hearing loss have hearing at birth but lose it around the time they become mobile. It is possible that such hearing loss, often called hereditary hearing loss, may be caused by perilymphatic fistulas that appear when the children begin to stand upright and thus experience fluctuations in the pressure of the inner ear fluid. Perhaps the weakness cannot sustain normal fluctuations in the pressure of the inner ear fluid, which causes the hearing loss or deafness.

Diagnosis of perilymphatic fistulas is a challenge and this condition may therefore persist undiagnosed or misdiagnosed. Several tests have been designed to detect such leakage of cochlear fluid. Perilymphatic fistulas also affect the vestibular system and may present with similar symptoms and signs as Ménière's disease. Observation of vestibular responses when a sudden change in air pressure in the ear canal is applied is commonly used to detect perilymphatic fistulas. The patient's eye movements are studied using either direct observation of eye movements or by using electrical recordings of eye movements (electronystagmography). The ECoG recordings in connection with changes in posture have in animal experiments shown some promising results as indicators of perilymphatic fistulas. The change in the ratio of the amplitudes of the summating potential and action potential elicited by click sounds or tone bursts (SP/AP) have been used as indicator of the presence of a fistula [25]. These methods are not precise indicators of fistulas and the patient's medical history must be taken into consideration.

It is possible to repair such leaks surgically, although it is similar to repairing a leaking boat from the inside and therefore not always successful, at least not the first time.

SUDDEN HEARING LOSS

Sudden hearing loss is not a disease, but because its symptoms are so characteristic it has often been treated as a particular entity. Sudden hearing loss is a sudden and often total loss of hearing, usually only in one ear. Perilymphatic fistulas may be one cause of sudden hearing loss. Its cause can rarely be identified, but viral infections or total interruption of the blood supply are commonly suggested. None of these have been proven and even suggestive evidence is rarely obtained. Hearing may return partly, but rarely totally. Sudden hearing loss is also discussed in Chapter 15.

Injuries to the Cochlea from Trauma

Injuries to the cochlea may be caused by trauma and skull fractures sometimes cause fractures of the cochlear bone, which usually causes total deafness. The hearing loss may be caused by massive leaks of inner ear fluid.

Auditory Nerve and Central Auditory Nervous System

ABSTRACT

1. Disorders of the central auditory nervous system are of two kinds, one that is associated with morphological changes and one that is not.
2. The most common disorder of the auditory nervous system that is associated with morphological changes is impaired neural conduction in the auditory nerve.
3. Auditory nerve lesions are, for the most part, caused by acoustic tumors but can also be caused by viral infections, vascular compression, and by surgically induced injuries.
4. Disorders of the auditory nerve cause hearing loss with greater impairment of speech discrimination than cochlear hearing loss of the same magnitude and the impairment cannot be predicted from the threshold elevation for pure tones. The audiogram often has irregular peaks and dips, the BAEP is abnormal, and the threshold of the acoustic middle ear reflex is elevated or absent.
5. Patients with disorders of the auditory nerve often have tinnitus (hearing meaningless sounds).

6. Cochlear nucleus implants offer help to people whose cochlear nerve is injured.
7. Disorders caused by lesions of brainstem structures are rare and auditory signs are complex.
8. Lesions of the auditory cerebral cortex often cause minimal threshold elevation and speech discrimination is often normal when tested using standard audiologic tests, but such lesions can be diagnosed by using low-redundancy speech and by imaging techniques.
9. Lesions of the auditory cerebral cortex often cause auditory hallucinations where a person hears meaningful sounds such as music or speech.
10. Disorders that are not associated with detectable morphologic changes are assumed to be a result of changes in synaptic efficacy, and such changes may be brought about by neural plasticity.
11. Changes in the function of the central auditory nervous system that result from neural plasticity are associated with a complex pattern of hearing impairment, often including hyperacusis and tinnitus and often affecting speech discrimination to a greater extent than expected from the threshold elevation.
12. Plastic changes in the nervous system are likely to be reversible or preventable by appropriate sound stimulation.

INTRODUCTION

Lesions to the auditory nervous system affect hearing in a more complex way than lesions to the ear. Diagnosis of such lesions therefore requires more sophisticated audiologic tests than diagnosis of lesions of the sound-conducting apparatus and the cochlea. The patients' own description of his/her hearing loss is important for proper diagnosis of disorders of the auditory nervous system. Detailed knowledge about the anatomy and the function of the auditory nervous system is necessary in order to make an accurate diagnosis of central auditory disorders.

Lesions to the auditory nerve are the most common cause of disorders of the auditory nervous system. Lesions to the auditory nerve may also affect the vestibular portion of the VIIIth cranial nerve and hearing deficits may thus be accompanied by symptoms from the vestibular (balance) system. Disorders that are associated with morphologically detectable pathologies of nuclei or fiber tracts of the ascending auditory pathway of the brainstem and the auditory cortex are extremely rare. They are usually associated with multiple symptoms and signs from other brain systems. Lesions of the auditory cerebral cortex often cause auditory hallucinations consisting of hearing meaningful sounds such as speech and music.

Usually, only lesions to the nervous system that are associated with detectable morphologic changes have been considered. There are, however, other kinds of changes in the function of the auditory nervous system which cannot be detected using the imaging techniques presently available. These changes in function are often a result of neural plasticity and they result in various kinds of pathologic signs such as tinnitus and hyperacusis. It is also possible that impaired speech discrimination can result from such plastic changes.

These kinds of pathologies have previously been largely disregarded, mainly because they have no morphologic correlates. Without such direct, visual evidence from imaging studies that the pathology existed, the disorder was often attributed to psychological causes and considered untreatable. There is increasing evidence, however, that such disorders have a true physiologic basis and can be treated successfully by appropriate sound stimulation [78] (see Chapter 16).

AUDITORY NERVE

The most common disease process that affects the auditory nerve is acoustic tumors. Other space-occupying lesions in the cerebellopontine angle are rare but any such lesion may cause symptoms and signs from the auditory nerve. Irritation or compression of the VIIIth cranial nerve from blood vessels may also cause symptoms such as tinnitus and hearing loss. Surgical injury to the auditory nerve from operations in the cerebellopontine angle may cause hearing loss or deafness together with tinnitus. Viral infections that affect the auditory nerve may cause hearing impairment.

It is assumed that normal speech discrimination depends on a high degree of temporal coherence. Normally, the conduction velocity of different auditory nerve fibers varies very little (see Chapter 5), ensuring a high degree of temporal coherence of nerve impulses that reach the cochlear nucleus. Mild injury to the auditory nerve makes nerve fibers conduct slower than normal and more severe injury can interrupt neural conduction in auditory nerve fibers. The reduced speech discrimination that is typical for injuries to the auditory nerve is assumed to be caused by impaired temporal coherence of nerve impulses that reach the cochlear nucleus. This may occur because the reduction in conduction velocity of auditory nerve fibers that occurs after injury is different for different nerve fibers.

AUDIOMETRIC SIGNS OF INJURY TO THE AUDITORY NERVE

The signs of hearing impairment from injury to the auditory nerve are more complex than those associated with cochlear injuries. Speech discrimination is reduced more than it would have been from similar hearing loss caused by cochlear injury or conduction impairment. The audiogram often has irregular shapes, with dips occurring at different frequencies (Fig. 15.1).

Anatomically, fibers from the apical portion of the auditory nerve are in the core of the nerve and fibers that are tuned to high frequencies are located

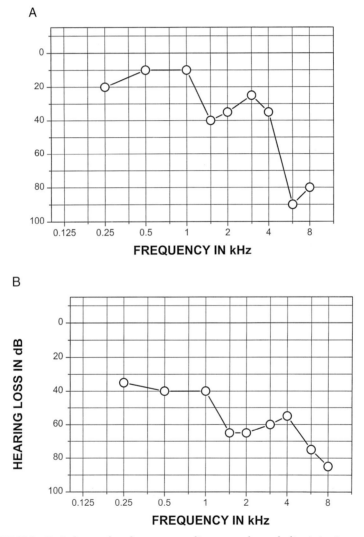

FIGURE 15.1   Typical examples of pure tone audiograms and speech discrimination scores of patients with acoustic tumors (courtesy M. B. Møller).

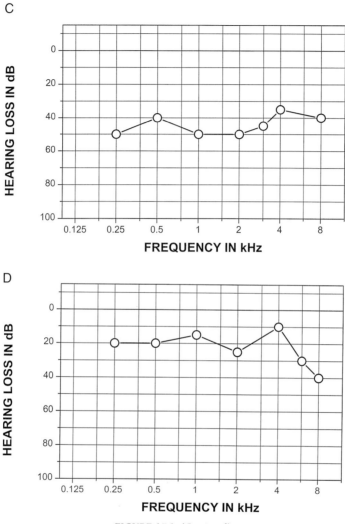

FIGURE 15.1 (*Continued*)

superficially, at least in animals (the cat) [146]. The anatomical arrangement of the auditory nerve is probably similar in humans, which may explain why injuries from compression of the auditory nerve, such as occur in acoustic tumors, mostly affect hearing at high frequencies. However, the fact that the auditory nerve is much longer in humans than in cats (2.5 cm versus about

0.8 cm [89, 90]) indicates that the anatomical arrangement of nerve fibers of the auditory nerve may be different in its intracranial course versus its peripheral course. This is probably the reason why lesions to the auditory nerve close to the brain stem, from close contact with blood vessels or from surgical trauma, cause hearing loss in the low- to mid-frequency range in humans (Fig. 15.2).

The threshold of the acoustic middle ear reflex is elevated and the growth of the reflex response is impaired in individuals with hearing loss from auditory nerve injuries. The acoustic reflex may even be absent in patients with signs of injury to the auditory nerve. This is puzzling, since mild injury to the auditory nerve is supposed to mainly affect the timing of the discharges and it might indicate that the acoustic middle ear reflex depends on coherence of the nerve activity that reaches the cochlear nucleus. The growth of the amplitude of the reflex response is reduced in patients with lesions of the auditory nerve and the maximally obtainable amplitude of the response is much less than it is normally (Fig. 15.3).

Temporal integration is affected by injuries to the cochlear nerve.[1] One way to obtain an estimate of temporal integration is to determine the threshold to tones of different duration, e.g., 20 ms vs 200 ms. The difference in threshold of tones of 20 and 200 ms duration is normally approximately 8 dB for tones of 1 and 4 kHz and it is less in individuals with acoustic tumors where the auditory nerve is injured (approximately 5–6 dB). Cochlear injuries also result in a slightly reduced temporal integration with a difference of 6–7 dB between the threshold of tones of 20 ms duration and 200 ms duration [139].

ELECTROPHYSIOLOGIC SIGNS OF INJURY TO THE AUDITORY NERVE

Signs of decreased conduction velocity in the auditory nerve can be demonstrated using various kinds of electrophysiological recordings. Of these methods, recording brainstem auditory evoked potentials (BAEP) is the most commonly used method for diagnosing disorders caused by injury to the auditory nerve. Recordings made directly from the exposed VIIIth cranial nerve are used in operations where the auditory nerve may be injured in surgical manipuations [109]. Such recordings show a prolonged latency of the compound action potential (CAP) because of the decrease in conduction velocity. The CAP also becomes broadened because different nerve fibers are affected differently (Fig. 15.4).

---

[1] The numerical value of a time constant for temporal integration is usually defined as the time it takes for a quantity to decay to an amplitude of $1/e$ (e is the base of the natural logarithm 2.72) of its original value. These definitions are taken from engineering terminology and assume an exponential decay or rise.

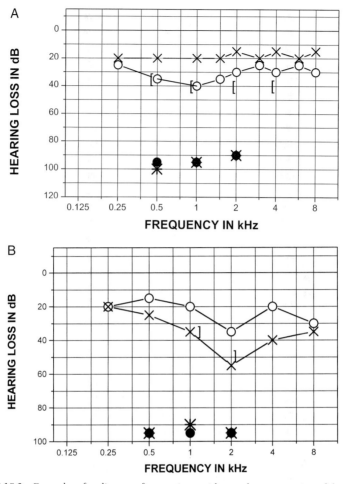

FIGURE 15.2 Examples of audiograms from patients with vascular compression of the auditory nerve. (A) Audiograms from a patient with vascular compression of the eighth cranial nerve near its entry into the brainstem showing a dip near 1500 Hz (circles) (modified from Møller, M. B. (1994). Audiological evaluation. *J. Clin. Neurophysiol.* **11**:309–318) and (B) from a patient with hemifacial spasm (crosses) showing a dip around 2000 Hz (from Møller, M. B., and Møller, A. R. (1985). Audiometric abnormalities in hemifacial spasm. *Audiology* **24**:396–405).

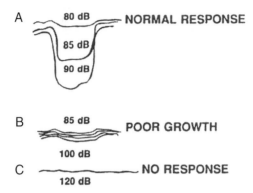

FIGURE 15.3   The growth of the acoustic middle ear reflex response. (A) In a patient without auditory nerve damage and (B) in a patient with an acoustic tumor (from Møller, M. B. (1994). Audiological evaluation. *J. Clin. Neurophysiol.* 11:309–318).

Injury to the intracranial portion of the auditory nerve causes the latency of all peaks of the BAEP to increase, except Peak I, which is generated by the most peripheral portion of the auditory nerve (see Chapter 11). Peaks II (generated by the central portion of the VIIIth nerve) and III of the BAEP may be obliterated and often only Peaks V and I are discernable in the BAEP of individuals whose auditory nerve is injured. Subtle injuries to the auditory nerve, such as from vascular compression, typically cause less than a 1-ms increase in latency. In patients with acoustic tumors, it is not unusual that the latency of Peak V is prolonged 3 ms or more. The normal conduction velocity of the auditory nerve is about 20 m/s [111]. The length of the auditory nerve is 2.5 cm, corresponding to a total conduction time of 1.25 ms. Prolongation of the conduction time by 3 ms implies a reduction of the conduction velocity of the auditory nerve to about 8 m/s if the change was uniform along its entire length. If only the intracranial portion of the auditory nerve is affected (length approximately 1 cm), the conduction velocity of that segment would decrease from 20 m/s to approximately 3 m/s.

CAUSES OF INJURY TO THE AUDITORY NERVE

As stated earlier in this chapter the auditory nerve may be affected by disease processes such as acoustic tumors and other space-occupying lesions of the

FIGURE 15.4   Typical CAPs recorded consecutively from the intracranial portion of the auditory nerve during an operation where the nerve was surgically manipulated. Møller, A. R. (1995). "Intraoperative Neurophysiologic Monitoring." Harwood, Luxembourg.

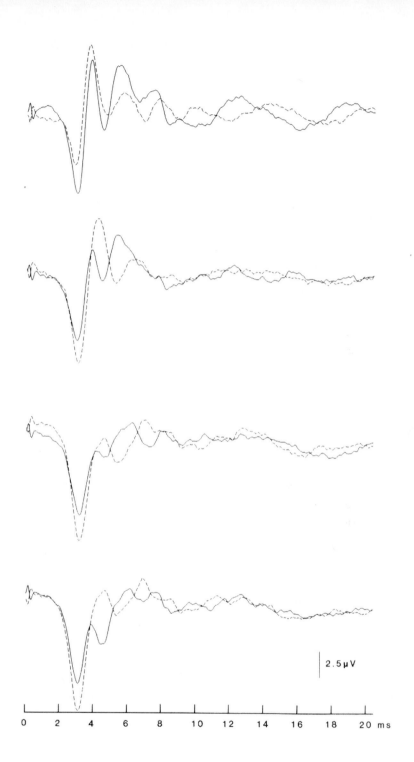

2.5 μV

0   2   4   6   8   10   12   14   16   18   20 ms

cerebellopontine angle. Virus infections may also affect neural conduction in the auditory nerve and close contact with a blood vessel can cause subtle changes in the function of the auditory nerve. The auditory nerve can be injured in operations in the cerebellopontine angle and head trauma may involve injuries to the auditory nerve. These disease processes are discussed in detail in this section.

ACOUSTIC TUMORS

Acoustic tumors are schwanomas that grow in the transition between peripheral (Schwann cell) myelin and central (oligodendrocyte) myelin (Obersteiner-Redlich, OR, zone). Acoustic tumors usually grow from the vestibular portion of the VIIIth cranial nerve and attempts have been made to give these tumors a more appropriate name such as vestibular schwanoma or vestibular neurilemoma. Recently, it has been decided that the official name should be "vestibular schwanoma," but these tumors are still commonly called acoustic tumors. We retain the old name "acoustic tumors" in this book.

> Acoustic tumors are benign tumors and they thus do not cause metastases. Acoustic tumors belong to a group of tumors known as skull base tumors, as they occur in or near the base of the skull. Acoustic tumors make up 40% of all intracranial tumors. In most cases, the tumors originate from the superior vestibular nerve but they can also originate from the inferior vestibular nerve or the auditory nerve. The OR zone of the VIIIth cranial nerve is located inside the internal auditory meatus.
>
> In a study from Denmark, the incidence of acoustic tumors was reported to be 0.78–0.94 per 100,000 [178]. These numbers are derived from diagnosed tumors and may thus be affected by the efficacy of diagnostic methods. There is also a geographical dependence, and since the incidence increases with age, the incidence depends on the longevity of a population. Acoustic tumors grow slowly [31]. The average growth rate of acoustic tumors is 0.2 cm per year, with large individual variation. Some tumors may grow rapidly or may decrease in size or even disappear.

The earliest symptoms of acoustic tumors are tinnitus and hearing loss in one ear, with a larger decrease in speech discrimination than seen with a similar hearing loss of cochlear origin. It may be surprising that vestibular symptoms are not common. The reason is that slowly decreasing vestibular function produces few or no symptoms because the other side's inner ear and other neural systems take over the function of the impaired vestibular system. This may be different when a tumor starts growing at a fast rate, especially if a person is in his or her sixties or older. Then the loss of vestibular function cannot be compensated for by the remaining vestibular system as it can in younger individuals. An elderly person who loses vestibular function on one side at that age typically experiences off-balance without vertigo.

The facial nerve travels in the internal auditory meatus together with the VIIIth cranial nerve, but acoustic tumors seldom cause noticeable signs from the facial nerve and impairment of facial function usually does not occur before the tumor is treated surgically. Injury to the facial nerve may then occur as a result of surgical manipulations in connection with removal of a tumor. While surgical removal of these tumors is the most common treatment, gamma radiation therapy ("gamma knife") is also used to treat such tumors.

Diagnosis of Acoustic Tumors

The presence of an acoustic tumor must be ruled out in individuals who have asymmetric hearing loss. While the most common early sign of acoustic tumors is tinnitus, only a few individuals with tinnitus have an acoustic tumor and many people have asymmetric hearing loss without a tumor. Recording BAEP is an effective test for acoustic tumors because the tumor affects neural conduction in the auditory nerve. Recently it has been regarded to be more appropriate to use MRI scanning but the effectiveness of BAEP using prolongation of the latency of Peak V is approximately as good as MRI scans. Selters and Brackman reported in 1977 [153] that BAEP had a high sensitivity when compensations for age-related changes in hearing threshold were made. More recently, Godey et al. [60] reported that the sensitivity of BAEP alone is 92% and, together with recordings of the acoustic middle ear reflex and caloric vestibular response, the sensitivity is 98%, with all the false-negative responses being in patients with tumors less than 1.8 cm in diameter [60]. These authors proposed BAEP and the acoustic middle ear reflex as first-line screening tests.

When MRI scans are compared with other diagnostic methods, it should be noted that commonly made estimates of the effectiveness of MRI scanning are subjected to misinterpretations. This is because MRIs are used both as a comparison between other methods and as the definitive proof of the presence of a tumor. Thus MRI scans are used as the standard with which all other tests are compared. Therefore, negative MRI scans are interpreted to mean that the patient does not have a tumor. However, negative MRI scans cannot be confirmed unless the patient is operated on because there is no other way to find out if the patient in fact has a tumor. If a patient with a negative MRI scan has a tumor and other tests indicate the presence of a tumor, these results are normally judged to be false positive. It may become obvious only many years later whether the MRI finding was a false negative. Most positive MRI scans are verified because most patients with positive MRI scans for an acoustic tumor are operated on, although some are now treated using radiation.

Thus, decisions to use MRI scans to rule out acoustic tumors should be reconsidered and, instead, the use of BAEP and acoustic reflex testing should be promoted as effective means to detect the presence of acoustic tumors, the cost of which is much less than MRI scans. The use of BAEP for diagnosis of acoustic tumors requires interpretation of the BAEP, and expertise for that is not always available.

OTHER SPACE-OCCUPYING LESIONS

Other space-occupying lesions of the cerebellopontine angle (CPA) are usually benign tumors such as meningiomas, cholesteatomas, neuromas of cranial nerves other than the VIIIth nerve, and arachnoidal cysts. Such lesions are rare, but they may grow to large sizes without causing much hearing loss. Malignant tumors in that region of the brain are extremely rare.

VASCULAR COMPRESSION OF THE AUDITORY NERVE

Close contact between the auditory nerve and a blood vessel (vascular compression) may cause tinnitus (see Chapter 16) and probably also hearing loss with decreased speech discrimination in some individuals [121].[2] A blood vessel in close contact with the auditory nerve can irritate the nerve and may give rise to abnormal neural activity and perhaps slight injury to the nerve. Over a long time such close contact with a blood vessel may cause changes in more centrally located structures of the ascending auditory pathway and this is believed to be the cause of symptoms such as tinnitus, hyperacusis, and distortion of sounds (see Chapter 16). In many patients where vascular compression of the auditory nerve is a cause of these symptoms the common complaint is that sounds are distorted or sound "metallic." In some patients, tinnitus may be relieved with microvascular decompression (MVD) operations, where the offending blood vessel is moved off the auditory nerve [76, 86, 121]. If such an operation is successful in alleviating tinnitus, it also often relieves the patient's hyperacusis and distortion of sounds. Speech discrimination may improve. This indicates that at least some of the effects of vascular compression on neural conduction in the auditory nerve that are caused by vascular compression are reversible.

---

[2] Vascular contact with a cranial nerve is known as "vascular compression" but there is evidence that the pathology associated with close vascular contact between a cranial nerve and a blood vessel does not depend on a mechanical action (compression) but rather that the mere contact (or irritation) causes the pathology.

The reason that close contact between a cranial nerve and a blood vessel has been assumed to "cause" diseases such as face pain [trigeminal neuralgia (TGN) or tic douloureux] and face spasm (hemifacial spasm, HFS) is that these diseases can be effectively cured by moving a blood vessel off the respective nerve in an operation known as microvascular decompression (MVD). It has also been shown that close contact between cranial nerves V and VII is rather common and occurs in as much as approximately 50% of individuals who do not have any symptoms from these cranial nerves, while the disorders that are associated with such vascular compression of these nerves (TGN and HFS, respectively) are extremely rare, with incidences of about 5 and 0.8 per 100,000 respectively [5, 80]. Vascular contact with the VIIIth cranial nerve is also common, although it is not known exactly how often that occurs. In fact, it is the experience from observations of many operations in the cerebellopontine angle in patients undergoing MVD operations for TGN and HFS that close vascular contact with the VIIIth cranial nerve is common in such patients without any associated vestibular or hearing symptoms.

The fact that vascular compression is common in asymptomatic individuals means that vascular contact is not sufficient to produce symptoms. The reason that vascular contact with a cranial nerve only rarely gives symptoms and signs from the respective cranial nerve could be that vascular compression varies in severity but a more plausible reason is that vascular compression is only one of several factors, all of which are necessary, but each one of which is not sufficient to cause symptoms and signs [108]. The fact that MVD operations for TGN and HFS have a high success rate (80–85%) indicates that vascular compression is necessary to cause symptoms. Removal of vascular compression can relieve symptoms despite the fact that other factors are still present because vascular compression is necessary for producing the symptoms. Vascular compression of a cranial nerve alone can thus not cause symptoms and signs [108].

Subtle injuries to the auditory nerve or irritation from close contact with a blood vessel are thus present in a large number of individuals, but only very few such persons show any symptoms. Detecting the presence of a blood vessel is therefore not sufficient to diagnose these disorders. Using MRI scans for that purpose has been attempted, but MRI scans are not effective in detecting the presence of close contact between vessels and cranial nerves. Recording BAEP and the acoustic middle ear reflex response can detect the effect of vascular contact with the auditory nerve because it is associated with slower neural conduction in the auditory nerve. Prolongation of the latency of Peak II in the BAEP (see Chapter 11) and delays of all subsequent peaks are thus signs of slight injury to the auditory nerve.

Small dips may be observed in the audiograms of patients with tinnitus that can be alleviated by MVD operations of the auditory nerve (Fig. 15.2A [117]). The audiograms of some patients with hemifacial spasm that is caused by vascular contact with the VIIth cranial nerve have similar dips [118] (Fig. 15.2B). The reason for this is assumed to be compression of the auditory nerve from the same vessel that was in contact with the facial nerve causing the patient's symptoms (HFS). These patients, however, did not have any symptoms from the auditory system and only the audiogram, taken as a part of the preoperative testing, revealed the involvement of the auditory nerve. The fact that these dips occurred in the mid-frequency range of hearing would indicate that nerve fibers originated from the middle portion of the basilar membrane and are located superficially in the auditory nerve. This would be different from what is seen in animals where high-frequency fibers are located superficially on the nerve [146].

This observation supports the findings discussed above that showed that vascular contact in itself does not cause symptoms and confirms that close contact between a blood vessel and the auditory nerve is only one of several factors that are necessary to cause symptoms such as tinnitus. This also means that tests that reveal contact between the auditory nerve and a blood vessel cannot alone provide the diagnosis of such disorders as tinnitus and hyperacusis and the case history must be taken into account to achieve a correct diagnosis of such disorders.

SURGICAL INJURY TO THE AUDITORY NERVE

Surgical injury to the auditory nerve is a relatively recent cause of hearing loss, tinnitus, and hyperacusis that began to appear when it became common to operate in the cerebellopontine angle for nontumor causes (such as vascular compression of cranial nerves to treat pain and spasm of the face). Hearing loss from such operations are, however, less frequent now than before because of advances in operative technique and the use of intraoperative monitoring of auditory evoked potentials [109, 112].

Surgical injuries can be caused either by compressing or by stretching the auditory nerve. Heat that spreads from the use of electrocoagulation to control bleeding can also injure the auditory nerve. Depending on the degree of compression, stretching, or heating, the injuries may consist of a slight decrease in conduction velocity, conduction block in some fibers, or, in the more severe situation, conduction block in all auditory nerve fibers. The acute effect on neural conduction may recover completely with time or partially or not at all depending on the severity of the injury. Compression probably mostly affects fibers that are located superficially in the nerve, whereas stretching is likely to affect all fibers. Surgically induced injuries to the auditory nerve caused by stretching of the nerve may affect all fibers of the auditory nerve [66], which explains why hearing loss from surgically induced injury often affects both low and high frequencies. Surgically induced injury to the auditory nerve typically causes a moderate change in the pure-tone audiogram and a marked impairment of speech discrimination (Fig. 15.5). In fact, moderate threshold elevation may be associated with total loss of speech discrimination. The effects of surgical injury to the auditory nerve at all degrees including total loss of hearing are often accompanied by tinnitus and hyperacusis (discussed in a Chapter 16).

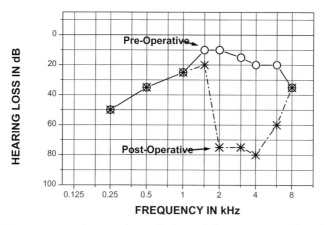

FIGURE 15.5 Pure-tone audiograms obtained before and after an operation where the auditory nerve was injured by surgical manipulations. The preoperative speech discrimination was 96% and the postoperative discrimination was 0%. The speech discrimination tests were obtained using recorded speech material 5 days after the operation. (From Møller A. R. Cranial nerve dysfunction syndromes: Pathophysiology of microvascular compression. In: *Neurosurgical Topics Book 13,* 'Surgery of Cranial Nerves of the Posterior Hossa,' Chapter 2. D. L. Barrow, ed. American Association of Neurological Surgeons, Park Ridge, Illinois, pp. 105–129, 1993.)

VIRAL INFECTIONS

Various forms of unspecific pathologies, such as viral infections, may affect the auditory nerve. Hearing loss can occur in connection with infections by the herpes virus, such as Ramsay Hunt syndrome [185], which often renders the patient deaf in one ear and may affect adjacent cranial nerves, such as the facial nerve, resulting in paralysis of facial muscles. Viral infections of the auditory nerve may occur without any other signs than hearing loss. To establish with any degree of certainty that a viral infection is the cause of hearing loss in an individual case is, however, difficult. Common childhood viral infections are suspected to cause hearing loss or deafness from inflammation of the auditory nerve and the cochlea with destruction of the organ of Corti, stria vascularis, and tectorial membrane. The cytomegalovirus (CMV) is usually harmless until some additional factors cause it to become activated, at which time it may cause sensorineural hearing loss. The CMV is now regarded to be one of the more common causes of early cochlear hearing loss. It is not known how a viral infection can render a nerve nonconductive in such a short time (a few hours to a day).

SUDDEN HEARING LOSS

"Sudden hearing loss" is a term often used to describe a sudden loss of hearing when no reason can be found. It may be a result of injury to the cochlea as discussed in Chapter 14. It has also been suggested that it may be a result of a conduction block in the auditory nerve and it has been suggested that inflammation of the auditory nerve, perhaps as result of viral infection, could cause sudden hearing loss. This would be similar to what is thought to occur to the facial nerve in the facial palsy that is known as Bell's palsy. Rarely is the vestibular nerve involved and rarely is the loss of hearing associated with tinnitus. Most individuals with sudden deafness recover completely without treatment and efforts have been made to predict the likelihood of recovery based on the signs of hearing loss. The many different treatments that have been tried have not proven to cause any noticeable change in the outcome. Treatment with anti-inflammatory drugs has been one of the most frequently tried treatments.

AGE-RELATED CHANGES IN THE AUDITORY NERVE

The distribution of fiber diameters of auditory nerve fibers widens with age [163] (see also Chapter 15). This causes a wider distribution of conduction velocities of auditory nerve fibers and thus a decreased temporal coherence of the auditory nerve impulses that arrives at the cochlear nucleus. These morphologic changes in the auditory nerve may be the reason why temporal processing in the auditory system deteriorates with age [167]). Since temporal coherence in the auditory nerve seems to be important for discrimination of complex sounds such as speech, age related changes in the auditory nerve may thus contribute to the deterioration of the ability to discriminate speech that often occurs with age.

The deterioration of temporal processing could thus be explained by a decrease in temporal coherence of nerve impulses that arrive at the cochlear nucleus caused by increased variations in neural conduction velocity of auditory nerve fibers.

BRAINSTEM AUDITORY PATHWAYS

Disorders that affect the brainstem nuclei of the auditory system and which are associated with detectable morphologic changes are extremely rare. Such disorders are tumors and vascular malformations and they usually give symptoms and signs from multiple systems. Rarely are the auditory signs present before signs occur from vital systems that are controlled from the brainstem. If a brainstem tumor manifests with auditory symptoms as the first sign, it usually does not take long before symptoms and signs from other systems appear. The changes that occur in the pure-tone audiograms from brainstem lesions are unpredictable. Changes in the BAEP may be the most distinct and the anatomical location of a lesion can often be determined by examining the BAEP and considering its neural generators.

HYPERBILIRUBINEMIA

Increased blood levels of bilirubin, hyperbilirubinemia, may occur when bilirubin, which is a breakdown product of the porphyrin ring of hemoglobin, is not broken down further in the liver by the enzyme glucuronyl transferase. Hyperbilirubinemia (jaundice) often occurs in newborns because the enzyme that breaks down bilirubin is poorly developed. Bilirubin is neurotoxic and causes bilirubinencephalopathy [2]. The pathologic sign of hyperbilirubinemia is kernicterus, meaning "yellow staining," which affects deep nuclei of the brain. Bilirubinencephalopathy affects the auditory nervous system specifically in addition to its general effect on the nervous system (i.e., causing retardation etc.). The brainstem auditory nuclei and, most noticeably, the cochlear nuclei are most affected [39, 44, 45, 58, 156]. The effect on the auditory system is characterized by high-frequency hearing loss that usually occurs bilaterally and symmetrically, with severely decreased speech discrimination. The interpeak latency (IPL) I-III of the BAEP is increased. The effect of hyperbilirubinemia is preventable, but it requires the condition to be detected early. Bilirubinemia has been studied extensively and an animal model of the disorder exists (Gunn rat model).

AUDITORY CORTICAL AREAS

Lesions that affect the auditory cortex are typically tumors of the temporal lobe, but bleeding, strokes, and trauma may also affect the auditory cortex. The symptoms are diffuse and often consist of auditory hallucinations in the form of hearing meaningful sounds such as music or speech. (It has been reported [53] that a stroke that affected the MGB caused similar auditory hallucinations, but such events are extremely rare.) This is unlike tinnitus, which involves hearing sounds that are not meaningful. (Tinnitus is regarded to originate in anatomically more peripheral levels of the ascending auditory nervous system, see Chapter 16).

  Injury to the primary auditory cortex does not usually cause any noticeable degree of threshold elevation, even for sounds presented to the contralateral ear that has the main projection to the affected auditory cortex and speech discrimination is usually normal when assessed using standard tests. Lesions of the auditory cortex may be detected by using low-redundancy speech [14] such as spectrally filtered speech, or speech that is presented at a very high rate or chopped speech. These three different ways to distort speech have a high degree of specificity to disorders of the central auditory nervous system [88]. Individuals with lesions of the temporal lobe affecting the auditory cortex have much lower discrimination scores for distorted speech when presented to the contralateral ear, while no abnormality can be detected when the speech was presented to the ear ipsilateral to the lesion (Fig. 15.6).

Low Redundancy Speech Tests

It was Bocca and co-workers [15] who first demonstrated that speech in which some of the redundancy was removed could detect auditory deficits in patients with central auditory lesions while the same individuals had normal or near-normal discrimination of undistorted speech. Bocca and his co-workers [15] showed that speech in which some redundancy was removed by spectral filtering could reveal impairments in the auditory system that could not be detected using normal speech material. Similar kinds of spectrally filtered speech were used by subsequent investigators to detect disorders that affect the auditory cortex in the temporal lobe, such as temporal lobe epilepsy as well as more general changes in the function of the nervous system, such as that of aging [83, 87].

Other ways of reducing the redundancy of speech, such as by periodically interrupting the speech (chopped speech), have been described. The effect of such alterations of speech on its intelligibility has been studied in individuals with normal hearing [13, 24, 104]. The efficiency of such low-redundancy speech in detecting central auditory lesions has been demonstrated [88]. Chopped and spectrally filtered speech have reduced redundancy because something has been removed from the speech, while time-compressed speech contains all components of the original speech, just presented in a shorter time than normal speech. Interruptions (10 per second) using a 1 : 1 ratio of speech to silence resulted in a marked reduction in speech discrimination in individuals with lesions to the central auditory nervous system, more so when the sound was presented to the contralateral ear than to the ipsilateral ear [88]. Normal hearing individuals could discriminate such interrupted speech at nearly 100% but a population of elderly individuals without specific disorders had reduced discrimination scores. When speech is speeded up by a factor of 2 (from normally 110–140 words per minute to about 250 words per minute) discrimination is reduced even by individuals with normal hearing [56].

It is interesting to note that rather extensive lesions of the auditory cortex allow normal speech discrimination in contrast to lesions of the auditory nerve where subtle injury causes severe deterioration of speech even under ideal circumstances such as those usually used in testing the speech discrimination for clinical purposes.

Now it is often regarded to be easier to use MRI scans for the diagnosis of cortical lesions but the high cost of MRI scans compared with speech tests makes low-redundancy speech tests an attractive alternative for diagnosis of temporal lobe lesions that affect the auditory cerebral cortex. It may also be worth noting that not all disorders of the nervous system, including the auditory nervous system, manifest themselves as detectable abnormalities in imaging

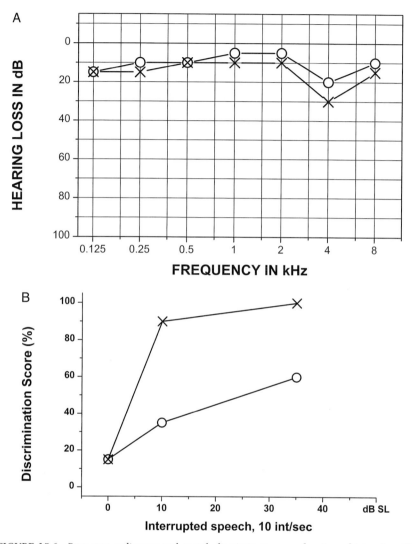

FIGURE 15.6 Pure-tone audiogram and speech discrimination as a function of intensity with which the speech was presented (in decibels above the patient's threshold, SL) for different kinds of low-redundancy speech. The patient had an astrocytoma in the left temporal lobe (modified from [88] with permission).

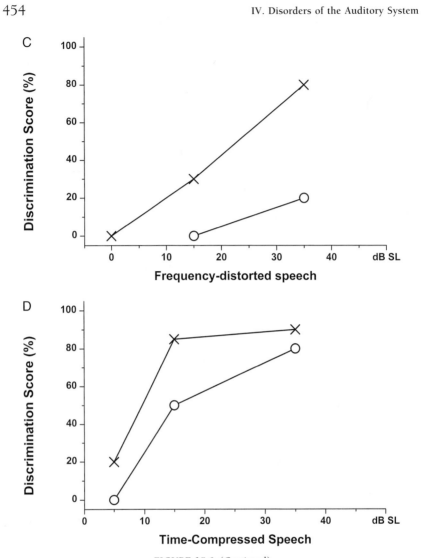

FIGURE 15.6 *(Continued)*

studies, which are limited to detecting changes in structure. Plastic changes in the nervous system cannot be detected with imagining methods such as the MRI. Perhaps tests such as the low-redundancy speech test should not have been totally abandoned.

EFFERENT SYSTEM

No specific disorders are known to affect the efferent auditory nervous system. However, due to its anatomical abundance it has been suggested that the efferent system is involved in several disorders, but little evidence has been presented to support such hypotheses. Because the cochlear efferent bundle travels together with the central portion of the vestibular nerve, it is severed in operations for vestibular neurectomy when done close to the brainstem. (Vestibular neurectomy is often used to treat patients with vestibular disturbances.) Scharf and co-workers [150] found little measurable changes in hearing in patients who had their vestibular nerve sectioned intracranially to treat vestibular disorders and thus had a severed olivocochlear bundle.

ROLE OF NEURAL PLASTICITY IN DISORDERS OF THE CENTRAL AUDITORY NERVOUS SYSTEM

The auditory nervous system possesses great abilities to change its function and there is increasing evidence that neural plasticity may be involved in such hyperactive hearing disorders as tinnitus and hyperacusis (Chapter 16). More recent studies indicate that neural plasticity may be involved in many more forms of hearing impairment than earlier believed. A variety of studies have shown that different brain functions, such as processing of sensory information and pain, and even motor functions, can change more or less permanently as a result of altered input or lack of input. This means that the brain is plastic to a much greater extent than previously believed. Earlier, it had been assumed that such plastic changes could only occur early in life during ontogenetic development of the central nervous system but it has become apparent that plastic changes can indeed occur in the adult nervous system, although to a lesser degree. The changes that occur are mainly a result of change in synaptic efficacy but outgrowths of new connections or degeneration of existing connections have also been demonstrated. The changes develop over time as a result of abnormal input, such as overstimulation, or deprivation of input, but novel stimulation can also cause such changes. The auditory nervous system thus possesses a great ability to change its function under certain circumstances, as does other sensory systems. These changes may reverse spontaneously or become permanent after the processes that initially caused therm have been eliminated.

Many studies have shown that changes in the function of the auditory nervous system can be induced by deprivation of input [57], overstimulation [6, 68, 192], or by input that is abnormal in one way or another. Cochlear damage may induce changes in the central auditory nervous system [81]. The changes reported in these early studies consist of increased sensitivity [57], broadening of tuning in cochlear nucleus units [68], or changes in the temporal pattern of responses from IC neurons. Few of these changes have detectable morphologic correlates and they thus represent a functional reorganization. The changes are possible because of the plasticity of the nervous system ("neural plasticity").

The changes in function of the central nervous system are assumed to be the result of changes in synaptic efficacy, change in the balance between inhibition and excitation, and degeneration of fibers and nerve cells. Decrease in inhibition or increase in excitation may cause hypersensitivity and hyperactivity. Several studies have shown that such changes in the central nervous system may result in pathologies of the auditory system that may cause tinnitus and hyperacusis (Chapter 16). Opening of dormant synapses can result in rerouting of information. The term "dormant synapses" was coined in 1977 by Wall [184] as an explanation of certain forms of pain. Changes of these kinds are known to cause certain types of pain, but it has been only relatively recently that neural plasticity has been implicated in disorders of the auditory system. Changes in the function of the cochlea can, however, also cause changes in the auditory nervous system that have morphologic correlates such as degeneration of axons [81].

CHANGES IN THE AUDITORY NERVOUS SYSTEM FROM NOISE EXPOSURE

It has been known for many years that exposure to loud noise causes hearing loss. Until recently it has been assumed that such noise-induced hearing loss (NIHL) was the result of injury to cochlear hair cells. While a large part of NIHL can indeed be explained by injury to the cochlea, it has more recently become evident that the amount of NIHL is affected by prior exposure to sound [27, 105, 160] and that exposure to loud noise can alter the function of parts of the auditory nervous system [144]. Thus hearing impairment from noise exposure is caused not only by alteration of the function of the cochlea; changes in the function of the auditory nervous system also contribute to acquired hearing impairment. These changes in the auditory nervous system are not related to detectable morphological changes but are brought about by plastic changes in function. This means that such changes most likely are reversible, which is in contrast to the morphological changes in the cochlea. There is considerable evidence that exposure to loud noise that causes relatively small changes in the response from the ear, the auditory nerve, and cochlear nucleus causes larger changes in the neural activity (spontaneous and driven) in the inferior colliculus, which are even greater in the auditory cerebral cortex. Some studies [81] have reported morphological changes in the auditory nervous system after noise exposure.

> Babigian *et al.* showed in 1975 [6] that there is a central component to auditory fatigue. These investigators showed that the response from the inferior colliculus decreased more than the response from the ear and the auditory nerve during a period of temporary threshold shift caused by prior sound stimulation. Syka and co-workers [135, 170] have shown that evoked potentials recorded from different places of the auditory nervous system are altered differently after exposure to loud noise. While the thresholds were elevated approximately the same amount in recordings from the auditory nerve, the inferior colliculus, and the auditory cortex, the responses to sound stimulation were altered in different ways at these three locations along the neural axis. Thus while the reduction in the amplitude of the evoked potentials was similar when recorded from the auditory nerve and the inferior colliculus, the response from the cortex increased at a steep rate as a function of sound intensity after noise exposure.
>
> Other investigators [144] found that noise exposure resulted in altered stimulus–response functions of neurons in the inferior colliculus. Mainly, the response increased at a steeper rate with increasing stimulus intensity after noise exposure, but these changes depended on the frequency that was tested and the spectrum of the noise to which the animals were exposed.

After noise exposure the responses from the auditory nerve and cochlear nucleus decrease and the threshold increases as a result of decreased cochlear function, but paradoxically, the response from the auditory cortex increases, indicating that the function of the auditory cortex changes as a result of noise exposure. An impairment of the amplitude compression in the cochlea might have resulted from the noise exposure (see Chapter 3), but that would have been expected to affect evoked potentials recorded from peripheral and central portions of the auditory nervous system equally. This paradoxical change in neural responses as well as in evoked potentials from the auditory cortex must be a result of changes in synaptic efficacy somewhere in the ascending auditory pathway, brought about by neural plasticity. The changes could be a result of deprivation of input, perhaps to neurons responding to high frequencies because of high frequency hearing loss. The changes in synaptic efficacy could also have been a result of the overstimulation during the noise exposure.

> Phantom limb syndrome that consists of sensations including pain that are referred to an amputated limb is assumed to be caused by reorganization of the central nervous system. Blocking the nerve that supplies the limb prior to the

amputation can eliminate the subsequent phantom limb symptoms [7], indicating that the massive stimulation of the somatic nervous system during the surgical procedure causes plastic changes in the function of the nervous system that cause these "phantom" sensations. It has been suggested that tinnitus may be a similar "phantom" sensation [77] (see Chapter 16).

The changes in function of the auditory nervous system from noise exposure that can de demonstrated by electrophysiologic methods are usually assumed to be a result of change in synaptic efficacy and a change in the balance between inhibition and excitation. Some investigators have proposed that a disinhibition may occur in the auditory cortex after exposure to loud noise. Morphologic studies by Morest and co-workers [81, 122, 123] have shown that injuries to cochlear hair cells cause degeneration of not only auditory nerve fibers but also cells in the cochlear nucleus and that transneural degeneration of axonal endings occurs in the superior olivary complex.

Whatever the cause of these plastic changes are, the increased neural activity in the cortex may explain why some people with NIHL have an abnormal perception of the loudness of sounds and experience normal sounds to be unpleasantly loud and even perceive loud sound as being painful. The abnormal function of the auditory cortex that these changes reflect may also explain why some people with NIHL have lower than expected speech discrimination.

Neural Components of Other Types of Hearing Loss

Recent studies have provided evidence that other forms of hearing loss that have traditionally been associated with injuries to cochlear hair cells, such as presbycusis and drug-induced hearing loss, have neural components that are similar to the reorganization of the central nervous system discussed in connection with NIHL [145].

Age-related changes may also occur in more centrally located parts of the auditory nervous system because the function of the nervous system may change as a result of the change in input from the cochlea caused by loss of hair cells. The changes that occur over time in the function of the cochlea may result in deprivation of input to neurons that are tuned to high frequencies because of loss of hair cells in the basal portion of the cochlea, which may reduce inhibition which, in turn, over time, may cause changes in the function of higher auditory centers [57] (see Chapter 16). There are other causes of decrease in inhibition in the auditory nervous system. It has been shown that GABA in the central nucleus of the inferior colliculus is lost with age, creating a deficit of an important inhibitory neurotransmitter [28]. There may thus be an age-related shift in the balance between inhibition and excitation in the central nervous system. All these factors may contribute to the hearing impairment that we normally call presbycusis. These complex changes in the ear and auditory nervous system are also assumed to be involved in the development of tinnitus, which often accompanies presbycusis (see Chapter 16).

Thus as more knowledge about age-related changes accumulates, it appears that presbycusis is more complex than normal age-related changes of cochlear hair cells. Some unexpected results of animal experiments have shown that the progression of age-related hearing loss can be slowed by sound stimulation [179]. This opens up the possibility for intervention in the progression of these changes in function.

Ototoxicity is normally associated with injury to the cochlea but some drugs affect neural processing of sounds. Thus salicylates and quinine, which affect the cochlea (see Chapter 14), also change the function of the auditory nervous system. Neural discharges in neurons in the secondary auditory cortex (AII) that receive their input from the nonclassical auditory system are affected by administration of both salicylates and quinine [42]. The spontaneous activity of neurons in the AII area increased while administration of these drugs caused a decrease in the spontaneous activity of neurons in primary auditory cortex (AI and AAF), which are part of the classical auditory system. These drugs are known to cause tinnitus and these findings are therefore significant in understanding the pathophysiology of tinnitus and hyperacusis.

Thus drugs that do not seem to affect the function of the cochlea may affect neural processing in the auditory system. Thus it has been shown that administration of scopolamine reduces discrimination of low-redundancy speech in young individuals with normal hearing [3].

It is likely that slowly decreasing hearing such as occurs in presbycusis may have similar effects on reorganizing the central auditory system, such as with NIHL and hearing loss from administration of ototoxic drugs. It may thus be neural plasticity that causes low discrimination of distorted speech in individuals with newly acquired cochlear injury, while individuals with similar levels of hearing, of conductive type have much higher discrimination scores, indicating that the auditory system can adapt to a poorly functioning cochlea [88]. It is also known that deprivation of auditory input that results from conductive hearing loss can affect auditory processing.

Evidence has been presented that exposure to some kinds of organized sounds, such as certain types of music, may be beneficial for the development of mental skills in the young individual and this has promoted the use of amplification in children with hearing loss and the use of cochlear implants in young children with severe hearing loss. A study in rats has shown that exposure *in utero* to music improves the rats' ability to complete a maze test in a shorter time with fewer errors than animals exposed to white noise or silence [140].

PATHOLOGIES THAT CAN AFFECT BINAURAL HEARING

Asymmetrical hearing loss of any kind may impair directional hearing and the perception of auditory space. Binaural hearing aids that are properly adjusted can restore some of these functions. The anterior commissure, a part of the corpus callosum, seems to be involved in perception of auditory space and there are some indications that this part of the brain may change with age so that it becomes more difficult to fuse the auditory input from the two ears into a single "image". Little is known, however, about that aspect of auditory function impairment.

CHAPTER 16

Tinnitus, Hyperacusis, and Phonophobia

ABSTRACT

1. Tinnitus is perception of sound that does not originate from a source outside the body.
2. Tinnitus is of two general types: (1) objective tinnitus and (2) subjective tinnitus.
3. Objective tinnitus is caused by a (physical) sound that is generated in the body, usually from flow of blood that has become turbulent. Objective tinnitus can usually be heard by an observer. Objective tinnitus is rare.
4. Subjective tinnitus is not caused by a (physical) sound and can therefore only be heard by the person who suffers from the tinnitus. Subjective tinnitus is common.
5. Subjective tinnitus has many forms and its severity varies from person to person. It can be divided into mild, moderate, and severe (disabling).
6. Severe subjective tinnitus is often accompanied by altered perception of normal sounds, which are often distorted, very loud (hyperacusis), unpleasant, and/or painful (phonophobia).

7. The neural activity that causes subjective tinnitus, hyperacusis, and phonophobia is not always evoked by the ear. The anatomical location of the physiologic abnormalities that cause the neural activity that is perceived as severe and disabling tinnitus is often located in the central nervous system. Hyperacusis and phonophobia is assumed to be caused by reorganization of the central auditory nervous system. The development of these conditions depends on neural plasticity, and it is often caused by overstimulation or deprivation of stimulation.

8. Tinnitus may be generated by neural activity in neurons other than those belonging to the classical auditory nervous system.

9. Phonophobia that often accompanies severe tinnitus may result from activation of the limbic system or other neural structures not normally activated by sound stimulation, but involved in emotional responses.

10. Many different pathologies affecting the ear and the auditory nervous system promote development of subjective tinnitus, hyperacusis, and phonophobia.

11. Severe tinnitus has many similarities to chronic pain.

INTRODUCTION

Tinnitus, hyperacusis, and phonophobia are abnormal perceptions that have few objective correlates. Tinnitus can be divided into two main types, objective tinnitus and subjective tinnitus. Objective tinnitus is caused by a physical sound generated within the body that stimulates the cochlea in a similar way as an external sound. Objective tinnitus can often be heard by an observer and it is often caused by blood flow that passes a constriction in an artery, causing the flow to become turbulent. Subjective tinnitus can only be heard by the individual who has the tinnitus. It does not involve a physical sound. This chapter deals only with subjective tinnitus.

Tinnitus consists of hearing meaningless sounds. The anatomical location of the structures that generate the neural activity that is perceived as a sound may be in the ear, the auditory nerve, or the auditory nervous system. Severe tinnitus often occurs together with altered perception of sounds. Sounds may be perceived as distorted, louder than normal (hyperacusis), or as being unpleasant or painful (phonophobia). Such altered perception of sound has received far less attention than tinnitus and yet hyperacusis, and in particular phonophobia, may be more annoying to the patient than tinnitus. The underlying physiologic abnormality of severe tinnitus and the abnormal perception of sounds are assumed to be identical or similar. However, some few disorders are associated with hyperacusis and phonophobia without tinnitus. This chapter deals with what is known at present about the physiology of tinnitus, hyperacusis, and phonophobia.

TINNITUS

Tinnitus is the perception of sound without any sound reaching the ear from outside the body. *Stedman's Medical Dictionary* defines tinnitus in the following way:

> Noises (ringing, whistling, booming, etc.) in the ears.
> The word tinnitus originates from [the] latin [word] tinnio, meaning a jingling.
> Tinnitus aurium: Sensation of sound in one or both ears usually associated with disease in the middle ear, the inner ear, or the central auditory apparatus. Syn: syrigmus, tympanophonia(1), tympanophony.
> Tinnitus cerebri: Subjective sensation of noise in head rather than ears.

Tinnitus can be intermittent or continuous in nature and its intensity can range from a just noticeable hissing sound to a roaring noise that affects all aspects of life. Tinnitus may be a high-frequency sound like that of crickets, a pure tone, or a broad-spectrum sound. Tinnitus is often different from any known sound. Some people hear intermittent noise, others hear continuous noise. Some hear their tinnitus as if it came from one ear, others hear it as if it came from inside of the head, thus bilaterally. Some people with tinnitus are slightly bothered by it, while others perceive their tinnitus as an unbearable annoyance that makes it impossible to sleep or to concentrate on intellectual tasks. Tinnitus can even cause people to commit suicide.

Tinnitus can have many different causes but it should be mentioned that the cause of tinnitus and in particular severe tinnitus in most cases is unknown. The most common known cause of tinnitus is noise exposure. Tinnitus is one of the three symptoms that define Ménière's disease and is a common symptom of acoustic tumors. Vascular compression of the auditory nerve can cause tinnitus; so can surgical injuries to the auditory nerve. Certain pharmacological substances may cause tinnitus; examples are salicylate (aspirin) and diuretics, such as furosemide. Ototoxic antibiotics (mainly aminoglycosides) may cause tinnitus and then possibly hearing loss. Many other substances can cause tinnitus. Most of these forms of tinnitus are reversible and the tinnitus subsides on its own, but tinnitus may be permanent depending on the magnitude of the exposure to noise or ototoxic substances and the length of exposure.

There are several reasons why tinnitus is thought to be caused by pathologies located in the ear. Tinnitus is perceived as a sound, which is one reason why it is referred to the ear. Another reason is that the main causes of tinnitus, noise exposure, ototoxic drugs, and Ménière's disease, all are associated with morphological changes in the ear. More recently, evidence that plastic changes in the central auditory nervous system can cause symptoms such as tinnitus and hyperacusis has accumulated. In fact it has been shown that many forms of cochlear pathology are associated with reorganization of the central nervous system (see Chapter 15). The changes in the central auditory nervous system cannot be detected by the imaging techniques we have available. Since the changes in the function of the central nervous system that are associated with tinnitus do not have any apparent morphologic abnormalities, these functional changes have escaped detection.

Insufficient understanding of the pathology of tinnitus and the absence of objective signs of tinnitus are problems inherent in trying to treat tinnitus. Like individuals with chronic pain, tinnitus patients often invoke a suspicion of malingering or of having psychological disturbances or psychiatric disorders. Tinnitus is therefore not only a problem for the patient but also for the physician, who does not know what to do to help the patient, who is clearly miserable. It may be tempting for the physician to state that "there is nothing wrong with you" because all test results are normal. We have to realize, however, that there are real disorders that are not associated with abnormal results of the tests we use at present. It would therefore be a more correct to state: "I do not know what is causing your tinnitus or how to treat it."

ASSESSMENT OF TINNITUS

Considerable efforts have been devoted to finding methods that can describe the character and intensity of an individual's tinnitus objectively, but since tinnitus has no objective signs it cannot be assessed by objective tests. Attempts to match the intensity of an individual's tinnitus to a (physical) sound have usually given the impression that the tinnitus is much weaker than the patient's perception of his or her tinnitus. Individuals who report that their tinnitus keeps them from sleeping or from concentrating on intellectual tasks often match their tinnitus to a physical sound of an intensity that is unbelievably low, often between 10 and 30 dB above threshold, thus sounds that would not be disturbing at all to a person without tinnitus.

Matching the sounds of a patient's tinnitus to an external sound has also generally been unsuccessful in confirming a patient's description of the character of his/her tinnitus and such tests cannot determine its severity. The results of comparing a patient's tinnitus with a large variety of synthesized sounds to gain information about the frequency and temporal pattern of their tinnitus have been disappointing. It is difficult for a person with tinnitus to describe the sounds he or she hears because tinnitus often does not resemble any normal sound. Only in a few individuals has it been possible to obtain a satisfactory match between the tinnitus and a real (synthesized) sound.

Hearing loss as it appears in a pure-tone audiogram has no apparent relationship to the severity of the tinnitus. Individuals with tinnitus may have hearing loss, but it is usually moderate and not commonly a complaint by individuals with tinnitus. The hearing loss typically is mild to moderate and sometimes the audiogram shows only small dips. Usually these small dips are only revealed when testing is done at half-octave frequencies. A few individuals with tinnitus have severe hearing loss and even total deafness can occur together with tinnitus. No known electrophysiologic test can quantify tinnitus and provide objective measures of its strength.

The only remaining option to classify tinnitus is to use the patient's own judgment about the severity of his/her tinnitus. Using individual patient's own perceptions of their tinnitus has resulted in a classification system of three broad groups: mild, moderate and, severe tinnitus [121, 141]. Mild tinnitus does not interfere noticeably with everyday life. Moderate tinnitus may cause some annoyance and may be perceived as unpleasant. Severe tinnitus affects a person's entire life in major ways, making it impossible to sleep and conduct intellectual work.

Thus, subjective tinnitus is an enigmatic disease that people suffer alone because it has no external signs of illness. The severity of tinnitus can only be determined from the perception of the person who has the tinnitus. Tinnitus thus has similarities with chronic pain. Rene Leriche, a French surgeon (1879–1955) has said about pain: "The only tolerable pain is someone else's pain," and that is true also for tinnitus.

It should be noted that tinnitus is often the first sign of an acoustic tumor. Individuals with one-sided tinnitus with or without hearing loss on one side (or asymmetric hearing loss) should always be examined for the possibility of an acoustic tumor. However, this rarely is the cause.

What Is the Anatomical Location of the Pathological Abnormalities That Generate Tinnitus?

The fact that tinnitus presents as a sensation of sound has led to the assumption that tinnitus is generated in the ear and that it involves the same neural system as is normally activated by a sound that reaches the ear. This is not always the case. Many studies have shown evidence that in many forms of tinnitus the ear is not always the anatomical location of the physiologic abnormality. Perhaps the strongest argument against the ear always being the location of the physiologic abnormalities that cause tinnitus is the fact that the auditory nerve can be severed (surgically) in some individuals with tinnitus without it alleviating their tinnitus. Thus, tinnitus may persist in patients with acoustic tumors, even after removal of the tumor, despite the fact that the auditory nerve has been severed during the operation [73]. This indicates that the injury to the auditory nerve from the tumor over time has caused changes in neural structures that are located more centrally. The finding that deaf people can have severe tinnitus and individuals with normal hearing without any signs of cochlear disorders can have severe tinnitus shows clearly that tinnitus can be generated in other places of the auditory system than in the ear.

> Intracranial recordings of auditory evoked potentials in patients with severe tinnitus during microvascular decompression (MVD) of the VIIIth cranial nerve have supported the hypothesis that the anatomical location of the physiological abnormality that causes severe tinnitus is not in the ear. Compound action potentials (CAP) recorded from the exposed intracranial portion of the auditory nerve and the vicinity of the cochlear nucleus showed no noticeable abnormalities in patients who had severe tinnitus [114]. Potentials generated by the termination of the lateral lemniscus in the inferior colliculus, however, had a small but not statistically significant abnormality. In this study results from tinnitus patients were compared with results from patients who had similar hearing loss as the tinnitus patients but no tinnitus.
>
> The patients studied had debilitating tinnitus and it seems surprising that patients with this degree of tinnitus did not have major abnormalities in their auditory evoked potentials. It would be expected that such strong tinnitus would be associated with a degree of altered responsiveness of the auditory nervous system and thus detectable by recordings of evoked potentials. Since the evoked response from the auditory nerve and the cochlear nucleus were not noticeably altered, it was assumed that the neural activity that caused the sensation of tinnitus was not generated in the ear, the auditory nerve, or the cochlear nucleus, but rather by more central structures of the auditory nervous system. The fact that several of these patients were cured of their tinnitus by moving an offending blood vessel off their auditory nerve (MVD) [114] supports the hypothesis that the changes in the function of these more centrally located auditory structures are caused by the abnormal input they receive from the auditory nerve because of the vascular irritation.

It has been shown that certain components of the magnetoencephalographic response to sound stimulation in individuals with tinnitus were abnormal [72]. The components that were abnormal were probably generated in the auditory cortex. However, other investigators [74] have had difficulties reproducing these results. They found no difference in the responses between tinnitus patients and those of individuals who did not have tinnitus.

Imaging studies in individuals who can voluntarily alter their tinnitus have supported the hypothesis that the neural activity that causes tinnitus is not generated in the ear. These studies show that the neural activity in the cerebral cortex that is related to tinnitus is not generated in the same way as sound-evoked activity [98]. Tinnitus activated the auditory cortex on only one side, whereas sounds activated the auditory cortex on both sides [98]. These investigators concluded that the neural activity associated with tinnitus did not originate in the ears. These findings are in good agreement with the results of studies that show that the nonclassical auditory nervous system is involved in tinnitus [113] and the hypothesis by Jastreboff, which states that tinnitus is a phantom sensation generated in the brain [77].

There is thus considerable evidence that some forms of tinnitus can be generated in the central auditory nervous system. There is also evidence that in some forms of tinnitus the anatomical location of the abnormal neural activity is in neural structures that are not normally activated by sound. Tinnitus associated with presbycusis and ototoxic antibiotics may be caused by pathologies of the cochlea. The fact that some individuals are relieved of their tinnitus by severing their auditory nerve [138] supports the assumption that in some individuals the cochlea is the anatomical location of the physiological abnormalities that generate the neural activity.

Also, animal studies support the hypothesis that the central auditory nervous system is involved in tinnitus. While the spontaneous activity in auditory nerve fibers is little affected by administration of salicylates, the spontaneous activity of neurons in the central nucleus of the ICC increases after administration of salicylate [79]. These findings support the hypothesis that the ascending auditory pathway can be involved in tinnitus.

Subjective tinnitus can thus be generated in the cochlea, the auditory nerve, or probably any part of the auditory nervous system (the ascending auditory pathway), but the anatomical location of the abnormality that causes tinnitus is seldom known. It is thus apparent that the neural activity that causes the sensation of tinnitus can be generated in different ways in different forms of tinnitus and the pathology that generates such abnormal neural activity is different for different forms of tinnitus. Recent research has provided some evidence regarding how pathologies of the central auditory nervous system may develop and what may cause them to develop.

Is Tinnitus Generated by the Same Neurons as Activated by Sound?

The fact that individuals with tinnitus often have difficulties in selecting sounds that are perceived in the same way as their tinnitus may indicate that neural circuits other than those normally activated by sound are involved in tinnitus. The fact that many individuals with tinnitus report that their tinnitus is unbearably strong while matching their tinnitus to sounds that are only 10–30 dB above the their hearing threshold also indicates that tinnitus is generated in parts of the central nervous system that do not normally process sounds.

Involvement of the Nonclassical Auditory Nervous System

Evidence has been presented that the nonclassical (or extralemniscal) ascending auditory nervous system is involved in some forms of severe tinnitus. The nonclassical auditory nervous system branches off from the classical auditory system, mainly at the level of the inferior colliculus, and connects to many parts of the brain such as the limbic system and association cortices (see

Chapter 5).[1] Little is known about the function of the adjunct or nonclassical auditory system or its role in normal hearing. Animal experiments show that neurons in the nonclassical auditory pathway respond in a much less specific way than neurons in the classical (lemniscal) system and neurons are broadly tuned in the nonclassical auditory system. This is why some investigators have named parts of the nonclassical pathways the "diffuse system." Many neurons of a part of the nonclassical system, known as the "polysensory system" (see Chapter 5), receive input from other sensory systems. The neurons of the nonclassical auditory system project to cortical areas other than the primary auditory system, such as the secondary (AII) area.

Interactions between the Auditory and Somatosensory Systems

The fact that some neurons in the nonclassical auditory nervous system (the polysensory system) receive input from the somatosensory system, in addition to auditory input, has been used to determine whether the nonclassical auditory system is involved in tinnitus. In a study of individuals with tinnitus [113] it was shown that in some of these individuals, the appearance of tinnitus could be changed by activating the somatosensory system by electrically stimulating the median nerve.

> In this study, the median nerve of 26 individuals with tinnitus was stimulated electrically. Ten of these patients experienced a change in their tinnitus (6 perceived a decrease and 4 an increase) during median nerve stimulation. Some individuals reported that their tinnitus became less unpleasant. That some of the individuals experienced a decrease in this tinnitus and some experienced an increase is in agreement with the fact that some of the neurons of the nonclassical auditory system receive inhibitory input from the somatosensory system while other neurons receive excitatory input. It is naturally possible that electrical stimulation of the median nerve also would affect the sensation of physical sounds in individuals who did not have tinnitus. However, a study of that showed only loud sounds that were perceived to be unpleasant, such as 40-pps clicks at a level of 105 dB PeSPL, were affected [113]. The participants in this study reported that median nerve stimulation caused a slight increase in the loudness of the sound (equivalent to approximately 2 dB [113]. The sensation of broad-band noise was affected less than that of loud clicks.

Other forms of abnormal interaction between the auditory and the somatosensory systems have been observed in some patients with tinnitus. Thus touching the face, moving the head, and changing gaze can change tinnitus in many individuals.

[1] The limbic system is a complex system of nuclei and connections consisting of brain structures such as the hippocampus, amygdala, and fornicate gyrus (part of the cingulate gyrus). These structures connect to other brain areas such as the septal area, the hypothalamus, and a part of the mesencephalic tegmentum. Through these systems the limbic system influences endocrine and autonomic motor systems and it affects motivational and mood states. It is also known as the visceral brain.

Effects of Drugs on the Nonclassical Auditory System

Recently, Eggermont and co-workers [42] found that administration of substances that cause tinnitus in humans, such as salicylate and quinine, cause an increased spontaneous firing of neurons in the secondary auditory cortical area (AII) in guinea pigs. Since that area receives its input from the nonclassical auditory system, the investigators concluded that tinnitus might result from increased spontaneous firing of neurons in the nonclassical auditory system. It is interesting to note that the same drugs decreased the spontaneous discharge rate in the cortical areas that are innervated by the classical ascending auditory system (AI and AAF). Using salicylate in animal experiments Chen and Jastreboff [32] had earlier shown that the spontaneous activity of neurons in the external nucleus of the inferior colliculus (ICX) increased after administration of salicylate, thus an indication of involvement of the nonclassical auditory system in tinnitus. The ICX, which receives its input from the central nucleus of the IC (ICC), is commonly regarded as an important link between the classical and the nonclassical auditory systems.

ROLE OF NEURAL PLASTICITY IN TINNITUS

Plasticity of the central auditory nervous system makes it possible for overstimulation or deprivation of input to cause functional changes that may result in tinnitus. Such induced changes in the function of the central auditory nervous system consist of altered synaptic efficacy and a shift in the balance between inhibition and excitation. Opening dormant synapses can result in rerouting information and a decrease in inhibition or an increase in excitation may cause hypersensitivity and hyperactivity. Gerken et al. [57] have demonstrated in animal experiments that deprivation of input to the central auditory nervous system can change its function so that the temporal integration of input changes radically. The central point of the comprehensive hypothesis regarding development of tinnitus presented by Jastreboff [77] is that changes may occur in the central nervous system as a result of sound deprivation, either from hearing loss or from lack of environmental sound stimulation. Based on that hypothesis Jastreboff and Hazel developed the tinnitus retraining therapy for tinnitus.[2] Similarities have recently been demonstrated between tinnitus and other hyperactive disorders believed to develop because of neural plasticity such as muscle spasms and chronic pain [110].

---

[2] The tinnitus retraining therapy has shown success in reducing the adverse effects of tinnitus. This method aims to "disconnect" the patient from the tinnitus while subjecting him or her to moderate levels of sounds to reverse the effect of sound deprivation on the function of the central nervous system.

Some studies also report on morphologic changes in the central auditory nervous system as a result of sound deprivation. These studies can be taken as evidence of neural plasticity.

> After the auditory nerve is severed, considerable morphologic changes develop in the cochlear nucleus [143]. These changes are most prominent in developing animals where cell size of the cochlear nucleus neurons and size of the cochlear nucleus are reduced. Changes in cells of nuclei of more centrally located structures of the ascending auditory pathway have also been demonstrated. Raising animals in a noise-free (sound-free) environment or reducing the sound input by occluding the ear canals causes similar changes in the nuclei of the ascending auditory pathway. In the newborn mouse Webster and Webster [186] showed that after 45 days of such sound deprivation, the cross-sectional areas of cells were reduced in the ventral cochlear nuclei, medial nuclei of the trapezoidal body, and central nuclei of the inferior colliculus, but no change was seen in the central part of the dorsal cochlear nucleus. These results were confirmed by other investigators using other animal species (rat and cat). Thus, Powell and Erulkar [137] found similar transneural degeneration in the ventral cochlear nuclei, ipsilateral superior olivary nuclei and the contralateral medial nuclei of the trapezoidal body and the contralateral lateral lemniscus. These studies show the greatest changes occur when sound deprivation occurs during postnatal life when the animal is immature, but even the mature nervous system has a considerable degree of plasticity.

Changes in Temporal Integration

Animal experiments have indicated that tinnitus may be associated with changes in temporal integration in the auditory nervous system. Temporal integration in the central auditory nervous system causes the threshold of hearing to decrease when the duration of a stimulus sound is increased and it causes long sounds to become louder than short ones of the same (physical) intensity. Normally, the time constant for temporal integration is about 100 ms for sounds above threshold and about 200 ms for sounds near threshold.[3] Less temporal integration means that the threshold of hearing decreases less with increasing duration of sounds.

Gerken et al. [57] have shown that temporal integration in the auditory nervous system is altered in animals (cats), with hearing losses of approximately 50 dB induced by noise exposure. These experiments were conducted by determining the (behavioral) threshold to electrical stimulation of the cochlear nucleus and the inferior colliculus using implanted electrodes. The threshold normally decreases exponentially with the number of impulses presented be-

---

[3] The numerical value of a time constant for temporal integration is usually defined as the time it takes for a quantity to decay to an amplitude of $1/e$ (e is the base of the natural logarithm). These definitions are taken from engineering terminology and assume an exponential decay or rise. In psychoacoustics, temporal integration is often expressed as the difference between threshold of tones of different duration, e.g., 20 and 200 ms (in decibels) [139].

cause of temporal integration. After hearing impairment the threshold was lower both for cochlear nucleus and inferior colliculus stimulation and it did not decrease with increasing the number of impulses, a sign of increased excitability and a change in temporal integration. Gerken *et al.* [57] concluded that the neural basis for temporal integration in the auditory nervous system is located in the inferior colliculus and that it can be affected by deprivation of auditory input. It is apparent from these experiments that the decrease in threshold as a result of sound deprivation is larger for short-duration sounds than for long-duration sounds.

Importance of Balance between Inhibition and Excitation

The inhibitory influence of neurons from other neural structures on neurons in the inferior colliculus is especially strong and it has been reported that the firing rate of some neurons may increase as much as 700% when the inhibition is released [134]. The inhibition is mediated by the neural transmitter gamma-amino butyric acid (GABA) and in these experiments the effect of GABA was reduced or eliminated by applying bicuculline, a $GABA_A$ antagonist, to the inferior colliculus. The effect of a decrease in GABAergic inhibition in the IC may therefore have a large effect and could produce hyperactivity in neurons in the IC and at more rostral structures of the ascending auditory pathway. The hypothesis that the inferior colliculus is involved in tinnitus has received considerable support recently.

> Specific components of the auditory evoked responses from the inferior colliculus are related to GABAergic inhibition [171]. In response to tone bursts the amplitude of these components decrease with increasing duration, indicating that the inhibition increases with increasing duration of the stimuli. After the animals were exposed to loud noise that presumably resulted in tinnitus, this specific component of the evoked potential decreased much less when the duration of the stimuli was increased, indicating a reduction in inhibition. The decrease of the response as a function of stimulus duration could be restored by administering agents that enhanced GABAergic inhibition, namely clonazepam, and one of the two forms of baclofen, $(-)$-baclofen, a known $GABA_B$ agonist [172].
>
> Benzodiazepines are generally known to bind to specific gamma-amino butyric acid ($GABA_A$) receptors and thus enhance the inhibitory effect of GABA, which in turn enhances inhibition in the central nervous system. Baclofen modulates the $GABA_B$ receptors. The fact that benzodiazepines can alleviate tinnitus in some individuals supports the hypothesis that GABAergic inhibition may be altered in some tinnitus patients and clearly points toward involvement of the central nervous system.

That the benzodiazepine clonazepam is an effective treatment of tinnitus in some patients but not in others supports the hypothesis that tinnitus is a group of diseases and not a single disease. It also supports the hypothesis that weakened GABAergic inhibition is involved in tinnitus. Clonazepam and diazepam have different effects on tinnitus and on inhibition in the inferior colliculus [172], which is the anatomical location that has been suspected to be involved in generation of at least some forms of tinnitus. Baclofen, a $GABA_B$ agonist, restores GABAergic inhibition in the inferior colliculus, but baclofen has not proven to be effective in treatment of tinnitus.

The factor of gender may have to do with the fact that female reproductive hormones can modulate GABAergic transmission [46]. The levels of these hormones vary throughout the menstrual cycle of women of reproductive age. It is possible that the resulting (cyclic) variation in the potency of some GABA receptors can facilitate recovery from the changes in the central nervous system that cause tinnitus. Although tinnitus and other hyperactive disorders may occur in young individuals, these disorders are more frequent in elderly individuals. This may be because GABA production decreases with age at a faster rate than does that of excitatory neural transmitters [28], which results in a reduction of inhibition.

Importance of the Temporal Pattern of Neural Discharges

Temporal coherence of the discharges of many nerve fibers may communicate information to the central nervous system about the presence of a sound and of its intensity [41, 106]. The central nervous system may use information about how many nerve fibers carry phase-locked neural activity to detect the presence of a sound and perhaps to determine the intensity of a sound. Low-frequency sounds cause neural activity in many nerve fibers to become phase locked to the waveform of the sound. When the neural activity in many nerve fibers becomes phase locked to the same sound, the activity of each such fiber also becomes phase locked to the other's neural activity. The number of nerve fibers that are activated by a sound and thus become phase locked to the sound is a function of the width of the spectrum of the sound and its intensity. This information may be used by the auditory nervous system to detect the presence of a sound and its intensity.

It has been hypothesized that slight injury to the auditory nerve could facilitate abnormal cross-talk between axons of the auditory nerve and cause pathologic phase locking of neural activity in groups of nerve fibers [106]. Such pathologic cross-transmission (ephaptic transmission) between nerve fibers could occur when the myelin sheath becomes damaged. Normally occurring spontaneous activity could thereby become phase locked, which would mimic the response to sounds that activate many nerve fibers and that would be interpreted by the central nervous system as a sound being present even in quiet surroundings, thus tinnitus. Eggermont [41] has elaborated on that hypothesis and extended it. He has suggested that cross-talk between the synapses of the hair cells of the cochlea might also cause the neural activity in many auditory nerve fibers to become correlated.

Abnormalities in the temporal pattern of the discharges in single auditory nerve fibers in animals has been demonstrated by several investigators. While this in itself may not be sufficient to explain severe tinnitus, it may promote (plastic) changes in more central structures of the auditory pathway. Schreiner and Snyder [151] showed that electrical (neurophonic) potentials recorded from the auditory nerve in animals that were treated with salicylate had a peculiar spectral component near 200 Hz that was assumed to be related to abnormal spontaneous neural activity in auditory nerve fibers caused by administration of the salicylate [38]. The 200-Hz

peak in the spectrum of the auditory nerve activity disappeared after administration
of lidocaine, a local anesthetic agent that is known to decrease tinnitus in some
patients. It is not known, however, what this 200-Hz peak in the spectrum of
auditory nerve activity represents, but in the absence of other objective signs of
tinnitus, it is appreciated as at least something that can be measured.

It has thus been found consistently that both morphologic and electrophysiologic changes may occur in several animal species after sound deprivation.
This emphasizes the importance of avoiding sound deprivation such as that
which would occur in children who have long-standing middle ear infections
or other disorders that impair sound conduction to the cochlea. It is particularly
important to avoid sound deprivation during development, when the nervous
system is immature and more pliant.

INVOLVEMENT OF THE SYMPATHETIC NERVOUS SYSTEM IN TINNITUS

Cochlear hair cells have an abundance of sympathetic innervation [35]. It is
conceivable that increased sympathetic activation can increase the sensitivity
of cochlear hair cells so that neural activity that resembles that generated by
a sound occurs in absence of sound. Stress activates the sympathetic nervous
system, which indicates that activation of the sympathetic nervous system can
aggravate tinnitus. Animal experiments have shown little effect of activation
of the sympathetic nervous system on auditory neural activity [93, 131], but
these studies were made in normal animals and it is possible that the sympathetic nervous system may affect the function of the auditory system when
specific pathologies are present.

The fact that sympathectomy or blockage of a cervical sympathetic ganglion
(the stellate ganglion) can alleviate tinnitus in some patients [55] supports
the hypothesis that the sympathetic nervous system is involved in some forms
of tinnitus. Deactivation of the sympathetic nervous system causes an increase
of blood flow in the ear by vasodilatation, which may improve hearing by
10–20 dB at low frequencies (125–500 Hz), but not at high frequencies (1000–
4000 Hz). It might be inferred that the beneficial effect of a sympathetic block
on tinnitus could be the result of restoring the blood supply to the ear but it
seems more likely that the origin of at least some of the effect is central. Other
investigators [1] have found that tinnitus in patients with Ménière's disease
could either increase or decrease because of sympathectomy.

HYPERACUSIS AND PHONOPHOBIA

Severe tinnitus is often accompanied by hyperacusis, which is a distorted
perception of sounds, and some tinnitus suffers perceive many sounds, even

sounds of moderate intensity, to be painful (phonophobia). Individuals with these forms of abnormal perception of sound often experience severe discomfort from sound, most pronounced for loud sounds, but often even sounds at physiologic levels give a feeling of discomfort. Hyperacusis and phonophobia greatly add to the discomfort of tinnitus and oftentimes a patient may consider their effects to be worse than the tinnitus. The terms hyperacusis and phonophobia are used to describe different forms of abnormal perception of sounds and reactions to sounds, while abnormal perception of loudness of sounds has also been labeled dysacusis [130].

HYPERACUSIS

According to *Stedman's Medical Dictionary* hyperacusis (hyperacusia) is:

> an abnormal acuteness of hearing due to increased irritability of the sensory neural mechanism. Syn: auditory hyperesthesia. Origin [hyper- + G. *akousis*, a hearing].

PHONOPHOBIA

According to *Stedman's Medical Dictionary* phonophobia is:

> Morbid fear of one's own voice, or of any sound. Origin [phono- + G. *phobos*, fear].

There are a few disorders that are associated with hyperacusis in the absence of tinnitus. Williams disease is one and Lyme disease is another. Autism is often associated with discomfort from loud sounds (phonophobia). Hyperacusis of loud sounds occurs after paralysis of the stapedius muscle or severance of the stapedius tendon that can occur during stapedectomy because the middle ear reflex no longer causes attenuation of loud sounds (see Chapter 12). People usually adapt to that condition after a short time. A form of hyperacusis, also known as recruitment of loudness, occurs in individuals with hearing loss from loss of outer hair cells. This is due to decreased impairment or loss of the function of the cochlear amplifier (see Chapter 3).

The abnormalities in processing of sound that cause hyperacusis and phonophobia are ascribed to functional changes in the processing of sound in the central auditory nervous system. While hyperacusis caused by absence of the acoustic middle ear reflex is not accompanied by tinnitus, hyperacusis of cochlear origin may be, and hyperacusis of central origin most always occurs together with tinnitus.

PATHOPHYSIOLOGY OF HYPERACUSIS AND PHONOPHOBIA

Little is known about the pathophysiology of abnormal loudness perception that may occur together with severe tinnitus, but it is often assumed that the mechanisms are similar. Reorganization of the central nervous system as is assumed to cause severe tinnitus has also been mentioned as a cause of hyperacusis and phonophobia. The pathophysiology of abnormal loudness perception that occurs together with the absence of the acoustic middle ear reflex and loss of outer hair cells of the cochlea are better understood.

HYPERACUSIS AND PHONOPHOBIA FROM CENTRAL CAUSES

While the word hyperacusis may accurately describe the abnormal loudness perception that individuals with absence of the acoustic middle ear reflex and cochlear hearing loss experience, the abnormal loudness perception that many individuals with tinnitus of central origin experience is different and not primarily a perception that sounds are too loud. Rather, such individuals often have an exaggerated perception of a wide variety of sounds. To persons with normal hearing, only certain exceptionally loud sounds are perceived as an annoyance. Therefore, the abnormal perception of sounds that many individuals with severe tinnitus experience is perhaps better described as an exaggerated annoyance from sounds rather than as increased loudness, as would be implied from the use of the word hyperacusis. It has been suggested that the word "phonophobia" be used to describe that kind of abnormal perception of sounds [130], which would make it compatible with "photophobia," which is often experienced in connection with head injuries and emphasizes that sensory stimuli may evoke abnormal emotional reactions. There is considerable evidence that deprivation of input to the central nervous system can promote a reorganization that may result in symptoms of dysfunction of sensory and motor system. Reorganization and change of synaptic efficacy in the auditory nervous system is likely to affect temporal integration. Changes in temporal integration may explain why some tinnitus individuals experience greater hypersensitivity to impulsive sounds and some of the unpleasantness of impulse sounds that such people experience (phonophobia). Reorganization of the auditory nervous system may also include opening of dormant synapses so that new functional connections are established, such as to the nonclassical auditory system. It has been suggested that the adverse perception of sound may result from activation of the nonclassical auditory pathway [113], which connects to the limbic system and thus makes sounds elicit fear and other emotional reactions, thus phonophobia.

Cochlear Origin of Hyperacusis

Recruitment of loudness describes an abnormal rapid rise of the sensation of loudness above an elevated hearing threshold due to loss of outer hair cells. As was discussed in Chapter 14, loss of outer hair cells impairs positive feedback ("motors") in the cochlea (see Chapter 3), which is associated with loss of sensitivity and broadening of tuning. The latter may contribute to the distorted perception of loudness because it implies that sounds activate more auditory nerve fibers than in the normally functioning ear because an increase of the sound intensity above the hearing threshold causes a rapidly increasing overlap between tuning of individual nerve fibers. This means that a small increase of sound intensity causes activation of many new nerve fibers.

In addition to loss of sensitivity, cochlear hearing loss is also associated with impairment of the automatic gain control that normally is present in the cochlea. This causes the spike rate versus sound intensity in individual auditory nerve fibers to become steeper than normal and could be a reason for the abnormal loudness perception that results from loss of outer hair cells. Another hypothesis emphasizes the importance of the temporal pattern of neural discharges in auditory nerve fibers and suggests that the degree of temporal coherence of nerve impulses in the auditory nerve determines the threshold as well as the growth of loudness [106]. Nevertheless, these models will most likely be revised considerably as more information about the function of the normal and the pathologic ear accumulates.

Hyperacusis from Absence of the Acoustic Middle Ear Reflex

Impaired function of the acoustic middle ear reflex is a cause of abnormal perception of loud sounds, thus hyperacusis. The acoustic middle ear reflex acts as an automatic gain control of the middle ear and thus reduces the input to the cochlea for sounds above the physiological sound levels (see Chapter 12). When the reflex is not functioning properly, abnormally high sound intensities may be conducted to the cochlea because the attenuation of loud sound by the action of the middle ear reflex that normally occurs is absent. The most common cause of absence of the acoustic middle ear reflex is a result of facial nerve dysfunction such as occurs in Bell's palsy. Severance of the stapedius tendon that occurs during stapedectomy also renders the acoustic middle ear reflex nonfunctional. People usually adapt to the absence of the middle ear reflex within a few weeks.

Hyperacusis (Phonophobia) that Occurs Together with Other Diseases

Besides Bell's palsy and cochlear hearing loss, hyperacusis ("phonophobia") often occurs in connection with head trauma, perilymphatic fistulae, endolymphatic hydrops, viral infections such as the Ramsay–Hunt syndrome, and Lyme disease. One genetic disorder, Williams–Beuren syndrome, is associated with a high incidence of hyperacusis or phonophobia [4, 20, 84].

> Williams–Beuren syndrome (WBS) is also known as infantile hypercalcemia. It is characterized by high blood levels of calcium, believed to be caused by hypersensitivity to vitamin D. Williams–Beuren syndrome produces multiple congenital anomalies, cardiovascular disorders, prenatal and postnatal growth retardation, facial abnormalities, and mental retardation including poor visuospatial skills but relatively preserved verbal skills, loquacity (talkativeness), hyperactivity, and hyperacusis. Reports of the incidence differ between investigators from 1 in 20,000 live births [20] to 1 in 50,000 [4], thus a noticeable incidence. As many as 95% of individuals with WBS have hyperacusis and react adversely to sounds of moderate intensity [84]. Individuals with WBS also have a high incidence of otitis media but their hyperacusis seems to be unrelated to that. It is not known if children with WBS react to sounds because of the physical nature of the sounds or because what they associate the sounds with is perceived as frightening.

SIMILARITIES BETWEEN SEVERE TINNITUS, HYPERACUSIS, PHONOPHOBIA, AND CHRONIC PAIN

Studying other diseases that have similarities with tinnitus might help in better understanding tinnitus, hyperacusis, and phonophobia. Chronic pain and severe tinnitus have many similarities. Tonndorf [174] was one of the first to formulate a distinct hypothesis connecting similarities between chronic pain and tinnitus. He alluded to the hypothesis of Melzack and Wall [103], which postulates that stimulation of large-diameter nerve fibers of peripheral nerves (A-fibers) can modulate pain impulses on their way to the brain (the gating hypothesis of pain).

In addition to chronic pain, such individuals often have an altered perception of normally innocuous somatosensory stimuli, which is known as allodynia. The reaction to painful stimuli is often exaggerated in individuals with chronic pain, which is known as hyperpathia. Another phenomenon that is commonly experienced by individuals with chronic pain is the "wind-up" phenomenon, which refers to the sensation of painful stimuli increasing when the stimulation is repeated. The "wind-up" phenomenon is a form of temporal summation.

All three phenomena have similarities with severe tinnitus. As we discussed above, many individuals with severe tinnitus have an altered perception of (physical) sounds, known as hyperacusis and phonophobia. Hyperacusis and phonophobia are similar to allodynia in that a person perceives some types of normal sounds as being painful. Strong sounds that individuals without tinnitus may perceive as being unpleasant are perceived as excessively painful by tinnitus patients, thus similar to hyperpathia. Repeating an unpleasant sound may be felt as excessively unpleasant to a person with severe tinnitus, thus similar to the wind-up phenomenon in chronic pain. Chronic pain often has an emotional component that is not present in acute pain, thus another similarity with severe tinnitus and phonophobia.

Other similarities between severe tinnitus and chronic pain are related to the way the abnormal neural activities that cause these sensations are generated. Thus, the anatomical location of the physiologic abnormality that causes chronic pain and severe tinnitus is not always the location to which the sensation of pain or tinnitus is referred. Phantom pain and tinnitus in deaf people or in people with severed auditory nerve provide the most convincing evidence of this [77]. Phantom pain or phantom sensations are often associated with amputated limbs. These patients localize the pain to somewhere within the amputated limb. Similarly, tinnitus is referred to the ear because it generates a similar sensation as sounds that reach the ear. In individuals in whom the auditory nerve is severed the abnormal neural activity that causes the tinnitus must be generated within the central auditory nervous system.

Contemporary Hypothesis about the Role of Neural Plasticity in Chronic Pain

Hypotheses about the development and maintenance of chronic pain assume that the somatosensory system maintains its plasticity into adulthood so that outside events may cause reorganization of the somatosensory nervous system. The prevailing hypothesis about the development of chronic pain postulates that certain neurons in the spinal cord, known as wide dynamic range (WDR) neurons, become sensitized so they respond to their normal input from large diameter sensory fibers with an abnormally high firing rate, which causes these WDR neurons to activate pain pathways. Sensitization of the somatosensory system may occur as a result of a reduction of inhibitory inflow from the periphery to those neurons; another cause is by sensitization of the mechanoreceptors that supply the input to those neurons.

The WDR neurons receive input from different kinds of mechanoreceptors in skin, joint, and muscle receptors and from pain and temperature receptors. Some of this input is excitatory and some is inhibitory. It is hypothesized that these WDR neurons may become hyperactive and hypersensitive either because of a decrease in the inhibitory input or as a result of an increased excitatory input. Normally the output of the WDR neurons travels in the ascending somatosensory pathway to the somatosensory cerebral cortex via the dorsal column nuclei and the thalamic relay neurons. When the WDR neurons become hyperactive, their output seems also to connect to neural circuits that normally serve the communication of pain sensations. This probably occurs as a result of opening dormant synapses.

Similar hypotheses have been presented for trigeminal neuralgia (TGN) (face pain) [52], where it has been assumed that the irritation of the trigeminal nerve root causes a persistent input to the trigeminal nucleus that over time makes the nucleus hyperactive because of increased excitability and/or a decrease of segmental inhibition. Typically, patients with TGN can trigger a pain attack by lightly touching the face, and not necessarily the exact region where the pain is. The activity that touch generates is communicated by large-diameter nerve fibers (A-fibers), and it has been assumed that a similar chain of events causes an elicitation of a pain attack as described above for pain related to the spinal cord.

Involvement of the Sympathetic Nervous System in Chronic Pain

When pain activates the sympathetic nervous system, a vicious cycle may be started, where sympathetic activity creates more sympathetic activity. That sequence of events is assumed to be responsible for such chronic pain conditions as causalgia and reflex sympathetic dystrophy (RSD). The mechanism of RSD seems to be an excessive liberation of norepinephrine from the sympathetic nerve terminals that are located in the skin near mechanoreceptors. Norepinephrine increases the sensitivity of mechanoreceptors and excess norepinephrine may cause these mechanoreceptors to initiate neural activity without any mechanical stimulation.

Thus, many studies agree that neural plasticity plays an important role for the development of both chronic pain and severe tinnitus, including hyperacusis and phonophobia. Earlier it was assumed that neural plasticity was limited to a young age but it has now become evident that the central nervous system maintains the ability to reorganize and change its function throughout adult life, although to a lesser extent than what is the case in younger people.

TREATMENT OF TINNITUS, HYPERACUSIS, AND PHONOPHOBIA

Many different treatments for tinnitus have been tried. The fact that administration of the local anesthetic lidocaine can totally abolish tinnitus in some individuals has been encouraging. Lidocaine is primarily thought of as a sodium channel blocker but it has numerous other effects and it has not been possible to determine which of these is effective in treating tinnitus. It was originally thought that lidocaine acted on cochlear hair cells but it may in fact be its effect on the central nervous system that suppresses tinnitus. As lidocaine can only be administrated intravenously, it is not a practical treatment for tinnitus. A similar drug, which can be administrated orally, tocainide, has not been shown to be effective and produces considerable side effects. Other medical treatments aim at restoring the balance between inhibition and excitation in the brain. These involve the use of benzodiazepines, GABA$_A$ receptor agonists. A GABA$_B$ agonist, baclofen, has also been tried, but with little practical success.

ELECTRICAL STIMULATION OF THE EAR

Electrical current (DC) passed through the cochlea can reduce tinnitus in some patients [29]. These investigators placed an electrode on the round window or the promontorium and passed a positive current through the cochlea. In one study [29] six of seven individuals with tinnitus obtained relief. It was assumed that the electrical current passed through the cochlea affected the hair cells so that the spontaneous activity in auditory nerve fibers decreased. However, the electrical current could also have affected the auditory nerve and it might have had its effect by stimulating the trigeminal nerve, thus activating the somatosensory system and thereby modulating the activity in the nonclassical pathways. Transcutaneous nerve stimulation seems to help only some individuals with tinnitus [152]. Despite these positive effects, electrical stimulation never became widely used in treatment of individuals with tinnitus.

In deaf people with tinnitus, electrical stimulation of the auditory nerve provided by a cochlear implant can relieve tinnitus [101, 159].

Surgical Treatment

There are mainly two kinds of surgical treatment of tinnitus: severance of the auditory nerve and intracranial microvascular decompression of the auditory nerve. These operations have been performed by only a few surgeons. Severing the auditory nerve can alleviate tinnitus in many patients with Ménière's disease. As early as 1941 the neurosurgeon Dandy reported relief of tinnitus in about 50% of patients with Ménière's disease after intracranially sectioning the VIIIth nerve cranial. Labyrinthectomy and translabyrinthine sectioning of the VIII cranial nerve has been done in patients with vertigo and tinnitus. Pulec reported that 70% of the patients were free of tinnitus after such operations [138]. Other surgeons have reported similar results. The results from sectioning of the VIII cranial nerve regarding tinnitus are generally better in patients with both vertigo and tinnitus [67, 73].

Sympathectomy or blockage of a cervical sympathetic ganglion (the stellate ganglion) has been used to treat tinnitus [55]. Microvascular decompression of the auditory portion of the VIIIth cranial nerve to treat tinnitus was suggested as a treatment 2 decades ago [76, 86, 119], but it is only recently that results of such treatment of many patients have been published [121]. Microvascular decompression of cranial nerves is an established treatment for disorders such as hemifacial spasm, trigeminal neuralgia, glossopharyngeal neuralgia, and certain forms of vertigo. Microvascular decompression of the VIIIth cranial nerve for tinnitus has a much lower success rate than that of trigeminal neuralgia and hemifacial spasm. The success rate of MVD for tinnitus is about 40% for total relief or significant relief [121]. The success rate of microvascular decompression for trigeminal neuralgia and hemifacial spasm is approximately twice of that [8, 9]. The success rate of microvascular decompression for tinnitus was very different for men and women. For men it was 29.3% relief, while for women it was 54.8%. The success rate of MVD for TN and HFS in men and women is, however, only slightly different, about 85%. The success rate for MVD as a cure for tinnitus was also related to how long time a patient had had tinnitus. Those who experienced total relief of their tinnitus or marked improvement had had their tinnitus for only 2.9–2.7 years respectively, but those who experienced only a slight improvement or no improvement at all had their tinnitus for an average of between 5.2 and 7.9 years. The diminished improvement may be due to plastic changes in the auditory system becoming permanent. These facts are important in selecting tinnitus patients for MVD. Patients who have MVD of their auditory nerve and who become free of tinnitus also were relieved of their hypersensitivity to sound, which indicates that tinnitus and hypersensitivity to sound may be caused by the same pathology [121]. The success rate for MVD is higher in patients with unilateral tinnitus than with bilateral tinnitus [180].

MULTIPLE FACTORS ARE NECESSARY TO CAUSE OF TINNITUS AND HYPERACUSIS

As has been pointed out earlier in this book, there is rarely a disease with only a single cause and most disorders are a result of multiple pathologies. Tinnitus is no exception and attempts to find *the* cause of tinnitus are therefore futile. For example, some forms of tinnitus can be cured by moving a blood vessel off the intracranial portion of the auditory nerve (microvascular decompression), but vascular compression of the auditory nerve is common, so it is incorrect to assume that vascular compression alone causes tinnitus. The fact that MVD operations can cure some forms of tinnitus indicates that vascular compression is a component in causing tinnitus. Further, the fact that many people have vascular compression but no tinnitus indicates that vascular compression is not sufficient to cause tinnitus. Vascular compression is thus only one of several factors that are necessary (but not sufficient) to cause tinnitus. This may also apply to other forms of tinnitus and, for that matter, to many other diseases. Thus, instead of attempting to find *the* cause of a certain form of tinnitus it may be more productive to try to identify a combination of factors, each of which may not cause any symptom alone. The inability to comprehend and deal with phenomena that depend on several causes may explain why it is common to find the diagnosis of "idiopathic tinnitus," i.e., "tinnitus of unknown origin."

If it is found that a certain disease is produced by two or more distinct abnormalities, it may mean that two or more physiological abnormalities are necessary to cause a specific disorder to become manifest. Attempts to exclude one is the natural reaction, yet it is probably more productive to assume that both are present. Thus, in the case of tinnitus, it might be more productive to try to find which two or more abnormalities are present, which of course, is a far more complex task.

Disorders that can be cured by microvascular decompression of a cranial nerve are examples of diseases that *only* become manifest when *multiple factors* are present. These disorders [hemifacial spasm, (HFS) and trigeminal neuralgia, (TGN)] are extremely rare, with incidences of 0.8 and 5 per 100,000 respectively [108], yet vascular compression is rather common, as evidenced by autopsies material. Vascular compression of the Vth and VIIth cranial nerves occurs frequently (in as much as 50%) in individuals who have no specific symptoms [100, 127, 168].

Thus it cannot be claimed that vascular compression *causes* these disorders. Evidence from treatment using microvascular decompression of TGN and HFS shows a success rate of about 85%; strong evidence that vascular compression plays a role in these disorders. Vascular compression must be *necessary* but not *sufficient* to cause a disease because many people have vascular compression without any signs of disease. Removal of one factor such as vascular compression can effectively cure the patient because that factor is necessary to cause the symptoms. This explains why TGN can be effectively cured by decompression of the Vth cranial nerve as well as by medicine. The other factor(s) that are *necessary* are unknown and they do not yield symptoms [108].

The fact that patients who were operated on for HFS by MVD of the VIIth cranial nerve often had a blood vessel in contact with the auditory nerve but did not have tinnitus may be an example of a situation where only one of several *necessary* factors were present for symptoms of tinnitus to become manifest. Attempts to understand a certain form of tinnitus on basis of a single pathophysiology are therefore not productive and abandoning the search for a single cause of tinnitus may resolve some current confusions. The finding that more than one treatment can effectively relieve the symptoms of tinnitus supports both the multifactor hypothesis and the assumption that tinnitus is a group of different diseases and complicates matters further.

It often happens that a drug that has shown promise in a pilot study or by individual physicians fails when subjected to the rigors of standard evaluation procedures, such as double-blind tests. Individuals with tinnitus are not a homogenous group with regard to pathology and one drug may be effective in some individuals but not in others. This can have serious implications in testing of the efficacy of drugs using double-blind studies. Thus, if a cohort of tinnitus patients that compose a test group have, for example, patients with three different pathologies, three different treatments might be needed to be effective. If any one of these treatments is tested alone on such an inhomogeneous group it may be impossible to obtain significant results, even in the situation where each of the treatments tested were effective in treating tinnitus of one particular kind. A physician can try different treatments in an individual patient and thus achieve good results. Unfortunately, the negative results of such double-blind studies may discourage the use of an effective treatment because of the great reliance on the results of double-blind studies.

If two (or more) factors are necessary to cause symptoms, a treatment that affects one factor may cure the disease. One treatment may eliminate one factor and another treatment may eliminate another factor. Combining two or more treatments that affect different "causes" of a disease may be the most effective therapy because they will have an additive effect and maybe even a synergistic effect. However, development of such combinatorial treatment is hampered by difficulties in testing efficacy and the deeply rooted conception that diseases have a single cause and that the effective treatment is a single drug.

SECTION IV REFERENCES

1. Adams, D. A., and Wilmot, T. J. (1982). Long-term results of sympathectomy. *J. Laryngol. Otol.* **96**:705–710.
2. Ahdab-Barmada, M., and Moossy, J. (1984). The neuropathology of kernicterus in the premature neonate: Diagnostic problems. *J. Neuropathol. Exp. Neurol.* **43**:45–56.
3. Antonelli, A. R., and Calearo, C. (1964). Drug effects on the auditory speech discrimination mechanisms. *Acta Otolaryngol. (Stockh.)* **58**:105.
4. Arnold, R., Yule, W., and Martin, N. (1985). The psychological characteristics of infantile hypercalccaemia: A preliminary investigation. *Dev. Med. Child Neurol.* **27**:49–59.
5. Auger, R. G., and Whisnant, J. P. (1990). Hemifacial spasm in Rochester and Olmsted County, Minnesota, 1960 to 1984. *Arch. Neurol.* **47**:1233–1234.
6. Babighian, G., Moushegian, G., and Rupert, A. L. (1975). Central auditory fatique. *Audiology* **14**:72–83.
7. Bach, S., Noreng, M. F., and Thellden, N. U. (1988). Phantom limb pain in amputees during the first 12 months following limb amputation, after preoperative lumbar epidural blockade. *Pain* **33**:297–301.
8. Barker, F. G., Jannetta, P. J., Bissonette, D. J., Shields, P. T., and Larkins, M. V. (1995). Microvascular decompression for hemifacial spasm. *J. Neurosurg.* **82**:201–210.
9. Barker, F. G., Jannetta, P. J., Bissonette, D. J., Larkins, M. V., and Jho, H. D. (1996). The long-term outcome of microvascular decompression for trigeminal neurologic. *N. Engl. J. Med.* **334**:1077–1083.
10. Beck, C., and Schmidt, C. L. (1978). 10 years of experience with intratympanically applied streptomycin (Gentamycin) in the therapy of Morbus Ménière. *Arch. Oto-Rhino-Laryngol.* **221**:149–152.
11. Békésy, von G. (1936/1960). *In* Békésy, von G. "Experiments in Hearing." McGraw–Hill, New York.
12. Bernstein, J. M. (1988). Middle ear mucosa: Histological, histochemical, immunochemical, and immunological aspects. *In* A. F. Jahn and J. Santos-Sacchi (Eds.), "Physiology of the Ear" Raven, (pp. 59–80). New York.
13. Bocca, E. (1958). Clinical aspects of cortical deafness. *Laryngoscope* **68**:301.
14. Bocca, E. (1965). Distorted speech tests. *In* B. A. Graham (Ed.), "Sensory-Neural Hearing Processes and Disorders." Little, Brown, Boston.
15. Bocca, E., Calearo, C., and Cassinari, V. (1954). A new method for testing hearing in temporal lobe tumours. *Acta Otolaryngol. (Stockh.)* **44**:219.

16. Borg, E. (1982a). Noise induced hearing loss in rats with renal hypertension. *Hear. Res.* 8:93–99.

17. Borg, E. (1982b). Noise induced hearing loss in normotensive and spontaneously hypertensive rats. *Hear. Res.* 8:117–130.

18. Borg, E., Canlon, B., and Engstrom, B. (1995). Noise-induced hearing loss: Literature review and experiments in rabbits. *Scand. Audiol.* 24 (Suppl. 40):1–147.

19. Borg, E., and Møller, A. R. (1978). Noise and blood pressure: Effects on lifelong exposure in the rat. *Acta Physiol. Scand.* 103:340–342.

20. van Borsel, J., Curfs, L. M. G., and Fryns, J. P. (1997). Hyperacusis in Williams Syndrome: A sample survey study. *Genet. Counsel.* 8:121–126.

21. Bunch, C. C. (1943). "Clinical Audiometry." Henry Kimpton, London.

22. Burns, W., and Robinson, D. W. (1970). "Hearing and Noise in Industry." Her Majesty's Stationery Office, London.

23. Caiazzo, A. J., and Tonndorf, J. (1977). Ear canal resonance and TTS. *J. Acoust. Soc. Am.* 61:S78.

24. Calearo, C., and Antonelli, A. R. (1963). "Cortical" hearing tests and cerebral dominance. *Acta Otolaryngol. (Stockh.)* 56:17.

25. Campbell, K. C. M., and Abbas, P. J. (1994). Electrocochleography with postural changes in perilymphatic fistula. *Annals Otol. Rhin. Laryngol.* 103:474–482.

26. Campbell, K. C. M., Harker, L. A., and Abbas, P. J. (1992). Interpretation of electro-cochleography in Ménière's disease normal subjects. *Ann. Otol. Rhin. Laryngol.* 101:496–500.

27. Canlon, B., Borg, E., and Flock, A. (1988). Protection against noise trauma by pre-exposure to a low level acoustic stimulus. *Hear. Res.* 34:197–200.

28. Caspary, D. M., Raza, A., Lawhorn Armour, B. A., Pippin, J., and Arneric, S. P. (1990). Immunocytochemical and neurochemical evidence for age-related loss of GABA in the inferior colliculus: Implications for neural presbycusis. *J. Neuronosci.* 10:2363–2372.

29. Cazals, Y., Negrevergne, M., and Aran, J. M. (1978). Electrical stimulation of the cochlea in man: Hearing induction and tinnitus suppression. *J. Am. Audiol. Soc.* 3:209–213.

30. Celestino, D., and Ralli, G. (1991). Incidence of Ménière's disease in Italy. *Am. J. Otol.* 12:135–138.

31. Charabi, S., Thomsen, J., Tos, M., Charabi, B., Mantoni, M., and Brgesen, S. E. (1998). Acoustic neuromavestibular schwannoma growth: Past, present and future. *Acta Otolaryngol. (Stockh.)* 118(3):327–332.

32. Chen, G., and Jastreboff, P. J. (1995). Salicylate-induced abnormal activity in the inferior colliculus of rats. *Hear. Res.* 82:158–178.

33. Cody, A. R., and Johnstone, B. M. (1980). Single auditory neuron response during acute acoustic trauma. *Hear. Res.* 3:3–16.

34. Dallos, P., and Cheatham, M. A. (1976). Compound action potential tuning curves. *J. Acoust. Soc. Am.* 59:591–597.

35. Densert, O. (1974). Adrenergic innervation in the rabbit cochlea. *Acta Otolaryngol. (Stockh.)* 78:345–346.

36. Dobie, R. A. (1993). "Medical–Legal Evaluation of Hearing Loss." van Nostrand–Reinhold, New York.

37. Dolan, T. R., Ades, H. W., Bredberg, G., and Neff, W. D. (1975). Inner ear damage and hearing loss after exposure to tones of high intensity. *Acta Otolaryngol. (Stockh.)* 80:343–352.

38. Dolan, D. F., Nuttall, A. L., and Avinash, G. (1990). Asynchronous neural activity recorded from the round window. *J. Acoust. Soc. Am.* 87:2621–2627.

39. Dublin, W. B. (1985). The cochlear nuclei pathology. *Otolaryngol. Head Neck Surg.* 93:448–463.

40. Eggermont, J. J. (1979). Summating potentials in Ménière's disease. *Arch. Oto-rhinolaryngol* **222**:63–75.

41. Eggermont, J. J. (1990). On the pathophysiology of tinnitus: A review and peripheral model. *Hear. Res.* **48**:111–124.

42. Eggermont, J. J., and Kenmochi, M. (1998). Salicylate and quinine selectively increase spontaneous firing rates in secondary auditory cortex. *Hear. Res.* **117**:149–160.

43. Ehret, G., and Romand, R. (Eds.) "The Central Auditory Pathway." Oxford Univ. Press, New York.

44. El Barbary, A. (1991a). Auditory nerve of the normal and jaundiced rat. I. Spontaneous discharge rate and cochlear nerve histology. *Hear. Res.* **54**:75–90.

45. El Barbary, A. (1991b). Auditory nerve of the normal and jaundiced rat. II. Frequency selectivity and two-tone rate suppression. *Hear. Res.* **54**:91–104.

46. Elkind-Hirsch, K. E., Stoner, W. R., Stach, B. A., and Jerger, J. F. (1992). Estrogen influences auditory brainstem responses during the normal menstrual cycle. *Hear. Res.* **60**:143–148.

47. Ernst, A., Snik, A. F. M., Mylanus, I. A. M., and Cremers, C. W. R. J. (1995). Noninvasive assessment of the intralabyrinthine pressure. *Arch. Otolaryngol. Head Neck Surg.* **121**:926–929.

48. Evans, E. F. (1975). Normal and abnormal functioning of the cochlear nerve. *Symp. Zool. Soc. Lond.* **37**:133–165.

49. Evans, E. F., and Harrison, R. V. (1976). Correlation between outer hair cell damage and deterioration of cochlear nerve tuning properties in the guinea pig. *J. Physiol. (London)* **256**:43–44.

50. Filipo R., *et al.*, in press.

51. Frazier, C. H. (1912). Intracranial division of the auditory nerve for persistent aural vertigo. *Surg. Gynecol. Obst.* **15**:524–529.

52. Fromm, G. H. (1991). Pathophysiology of trigeminal neuralgia. *In* G. H. Fromm and B. J. Sessle (Eds.), "Trigeminal Neuralgia" (pp. 105–130). Butterworth–Heinemann, Boston.

53. Fukutake, T., and Hattori, T. (1998). Auditory illusions caused by a small lesion in the right geniculate body. *Am. Acad. Neurol* **51**:1469–1471.

54. Garetz, S. L., and Schacht, J. (1996). Ototoxicity: Of mice and men. *In* T. R. Van De Water, A. N. Popper, and R. R. Fay (Eds.), "Clinical Aspects of Hearing" (pp. 116–154). Springer-Verlag, New York.

55. Garnet Passe, E. R. (1951). Sympathectomy in relation to Ménière's disease, nerve deafness and tinnitus: A report of 110 cases. *Proc. Roy. Soc. Med.* **44**:760–772.

56. Garvey, W. D. (1953). The intelligibility of speeded speech. *J. Exp. Psychol.* **45**:102.

57. Gerken, G. M., Solecki, J. M., and Boettcher, F. A. (1991). Temporal integration of electrical stimulation of auditory nuclei in normal hearing and hearing-impaired cat. *Hear. Res.* **53**:101–112.

58. Gerrard, J. (1952). Nuclear jaundice and deafness. *J. Laryngol. Otol.* **66**:39–47.

59. Glasscock, M. C., Thedinger, B. A., and Cueva, R. A. (1991). An analysis of the retrolabyrinthine vs the retrosigmoid vestibular nerve section. *Otolaryngol. Head Neck Surg.* **104**:88–95.

60. Godey, B., Morandi, X., Beust, L., Brassier, G., and Bourdiniere, J. (1998). Sensitivity of auditory brainstem response in acoustic neuroma screening. *Acta Otolaryngol. (Stockh.)* **118**:501–504.

61. Golding-Wood, P. H. (1973). Cervical sympathectomy in Ménière's disease. *Arch. Otolaryngol.* **97**:391–394.

62. Goldstein, J. (1978). Mechanisms of signal analysis and pattern perception in periodicity pitch. *Audiology* **17**:421–445.

108. Møller, A. R. (1993). Cranial nerve dysfunction syndromes: Pathophysiology of microvascular compression. *In* D. L. Barrow (Ed.), "Neurosurgical Topics: Surgery of Cranial Nerves of the Posterior Fossa" (pp. 105–129). American Association of Neurological Surgeons, Park Ridge, IL.
109. Møller, A. R. (1995). "Intraoperative Neurophysiologic Monitoring." Harwood, Luxembourg.
110. Møller, A. R. (1997). Similarities between chronic pain and tinnitus. *Am. J. Otol.* 18:577–585.
111. Møller, A. R., Colletti, V., and Fiorino, F. G. (1994). Neural conduction velocity of the human auditory nerve: Bipolar recordings from the exposed intracranial portion of the eighth nerve during vestibular nerve section. *Electroencephalogr. Clin. Neurophysiol.* 92:316–320.
112. Møller, A. R., and Møller, M. B. (1989). Does intraoperative monitoring of auditory evoked potentials reduce incidence of hearing loss as a complication of microvascular decompression of cranial nerves? *Neurosurgery* 24:257–263.
113. Møller, A. R., Møller, M. B., and Yokota, M. (1992). Some forms of tinnitus may involve the extralemniscal auditory pathway. *Laryngoscope* 102:1165–1171.
114. Møller, A. R., Møller, M. B., Jannetta, P. J., and Jho, H. D. (1992). Compound action potentials recorded from the exposed eighth nerve in patients with intractable tinnitus. *Laryngoscope* 102:187–197.
115. Møller, M. B. (1981). Hearing in 70- and 75-year old people: Results from a cross-sectional and longitudinal population study. *Am. J. Otolaryngol.* 2:22–29.
116. Møller, M. B. (1990). Results of microvascular decompression (MVD) of the eighth nerve as treatment for disabling positional vertigo (DPV). *Ann. Otol. Rhinol. Laryngol.* 99:724–729.
117. Møller, M. B. (1994). Audiological evaluation. *J. Clin. Neurophysiol.* 11:309–318.
118. Møller, M. B., and Møller, A. R. (1985). Audiometric abnormalities in hemifacial spasm. *Audiology* 24:396–405.
119. Møller, M. B., Møller, A. R., Jannetta, P. J., and Sekhar, L. N. (1986). Diagnosis and surgical treatment of disabling positional vertigo. *J. Neurosurg.* 64:21–28.
120. Møller, M. B., Møller, A. R., Jannetta, P. J., and Jho, H. D. (1992). Results of microvascular decompression (MVD) surgery in patients with disabling positional vertigo (DPV). *In* "Proceedings of the 2nd European Congress of Oto-Rhino-Laryngology and Cervico-Facial Surgery (EUFOS) (pp. 429–432). Monduzzi Editorie S.p.A., Bologna, Italy.
121. Møller, M. B., Møller, A. R., Jannetta, P. J., and Jho, H. D. (1993). Vascular decompression surgery for severe tinnitus: Selection criteria and results. *Laryngoscope* 103:421–427.
122. Morest, D. K., Ard, M. D., and Yurgelun-Todd, D. (1979). Degeneration in the central auditory pathways after acoustic deprivation or over-stimulation in the cat. *Anat. Rec.* 193:750.
123. Morest, D. K., and Bohne, B. A. (1983). Noise-induced degeneration in the brain and representation of inner and outer hair cells. *Hear. Res.* 9:145–152.
124. Myers, E. N., and Bernstein, J. M. (1965). Salicylate ototoxicity. *Arch. Otolaryngol.* 82:483–493.
125. Nigam, A., and Samuel, P. R. (1994). Hyperacusis and Williams syndrome. *J. Laryngol. Otol.* 108:494–496.
126. Nilsson, R., and Borg, E. (1983). Noise-induced hearing loss in shipyard workers with unilateral conduction hearing loss. *Scand. Audiol.* 12:135.
127. Ouaknine, G. E. (1981). Microsurgical anatomy of the arterial loops in the ponto-cerebellar angle and internal acoustic meatus. *In* M. Samii and P. J. Jannetta (Eds.), "The Cranial Nerves" (pp. 378–390). Springer-Verlag, Heidelberg.
128. Payne, M. C., and Githler, F. J. (1951). Effects of perforations of the tympanic membrane on cochlear potentials. *Arch. Otolaryngol.* 54:666–674.
129. Peake, W. T., Rosowski, J. J., and Lynch, T. J., III (1992). Middle-ear transmission: Acoustic versus ossicular coupling in cat and human. *Hear. Res.* 57:245–268.

130. Phillips, D. P., and Carr, M. M. (1998). Disturbances of loudness perception. *J. Am. Acad. Audiol.* 9:371–379.
131. Pickles, J. O. (1979). An investigation of sympathetic effects on hearing. *Acta Oto-Laryngol.* 87:69–71.
132. Pierson, L. L., Gerhardt, K. J., Rodriguez, G. P., and Yanke, R. B. (1994). Relationship between outer ear resonance and permanent noise induced hearing loss. *Am. J. Otol.* 15:37–40.
133. Pierson, M. G., and Møller, A. R. (1981). Prophylaxis of kanamycin-induced ototoxicity by a radioprotectant. *Hear. Res.* 4:79–88.
134. Pollak, G. D., and Park, T. J. (1993). The effects of GABAergic inhibition on monaural response properties of neurons in the mustache bat's inferior colliculus. *Hear. Res.* 65:99–117.
135. Popelar, J., Syka, J., and Berndt, H. (1987). Effect of noise on auditory evoked responses in awake guinea pigs. *Hear. Res.* 26:239–248.
136. Portmann, G. (1927). The saccus endolymphaticus and an operation for draining the same for the relief of vertigo. *J. Laryng. Otol.* 42:809.
137. Powell, T. S., and Erulkar, S. D. (1962). Transneuronal cell degeneration in the auditory relay nuclei of the cat. *J. Anat. (Lond.)* 96:249–268.
138. Pulec, J. L. (1984). Tinnitus: Surgical therapy. *Am. J. Otology* 5:479–480.
139. Quaranta, A., Sallustio, V., and Scaringi, A. (2000). Cochlear function in ears with vestibular schwannomas. *In* M. Sanna. "Third International Conference on Acoustic Neurinoma and other CPA Tumors" (in press). Monduzzi Editore, Bolongna, Italy.
140. Raucher, F. H., Robinson, K. D., and Jens, J. J. (1998). Improved maze learning through early music exposure in rats. *Neurol. Res.* 20:427–432.
141. Reed, G. F. (1960). Audiometric study of 200 cases of subjective tinnitus. *Arch. Otolaryngol.* 71:94–104.
142. Rosowski, J. J. (1991). The effects of external-and middle-ear filtering on auditory threshold and noise-induced hearing loss. *J. Acoust. Am.* 90:124–135.
143. Ruben, R. J., Hudson, W., and Chiong, A. (1962). Anatomical and physiological effects of chronic section of the eighth cranial nerve in cat. *Acta Otolaryngol.* 55:473–484.
144. Salvi, R. J., Wang, J., and Powers, N. (1996). Rapid functional reorganization in the inferior colliculus after acute cochlear damage. *In* R. J. Salvi, D. Henderson, F. Fiorino, and C. Colletti (Eds.), "Auditory Plasticity and Regeneration" (pp. 275–296). Thieme Medical, New York.
145. Salvi, R. J., Wang, J., Lockwood, A. H., Burkhard, R., and Ding, D. (1999). Noise and drug induced cochlear damage leads to functional reorganization of the central auditory system. *Noise Health* 2:28–42.
146. Sando, I. (1965). The anatomical interrelationships of the cochlear nerve fibers. *Acta Otolaryngol. (Stockh.)* 59:417–436.
147. Sass, K. (1998). Sensitivity and specificity of transtympanic electrocochleography in Ménière's disease. *Acta Oto-laryngol. (Stockh.)* 118:150–156.
148. Sass, K., Densert, B., and Arlinger, S. (1998). Recording techniques for transtympanic electrocochleography in clinical practice. *Acta Otolaryngol. (Stockh.)* 118:17–25.
149. Sataloff, R. T., and Sataloff, J. (1993). "Hearing Loss." Marcel Dekker, New York.
150. Scharf, B., Magnan, J., Collet, L., Ulmer, E., and Chays, A. (1994). On the role of the olivocochlear bundle in hearing: A case study. *Hear. Res.* 75:11–26.
151. Schreiner, C. E., and Snyder, R. L. (1987). A physiological animal model of peripheral tinnitus. *In* H. Feldmann (Ed.), "Third International Tinnitus Seminar" (pp. 100–106). Harsch-Verlag, Karlsruhe.
152. Schulman, A., Tonndorf, J., and Goldstein, B. (1985). Electrical tinnitus control. *Acta Otolaryngol. (Stockh.)* 99:318–325.
153. Selters, W. A., and Brackmann, D. E. (1977). Acoustic tumor detection with brainstem electric response audiometry. *Arch. Otolaryngol.* 103:181–187.

154. Sha, S. H., and Schacht, J. (1999). Stimulation of free radical formation by aminoglycoside antibiotics. *Hear. Res.* **128**:112–118.

155. Shambaugh, G. E. (1966). Surgery of the endolymphatic sac. *Arch. Otol.* **83**:302.

156. Shapiro, S. M. (1991). Binaural effects in brainstem auditory evoked potentials of jaundiced Gunn rats. *Hear. Res.* **53**:41–48.

157. Shea, J. (1958). Fenestration of the oval window. *Ann. Otol. Rhinol. Laryngol.* **67**:932–951.

158. Silverstein, H., Wanamaker, H. H., and Rosenberg, S. I. (1994). Vestibular neurectomy. *In* R. K. Jackler and D. E. Brackmann (Eds.), "Neurotology" (pp. 945–954). Mosby, St. Louis.

159. Sininger, Y. S., Mobley, J. P., House, W., and Nielsen, D. (1987). Intra-cochlear electrical stimulation for tinnitus suppression in a patient with near-normal hearing. *In* H. Feldmann (Ed.), "Proceedings of the III International Tinnitus Seminar." Harsch-Verlag, Karlsruhe.

160. Skellett, R. A., Cullen, J. K., Jr., Fallon, M., and Bobbin, R. P. (1998). Conditioning the auditory system with continuous vs. interrupted noise of equal acoustic energy: Is either exposure more protective? *Hear. Res.* **116**:21–32.

161. Spandow, O., Anniko, M., and Møller, A. R. (1988). The round window as access route for agents injurious to the inner ear. *Am. J. Otolaryngol.* **9**:327–335.

162. Spoendlin, H. (1986). Anatomical changes following various forms of noise exposure. *In* D. Henderson, R. P. Hamernik, D. S. Dosanjh, and J. H. Mills (Eds.), "Effects of Noise on Hearing" (pp. 69–90). Raven, New York.

163. Spoendlin, H., and Schrott, A. (1989). Analysis of the human auditory nerve. *Hear. Res.* **43**:25–38.

164. Spoor, A. (1967). Presbycusis values in relation to noise-induced hearing loss. *Int. Audiol.* **6**:48–57.

165. Stahle, J., Ahrenberg, K., and Stahle, C. (1978). Incidence of Ménière's disease. *Arch. Otolaryngol.* **104**:99–103.

166. Stangerup, S. E., and Tos, M. (1985). The etiologic role of acute suppurative otitis media in chronic secretory otitis. *Am. J. Otol.* **6**(2):126–131.

167. Strouse, A., Ashmead, D. A., Ohde, R. N., and Grantham, W. (1998). Temporal processing in the aging auditory system. *J. Acoust. Soc. Am.* **104**:2385–2399.

168. Sunderland, S. (1981). Cranial nerve injury: Structural and pathophysiological considerations and a classification of nerve injury. *In* M. Samii and P. J. Jannetta (Eds.), "The Cranial Nerves" (pp. 16–26). Springer-Verlag, Heidelberg.

169. Syka, J., and Popelar, J. (1982). Noise impairment in the guinea pig. I. Changes in electrical evoked activity along the auditory pathway. *Hear. Res.* **8**:263–272.

170. Syka, J., Rybalko, N., and Popelar, J. (1994). Enhancement of the auditory cortex evoked responses in awake guinea pigs after noise exposure. *Hear. Res.* **78**:158–168.

171. Szczepaniak, W. S., and Møller, A. R. (1995). Evidence of decreased GABAergic influence on temporal integration in the inferior colliculus following acute noise exposure: A study of evoked potentials in the rat. *Neurosci. Lett.* **196**:77–80.

172. Szczepaniak, W. S., and Møller, A. R. (1996a). Effects of (−)-baclofen, clonazepam, and diazepam on tone exposure-induced hyperexcitability of the inferior colliculus in the rat: Possible therapeutic implications for pharmacological management of tinnitus and hyperacusis. *Hear. Res.* **97**:46–53.

173. Szczepaniak, W. S., and Møller, A. R. (1996b). Evidence of neuronal plasticity within the inferior colliculus after noise exposure: A study of evoked potentials in the rat. *Electroencephalogr. Clin. Neurophysiol.* **100**:158–164.

174. Tonndorf, J. (1987). The analogy between tinnitus and pain: A suggestion for a physiological basis of chronic tinnitus. *Hear. Res.* **28**:271–275.

175. Tos, M., and Bak-Pedersen, P. (1972). The pathogenesis of chronic secretory otitis media. *Arch. Otolaryngol.* **95**:511–521.

176. Tos, M., Stangerup, S. E., and Andreassen, U. K. (1985). Size of the mastoid air cells and otitis media. *Ann. Otol. Rhinol. Laryngol.* **94**:386–392.

177. Tos, M., and Stangerup, S. E. (1985). Secretory otitis and pneumatization of the mastoid process: Sexual differences in the size of mastoid cell system. *Am. J. Otolaryngol.* **6**(3):199–205.

178. Tos, M., Thomsen, J., and Charabi, S. (1992). Incidence of acoustic neuromas. *Ear Nose Throat J.* **71**(9):391–393.

179. Turner, J. G., and Willott, J. F. (1998). Exposure to an augmented acoustic environment alters auditory function in hearing-impaired DBA2J mice. *Hear. Res.* **118**:101–113.

180. Vasama, J.-P., I. Møller, M. B., and Møller, A. R. (1998). Microvascular decompression of the cochlear nerve in patients with severe tinnitus: Preoperative findings and operative outcome in 22 patients. *Neurol. Res.* **20**:242–248.

181. Vernon, J. A., and Fenwick, J. A. (1985). Attempts to suppress tinnitus with transcutaneous electrical stimulation. *Otolaryngol. Head Neck Surg.* **93**:385–389.

182. Voss, S. E., Rosowski, J. J., and Peake, W. T. (1996). Is the pressure difference between the oval and round windows the effective acoustic stimulus for the cochlea? *J. Acoust. Soc. Am.* **100**(3):1602–1616.

183. Wable, J., Collet, L., and Chery Croze, S. (1996). Age-related changes in perilymphatic pressure: preliminary results. *In* A. Ernst, R. Marchbanks, and M. Samii (Eds.), "Intracranial and Intralabyrinthine Fluids" (pp. 191–198). Springer-Verlag, Berlin.

184. Wall, P. D. (1977). The presence of ineffective synapses and circumstances which unmask them. *Phil. Trans. R. Soc. Lond.* **278**:361–372.

185. Wayman, D. M., Pham, H. N., Byl, F. M., and Adour, K. K. (1990). Audiologic manifestations of Ramsay Hunt syndrome. *J. Laryngol. Otol.* **104**:104–108.

186. Webster, D. B., and Webster, M. (1977). Neonatal sound deprivation affects brain stem auditory nuclei. *Arch. Otolaryngol.* **103**:392–396.

187. Wever, E. G., and Lawrence, M. (1950). The acoustic pathways to the cochlea. *J. Acoust. Soc. Am.* **22**:460–467.

188. Wever, E. G., and Lawrence, M. (1954). "Physiological Acoustics." Princeton Univ. Press, Princeton, NJ.

189. Wever, E. G., Lawrence, M., and Smith, K. R. (1948). The middle ear in sound conduction. *Arch. Otolaryngol.* **48**:19–35.

190. Whitworth, C., Morris, C., Scott, V., and Rybak, L. P. (1993). Dose–response relationships for furosemide ototoxicity in rat. *Hear. Res.* **71**:202–207.

191. Willott, J. F. (1991). "Aging and the Auditory System: Anatomy, Physiology, and Psychophysics." Singular, San Diego.

192. Willott, J. F., and Lu, S. M. (1981). Noise-induced hearing loss can alter neural coding and increase excitability in the central nervous system. *Science* **216**:1331–1332.

193. Wladislavorsky-Wasserman, P., Facer, G. W., Mokri, B., and Kurland, L. T. (1984). Ménière's Disease: A 30 year epidimiologic and clinical study in Rochester, MN, 1951–1980. *Laryngoscope* **94**:1098–1102.

194. Yamasoba, T., Schacht, J., Shoji, F., and Miller, J. M. (1999). Attenuation of cochlear damage from noise trauma by an iron chelator, a free radical scavenger and glial cell line-derived neurotrophic factor *in vivo*. *Brain Res.* **815**(2):317–325.

195. Zakrisson, J. E., Borg, E., Diamant, H., and Møller, A. R. (1975). Auditory fatigue in patients with stapedius muscle paralysis. *Acta Otolaryngol.* (*Stockh.*) **79**:228–232.

196. Zwicker, E. (1974). On a psychoacoustical equivalent of tuning curves. *In* E. Zwicker and E. Terhardt (Eds.), "Facts and Models in Hearing" (pp.132–141). Springer-Verlag, Berlin.

INDEX